GEOTHERMAL ENERGY

GEOTHERMAL ENERGY

Edward R. Berman

NOYES DATA CORPORATION
Park Ridge, New Jersey London, England
1975

Library of Congress Catalog Card Number: 74-24721
ISBN: 0-8155-0563-9
Printed in the United States

Published in the United States of America by
Noyes Data Corporation
Noyes Building, Park Ridge, New Jersey 07656

FOREWORD

This Energy Technology Review is based on studies conducted by industrial and engineering firms or university research teams under the auspices of various governments and governmental agencies.

It describes in detail the nature of the geothermal resources, their extent, and the currently available technology by which these natural sources of energy can be exploited and utilized to the greatest advantage.

In this book are condensed vital data that are scattered and difficult to pull together. Important power schemes are described and illustrated by actual case histories. This condensed information will provide a sound background for action toward the realization of energy from geothermal sources. In view of the rise in price of oil and the intensive search for new sources of energy, geothermal power will continue to be of interest as an alternate source of energy.

Advanced composition and production methods developed by Noyes Data are employed to bring new durably bound books to the reader in a minimum of time. Special techniques are used to close the gap between "manuscript" and "completed book." Industrial technology is progressing so rapidly that time-honored, conventional typesetting, binding and shipping methods are no longer suitable. Delays in the conventional book publishing cycle have been bypassed to provide the user with an effective and convenient means of reviewing up-to-date information in depth.

The Table of Contents is organized in such a way as to serve as a subject index and provides easy access to the information contained in this book.

CONTENTS AND SUBJECT INDEX

INTRODUCTION

The first large scale geothermal power plant was built at Larderello, Italy in 1904. New Zealand began developing its geothermal fields in the 1950s. Geothermal power plants are in operation in Japan, Mexico and the USSR. In the United States, the first commercial geothermal power plant was built at The Geysers, in California in 1960.

Geothermal fields may be classified into three types—dry steam, hot water and dry rock reservoirs. The hot water fields may be further subdivided into those containing superheated water (wet steam) and reservoirs of lower temperatures. Larderello and The Geysers are examples of dry steam fields; the Wairakei, New Zealand fields are of the wet steam type. Dry rock reservoirs have not yet been exploited commercially but research is underway with both conventional rock fracturing methods and use of nuclear explosives (the Plowshare concept) under consideration.

In addition to electric power, geothermal energy has been used for space heating and cooling, notably in Iceland and the USSR. Geothermal energy has been used to heat greenhouses where vegetables and fruits are raised in areas which have short outdoor growing seasons. The USSR has many multiple use projects. At Kamchatka, for example, hot wastewater from the power plant is piped under highways to melt snow.

Possible by-products of geothermal power plants are natural gas, minerals and desalinated water.

Most experts agree that geothermal energy compares favorably with other sources of energy in regard to both pollution and cost.

GEOTHERMAL ENERGY AS AN ALTERNATE RESOURCE

The following information is derived from LA-5289-MS issued in May 1973 entitled *Geothermal Energy* by Morton C. Smith of Los Alamos Scientific Laboratory of the University of California.

The total volume of the earth is about 260 billion cubic miles, and all but a thin crust at its outermost surface is very hot. Exactly how hot is uncertain, but the evidence from lava flows—and, less spectacularly, from hot springs and geysers and fumaroles—is that it is certainly hot enough to supply most of mankind's needs for energy. In fact, when the volume of the earth's interior and the specific heat of rock are taken into account, it becomes evident that this geothermal heat is by far the largest reservoir of energy that is now available to man. Further, the energy which it contains is inherently clean energy. It already exists as heat, and it is therefore not necessary to burn a fuel or operate a reactor to produce that heat.

In a few fortunate places nature delivers geothermal energy to the earth's surface in the form of steam or of superheated water that flashes to steam, and this natural steam can be used directly to generate electricity. Elsewhere the water from springs or wells is at least hot enough to heat homes and greenhouses, and to supply energy for low-temperature chemical processes. In general however, and particularly considering the magnitude of the resource that it represents, the present worldwide usage of geothermal energy is in fact very small.

The ideas and most of the technology required to correct this now exist. It appears quite possible that within a few years geothermal energy can be made broadly available at a cost in money, effort, and environmental deterioration that men can afford. If so, this oldest form of energy will contribute significantly to solving two of the world's newest problems: a vanishing fuel supply, and an overwhelming burden of pollution.

Where natural steam or superheated water that flashes to steam is produced, the steam is simply passed through a screen to remove entrained particles, fed to a turbine, and thus used directly to generate electricity. The "fuel cost" of such a power plant is represented by the capital charges against the production wells and steam-collection system, and is low. The generating plant is also relatively inexpensive, since it requires no fuel-handling facilities, boiler-plant, or smoke-stack. Accordingly, in this convenient natural form, geothermal energy is a very economical source of electricity. It is unfortunate that occurrences of natural steam, or of water sufficiently superheated to produce it in useful amounts, so far appear to be relatively rare.

Much more common are natural reservoirs of hot water at temperatures below boiling, which are the sources of most of the world's many hot springs. These have been used for thousands of years for recreational and religious and medicinal purposes, and for domestic heating. They are still used, very economically, for heating homes, greenhouses, resort hotels, and occasionally factories, mines, and even whole cities. In general they are an inexpensive source of low-grade heat, and their use has the great advantages of conserving higher-grade energy sources for more demanding uses, and of minimizing pollution locally.

Natural hot water at intermediate temperatures—significantly above boiling, but not quite hot enough to be used economically as a steam source for a conventional generating plant—occurs in large amounts in several parts of the world, notably in the Imperial Valley of California. The technology required to utilize this large geothermal resource is developing rapidly, and demonstration plants which will desalinate brines and generate electricity in turbines driven by the vapors of low-boiling organic liquids are in the testing state. To this point, however, natural hot water in the intermediate temperature range is in general used only locally, and again principally for space-heating.

No demonstrated system yet exists for extracting energy from the hot "dry" rock which composes so much of the earth's crust but which—for lack either of permeability or of a ground-water supply—does not spontaneously yield useful amounts of either steam or hot water. However this energy reservoir is so vast, so broadly distributed and in many accessible places so hot, that it is now at last being given the serious technical attention that it deserves.

HISTORICAL BACKGROUND

Local use of natural geothermal water for various heating purposes is older than history, but has expanded notably in recent years—particularly in Iceland, Hungary, and the USSR. The first large-scale commercial use of natural steam was at Larderello, Italy, where it has been used to generate electricity since 1904. Development of the Wairakei geothermal area in New Zealand, where natural superheated water is flashed to steam, began in about 1950. In the United States the only commercial production of geothermal power is at The Geysers in California, where generation of electricity from natural steam began in 1960 and has since expanded steadily.

A major geothermal power plant based on superheated water is now being constructed at Cerro-Prieto in northwest Mexico, and there are very small geothermal generating plants operating elsewhere in Mexico, New Zealand, and Italy, and in Japan, the USSR, and Iceland. At this time, however, the total installed generating capacity of the world deriving its energy from geothermal sources is less than 1000 Mw electrical, and thus is less than the capacity of one large fuel-fired or nuclear power plant. All but a minute fraction of this capacity is based on the use either of natural dry steam or of wet steam flashed from naturally superheated water. (See Table 1.1.)

TABLE 1.1: WORLD GEOTHERMAL POWER CAPACITY 1972

		----Electrical Capacity, Mw----	
Country	**Field**	**Operating**	**Under Construction**
Italy	Larderello	358.6	
	Mt. Amiata	25.5	
USA	The Geysers	192	220
New Zealand	Wairakei	160	
	Kawerau	10	
Japan	Matsukawa	20	
	Otake	13	
Mexico	Pathe	3.5	
	Cerro Prieto		75
USSR	Pauzhetka	5	
	Paratunka*	0.7	
Iceland	Namafjall	2.5	
		790.8	295

*Freon plant

Source: PB 224,882

There is now active exploration for or development of natural steam or superheated water fields in many parts of the world, and there is no real doubt that new geothermal sources of this type will be found and developed over the years. When they are, they will be economical sources of energy, and, locally, they may be of great importance. However, there is no evident reason to suppose that they will ever be found in very large numbers or that, individually, they will be much larger than are the similar fields now being exploited. Accordingly, if only "existing technology" is considered, there is an honest basis for the commonly stated conclusion that geothermal sources can never satisfy a significantly large fraction of the country's or the world's total energy requirement. If, on the other hand, consideration is given to applications of present technology to new kinds of geothermal energy developments and to the probability that entirely new technologies may soon appear, then the forecast of geothermal energy's future contribution becomes much more optimistic.

SYSTEM CHARACTERISTICS

There are still mysteries about wet geothermal reservoirs related, for example, to the natural flow of water into them, convective circulation within them, and many details of their geology and heat-flow. However, their general nature is reasonably well understood, and is illustrated schematically in Figure 1.1.

FIGURE 1.1: SCHEMATIC DIAGRAM OF A WET GEOTHERMAL RESERVOIR

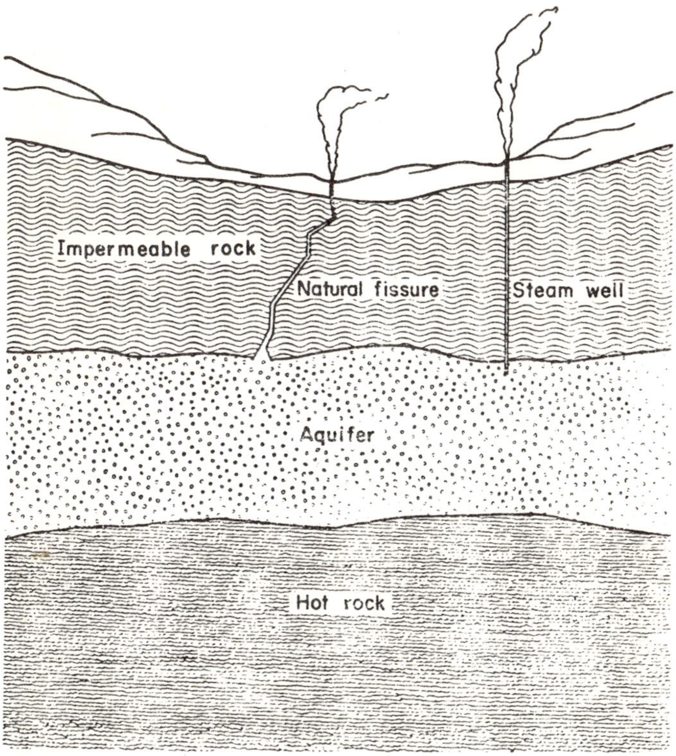

Source: LA-5289-MS

Apparently three lithologic features must exist together to create and maintain a wet geothermal field: (1) A large supply of usefully hot rock at depth, which of course is the major heat source; (2) An aquifer or permeable formation above it, through which ground-water reaches the hot rock, and within which the heated fluid—steam, hot water, or a mixture of the two—circulates convectively and transports heat to a higher level; (3) An impermeable cap above the aquifer, to prevent the large-scale loss of fluid at the surface which otherwise, over the millenia, would have cooled the system sufficiently to make it uninteresting as an energy source. This cap appears often to have formed spontaneously through plugging of an initially permeable formation by minerals dissolved at depth and reprecipitated in cooler rock near the surface.

When the fluid circulating in at least the upper part of the aquifer is steam, the natural system is generally described as a vapor-dominated geothermal reservoir. When temperatures are lower, pressures higher, or the concentration of dissolved minerals greater, so that the circulating fluid is water or brine, then it is a liquid-dominated reservoir (L.J. Muffler and D.E. White, "Geothermal Energy," *The Science Teacher,* 39, 3, March 1972.)

The surface evidence that a wet geothermal reservoir exists at depth is usually the natural occurrence of hot springs, geysers, or fumaroles. However, it is now believed that many such reservoirs may exist with no evidence at the surface that they are present. Prospecting for these hidden reservoirs will evidently be difficult.

It is apparent that thermal energy is present in the hot rock even when no permeable reservoir or circulating fluid exists above it to transport the heat to the earth's surface. Frequently, e.g., in Hawaii and other regions of recent vulcanism, extremely hot rock is quite close to the earth's surface, and its energy content is impressive. Thus, as is suggested by Figure 1.2, cooling 40 cubic miles of rock by 360°F would release 0.07Q of energy, which was just the energy requirement of the United States for all purposes in 1970. In this form the energy would of course not be suitable for many purposes, such as smelting iron ores and powering aircraft. It could, however, be used for many purposes, to conserve higher-grade energy sources for uses that geothermal energy itself cannot serve. And, while 40 cubic miles is a lot of rock, it is infinitesimal compared to the total volume of the earth's interior, or even to that relatively small fraction of the earth's interior that can now be reached by conventional drilling methods.

FIGURE 1.2: THE ENERGY EXTRACTED FROM 40 CUBIC MILES OF ROCK COOLED 200°C (360°F)

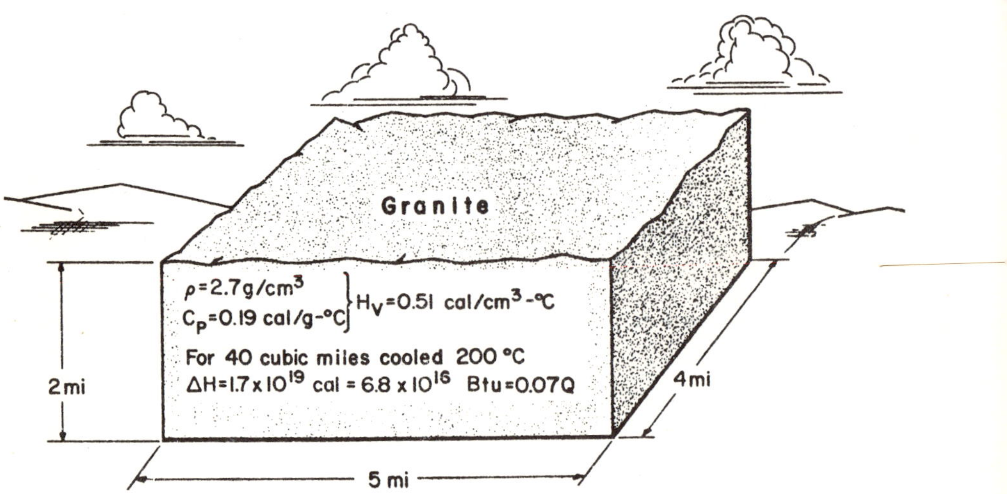

Source: LA-5289-MS

The technology for producing and using either wet or dry steam for generating electricity is already highly developed, and is still improving steadily. That for lower-temperature waters and for concentrated brines is being developed actively. The investigation of methods for extracting useful energy from the much abundant dry geothermal reservoirs has just begun.

General

Economically, the most attractive geothermal reservoirs are the vapor dominated ones, which produce dry steam—i.e., superheated steam, carrying with it minor amounts of other gases, but little or no water. Aside from mechanical removal of any dirt and rock carried out of the well by the steam, no treatment of it is ordinarily required, and in general the steam is piped directly to a turbine and used to generate electricity. Larderello in Italy, The Geysers in California, and Matsukawa in Japan are the principal dry-steam fields now being exploited commercially.

Two small dry-steam fields in the Monte Amiata region of Italy are marginally commercial, and several areas in Africa and in Central and South America now being investigated by the United Nations may be of this type. As new dry-steam fields are discovered, industry and/or government will be anxious to develop and manage them.

Considerably more common than dry-steam fields are the liquid-dominated or superheated-water fields, from which wet steam is produced. In these the temperature at depth is above the normal boiling point of water, but the water or brine in the reservoir is kept from boiling by the pressure on it, which is greater than atmospheric. As the fluid is brought out of the reservoir its pressure is reduced, boiling occurs, and a mixture of steam and hot water is produced from the well. Normally about 20% of the fluid is converted to saturated (wet) steam, and only this steam is used to generate electricity. The other 80% of the fluid, hot water at the boiling point, is usually wasted.

Wet-steam fields of this type are now being exploited at Wairakei and Broadlands in New Zealand and in smaller plants in Japan, Mexico, the USSR, and Iceland. Much of the hot brine in the Imperial Valley of California and at Cerro Prieto in northwestern Mexico is in this class.

Still more common are natural reservoirs of hot water at temperatures below the boiling point. One small power plant on the Kamchatka peninsula is now generating electricity by using natural hot water at only 178°F to vaporize Freon, and using the Freon vapor to drive a turbine. A similar two-fluid system, using isobutane instead of Freon, is being developed in the Imperial Valley. Otherwise, natural hot water at temperatures below boiling has so far been used only for space-heating and low-grade process heat.

Most common of all are dry geothermal reservoirs, which yield no steam or hot water either because they are relatively impermeable to water or because the ground-water circulation does not reach them. In principle, dry hot rock at a temperature high enough to be useful as an energy source could be reached from any point on earth's surface simply by drilling a deep enough hole. In practice, the difficulty and expense of drilling increase exponentially with hole depth. In most types of geology, holes are now drilled more or less routinely to depths of the order of 20,000 feet, but each increment of depth below this becomes more of an adventure.

Practically, then, with existing drilling technology, the dry geothermal reservoirs of interest are those within about 20,000 feet of the earth's surface. These vary widely in temperature and accessibility, the extremes being hot lava pools essentially at the surface with temperatures above 2000°F, and regions of extensive, geologically recent glaciation, where even at 20,000 feet the rock temperature may be less than 300°F. No demonstrated method exists for extracting energy economically from hot dry rock in the earth's crust. Several proposed systems for doing so are discussed below.

Availability

Dry steam fields with reserves sufficient to justify commercial development are now known to exist in less than a dozen places in the world, generally in regions of recent vulcanism. Systematic prospecting may discover a few dozen more, but the geological and thermal situation that produces a vapor-dominated geothermal reservoir is evidently rare, and the individual vapor-dominated reservoirs in general do not appear to be large.

Liquid-dominated systems capable of producing wet steam in commercial amounts appear to be both larger and more common than vapor-dominated systems. They occur in many parts of the world, and represent a very large and relatively accessible energy resource.

Natural hot water at lower temperatures is of course still more common and more readily available, and dry hot rock is the most common and accessible geothermal resource of all.

Energy Estimates

Estimates of the amount of energy potentially available from vapor-dominated and liquid-dominated geothermal reservoirs of types now being exploited commercially vary widely with the degree of optimism of the person extrapolating the boundaries and long-term productivities of existing fields, predicting discoveries of new ones, and anticipating technologies which may make them more broadly useful.

One probably conservative estimate is that—with present technologly—the identified, recoverable resource of natural steam in the United States is sufficient to support only 1000 Mw of generating capacity for 50 years, but that discoveries of new steam fields will probably make it possible eventually to support another 3000 to 6000 Mw of generating capacity, again for 50 years. (PB 224,882, D.L. Peck, et al, *Assessment of Geothermal Energy Resources,* June, 1972.)

Other estimates can be found which would increase the probable magnitude of the nation's immediately useful reservoirs of natural steam by factors as large as 5 to 10. However, even the most optimistic of these estimates suggests that, if only existing technology is used to exploit it, the geothermal resource is indeed small relative to the total energy needs of the United States.

The degree to which the magnitude of this resource can be increased simply by improvements in technology is the subject of much speculation. To most specialists in the field, it would probably seem reasonable to assume that, during the next 5 to 10 years, there will be:

(1) Significant improvements in techniques of prospecting for immediately usable geothermal reservoirs, so that more of them will be discovered than are now believed to exist.

(2) Development of improved drilling equipment and of totally new drilling systems so that exploration, development, and production costs will decrease for reservoirs at normal depths, and much deeper reservoirs will become accessible and economical energy sources.

(3) The chemical problems associated with relatively concentrated hot brines, such as those of the Imperial Valley, will be brought under control, vastly increasing the supply of superheated water that is directly useful for generating electricity.

(4) Two-fluid generating systems, involving transfer of heat from natural hot water to a lower-boiling organic liquid, will have been demonstrated convincingly, making lower-temperature geothermal reservoirs valuable for generating electricity.

(5) Non-electrical uses of the heat from natural hot water (and from power-plant coolants and condensates) will increase in such applications as domestic and industrial heating and air-conditioning, desalination of water, distillation of domestic and industrial wastes, recovery of minerals, and physical and chemical processing.

(6) Increased understanding of the geology, hydrology, heat- and fluid-flow, chemical and mechanical behavior, and artificial stimulation of geothermal reservoirs of all types will significantly increase their productivity.

The magnitude of the useful geothermal resource will increase with the degree to which any of these technological advances is achieved, and a resulting total increase in the resource by a factor of 20 in the next 10 years would not be unexpected. Thus Peck et al estimate that such advances may increase the geothermal resource of the United States sufficiently so that it can support a generating capacity of about 440,000 Mw for 50 years, in addition to its nonelectrical uses. Again, estimates 5 to 10 times this large can be found elsewhere in the literature, but the Peck estimate alone exceeds the total presently installed generating capacity of the US.

The further technological developments required to make possible the economical extraction of energy from dry hot rock would enormously increase the magnitude of the geothermal resource, The extent of the projected increase depends principally upon the assumptions made concerning the types and extents of geological structures in which suitable energy systems can be created, the minimum rock temperature that would be useful, the maximum depth to which the system might economically be extended, the costs associated with geochemical and other systems problems, and the anticipated life of the system.

It seems probable that rock temperatures as low as 300°F will be useful, that depths to about 20,000 feet can be reached more or less routinely, that systems problems and costs will be no greater than for the utilization of hot brines, and that systems can be developed having lifetimes not less than 20 years. With such assumptions, the estimated generating capacity that might be supported by dry geothermal reservoirs in the United States becomes many millions of megawatts for at least a century. If even a fraction of one percent of this energy can in fact be extracted economically, it will satisfy a major part of the nation's total energy requirements. The incentive to develop the required technology is therefore very great.

Output Form

The direct output of any geothermal energy system is heat, normally contained in steam or hot water. If its temperature level is high enough, this heat can be used to drive a turbine, a screw-expander, or some other type of heat-engine, producing mechnical energy that can be used to generate electricity or do other useful work. At similar or lower temperatures the heat can be used directly for space-heating, air-conditioning, distillation and chemical processes, and a multitude of other uses. Multiple-use systems, in which the higher-temperature heat is used to generate electricity and the rest is used directly as domestic or industrial heat, are particularly attractive.

Efficiency Range

While the temperatures of natural geothermal fluids vary over wide limits, in general they are below 400°F. This is relatively low for driving a steam turbine, and normal generating efficiency is correspondingly low. Power-conversion efficiency may be in the vicinity of 20% for dry steam, but is likely to be 15% or less for a hot-water system. Efficiencies will be increased significantly if economical two-fluid generating systems are developed, to be used either alone or as a bottoming cycle in conjunction with a steam topping cycle. Efficiencies will also be higher if water temperatures can be increased, which appears possible with man-made dry-rock systems.

In applications other than the generation of electricity, where part or all of the geothermal energy is used directly as heat, much higher efficiencies are possible than when the heat is used only for generating electricity.

Size Limitation

With existing technology, the natural unit size of a geothermal power plant is small—of the order of 50 to 400 Mw of electrical generating capacity. Principally this is because of the costs and inefficiencies associated with piping low-temperature steam over long distances. Typically the power plant is built to handle the steam production of about 10 to 30 wells in its immediate vicinity. Then, instead of extending the collection system and enlarging the existing plant, another small generating plant is built to use the steam from the next group of wells that is developed. However, the disadvantages associated with small plant size are less for geothermal power than for most other energy systems.

Because of the low temperature and pressure of the steam used, turbines are large relative to generating capacity, and their most economical individual size probably corresponds to about 50 Mw electrical or less. The associated plant equipment is small and simple, since no fuel or ash handling, combustion system, or smokestacks are required. Accordingly, the cost of constructing a new, small plant is not much greater than that of enlarging an old one, and—above a plant capacity of perhaps 100 to 150 Mw—the advantage of scaling up in plant size is not likely to balance the increased costs and heat losses of an extended piping system.

This picture may change somewhat as hot-water and pressurized dry-rock systems are developed. Energy output per well may be much larger when hot water instead of steam is produced because, for a given pressure drop and temperature, a much greater mass of liquid than of vapor is brought up the pipe, delivering a correspondingly greater amount of energy to the surface. Further, unlike steam, hot water at moderate pressure and temperature can be piped economically and efficiently over distances of many miles.

Finally because the organic vapors produced in a two-fluid system are much denser than steam, the turbines required for a given generating capacity are much smaller. Therefore, as the technology of hot-water and dry-rock systems is developed it may prove economical to use larger collection systems than are now used for natural steam. Combined with greater thermal output per well this would result in fewer power plants and larger generating capacity per plant, although if two-fluid systems are used the physical size of the plant may not increase.

Unit Size

The output of a single geothermal well varies widely with the nature, temperature, and quantity of fluid produced from it, and—for geologic reasons that are not usually understood—may differ greatly between adjacent wells in the same field. In general, a single steam well capable of supporting 20 Mw of generating capacity is unusual, and the average production per well corresponds to less than 10 Mw. Because of the greater mass flow possible when hot water is produced instead of steam, the average energy output per well may be significantly higher for hot-water and dry-rock systems than for natural steam. This, however, obviously depends upon the temperature at which the water can be delivered to the surface plant.

Pollution Products

Natural steam brings with it to the surface a proportion of noncondensible gas that varies usually between about 0.5 and about 5%. Usually this gas is principally carbon dioxide, which is inoffensive. However, it also usually contains some proportion of hydrogen sulfide, in particular, and minor amounts of such other species as methane and ammonia. The gas is commonly discharged to the atmosphere either directly or upon release from the condensate used in wet cooling towers. Hydrogen sulfide has a very offensive odor and, if permitted to collect in low spots or confined area, can be dangerous to human and animal life.

Trace amounts of boron, arsenic and other volatile compounds are commonly found in steam condensates which, if discharged into surface waters or permitted to escape as wind-blown droplets, can be damaging to plant life and may enter the food chain. If, as has been common, the condensate is discharged at an elevated temperature, then it also is a source of thermal pollution.

In the production of wet steam from liquid-dominated geothermal reservoirs, usually about 20% of the mass-flow from the well is steam, which carries with it normal amounts of the pollutants discussed above. The other approximately 80% is hot water which may contain higher concentrations of trace elements than does the steam, and typically has much higher contents of various dissolved minerals—especially carbonates and silica. If discharged into surface waters, this geothermal water can be damaging to fish, animal, and plant life, and because of its larger volume it is a much more serious source of both chemical and thermal pollution than is the steam condensate.

In two-fluid systems, the primary fluid produced by the well will in general be passed through a heat exchanger and then reinjected into the underground reservoir through a second well. Since this circulation can be completely contained, such a system can in principle be completely nonpolluting. There will, however, be dissolved gases and minerals which will come out of solution as the pressure and temperature of the fluid are reduced, and these will present a disposal problem that may not always be solved by reinjection. Nevertheless, pollution problems should be mauch easier to control in hot-water and dry-rock systems than in natural steam systems.

Natural steam systems also have an acoustic-pollution problem which is difficult to control. Superheated water flashing to steam produces an intense noise which can be muffled only by causing the flashing to occur in an acoustically designed chamber. Because shutting in a well or changing the production rate from it suddenly or drastically is very likely to damage the well, it is common to vent either wet or dry steam to the atmosphere continously over quite long periods, during testing of the well and its associated equipment, shut-downs elsewhere in the system, and the drilling of enough adjacent wells to supply a new power plant. This too produces an ear-splitting roar.

Noise-suppressers capable of reducing the sound level to an acceptable value are available, but those which are effective are also expensive and are likely to wear out rapidly from abrasion by dirt, rocks, water drops, and precipitated minerals carried by the steam. Acoustic pollution is a continuing problem in natural steam fields.

Because geothermal energy is normally produced at relatively low temperature, its use in heat machines is relatively inefficient. Therefore, a geothermal power plant has a major waste-heat problem, which of course it shares with all other low-temperature power systems. The typically small size of a geothermal plant permits thermal pollution to be distributed, which mitigates the problem to some degree but of course does not solve it.

Geothermal energy, then, presents a variety of pollution problems, some of which it shares with other prower systems. In general the problems of geothermal systems appear to be small compared with those of fuel-fired and nuclear plants, and to have straightforward engineering solutions. From an environmental standpoint, geothermal energy systems are probably the least objectionable of any power systems in existence or so far proposed.

It is also worth noting that geothermal energy is usually developed in regions where there already is abundant surface evidence of the presence of an underground thermal reservoir—in the form of fumaroles, geysers, mineral hot springs, and the odors, noises, mineral deposits, and stream contamination associated with them. Aside from its visual impact, controlled development of a well-engineered geothermal energy system is in fact very likely to reduce the level of local pollution to below its original, natural value.

Problems

The problems of locating and evaluating wet natural steam fields are similar to those of prospecting for oil and natural gas, and are being attacked with similar geological and geophysical methods supplemented by geochemical, hydrologic, and heat-flow studies. However natural steam and hot-water reservoirs are even less well understood than are oil and gas reservoirs, and the correlation of geological structure and of any ground or aerial observation with the presence of underground steam or hot water remains very uncertain. Drilling is still the final and conclusive prospecting method and is still very expensive.

The initial prospect drilling is commonly to depths of only a few hundred feet, to examine the local geothermal gradient, hydrology, and lithology. The sites for this drilling are usually selected on the basis of surface evidence of hydrothermal activity, supplemented by geochemical studies of natural waters and condensates, deep electrical resistivity measurements, microseismic observations, and other geophysical techniques. Exploration methods for geothermal energy are developing rapidly, but further development is needed.

Drilling is still essential in prospecting for geothermal energy, evaluating discoveries, and developing and producing the field. Except that high temperature and hot fluids may be encountered, prospect drilling is now relatively routine, and would benefit principally from development of equipment and techniques that would reduce the costs of small-diameter exploration holes and permit them to be used without casings.

Drilling of production wells has special problems associated with generally incompetent formations, high temperatures and pressures, and large flows of corrosive fluids.

Production wells are now drilled more or less routinely in existing steam and hot-water fields, but the drilling is difficult, expensive, and dangerous, and more often than not the final hole depth is decided by an equipment failure rather than by a development plan. New drilling techniques and improved equipment could contribute greatly to geothermal exploration, development, and production.

Down-hole geologic, geophysical, and hydrologic investigations are an essential part of a sound exploration and development program. Much of the existing hole-logging equipment was developed for oil-field use, and is not designed for the temperatures encountered in geothermal wells. This severely limits the down-hole observations that can be made and the reservoir information that can be collected. Improved high-temperature logging equipment and techniques are now being developed, and it is important that this development continue.

Since wet geothermal reservoirs are not yet well understood, the analyses and models that are the bases of sound reservoir engineering are not yet fully developed. Well spacings and arrangements, drilling sequences, production rates, and reinjection procedures are still guided more by intuition and local experience than by real understanding, and are probably not optimum. Good reservoir models, and methods of producing them, are still lacking in the geothermal industry, and must be developed.

Geochemical problems and the pollutants involved in them have been considered above. An important trend in the industry is to dispose of steam condensates and excess water, together with as much as possible of their dissolved gases and minerals, by reinjecting them into the producing formation through unproductive wells or holes drilled specifically for that purpose. This returns most of the chemical pollutants to their source, avoids thermal pollution of surface waters, replenishes the fluid supply in the reservoir, and reduces the danger of surface subsidence caused by fluid withdrawal from the reservoir. However it is expensive, and involves the risk of plugging the formation in the vicinity of the injection hole with particles or precipitates from the injected water.

Corrosion, scaling and plugging of drilling equipment, borehole casing, and surface plumbing are often serious problems with the hot, mineral-laden solutions produced from geothermal reservoirs. Until chemical methods of controlling these problems are developed, they will be handled principally by using expensive corrosion-resistant materials for surface-plant construction and by shutting down occasionally to rework wells and to clean out or replace surface plumbing.

Man-made, dry-rock geothermal-energy systems, as they are developed, will involve several of the problems of wet systems plus some peculiar ones of their own. Prospecting for them should be relatively simple, since only high temperature and a desirable lithology are required. Exploratory, development, and production drilling should also be generally less difficult than in wet geothermal reservoirs, since only high temperature—which can be controlled by circulation of adequate coolant—will normally be encountered, and not strong flows of hot, high-pressure, corrosive water or steam.

The major problems will be the engineering ones associated with extraction of energy from hot dry rock in a useful form, at a usefully high rate, and at a cost competitive with other energy systems. The principal basis for these problems is the fact that rocks in general are very poor conductors of heat.

Unless heat is carried into a drill hole by the mass flow of water or steam, it must be brought in by conduction through the rock, and this is very slow. The common idea of simply drilling a hole and lowering a heat-exchanger into it to extract energy from the earth is impractical except in the rare places where there is a strong horizontal flow of hot water through the formation containing the hole. In general, when heat enters the hole only by conduction through the rock, it does so at a rate so low that—even if it could be brought out of the hole at perfect efficiency and zero cost—its value would not be sufficient to amortize the cost of drilling the well. Drilling a larger hole would not materially change this situation, since drilling cost increases with hole diameter at about the same rate as does the surface area of the hole produced.

To extract energy economically and at a usefully high rate from hot, dry rock evidently requires that a large heat-transfer area be created in the rock by some means less expensive than drilling a large number of holes, and that a large volume of fluid be circulated over this new surface to extract energy from it. The fluid with its contained heat must then be recovered and brought to the surface, where in general the thermal energy will be extracted in a heat-exchanger, and the cool fluid returned to the underground circulation system to extract more heat. There have been several suggestions for developing energy systems of this general type.

In arid regions or where porous or fractured formations are overlain by impermeable strata, there is undoubtedly hot rock which does not produce hot water or steam simply because no water reaches it. Here it may not be necessary to produce new surface, but only to introduce water, permit it to percolate through the formation, and then recover it as superheated water or steam. A water-flooding technique, such as that used in secondary recovery of petroleum, should permit very large amounts of heat to be recovered from such a formation. This would involve drilling a systematic array of holes, injecting cool water into some of them, and recovering hot water or steam from the others.

Since hot water is less viscous than cold water, the cool water should sweep essentially all of the heated water ahead of it toward the recovery holes, and energy extraction should be very efficient. Such a system would involve a high drilling cost and a long development period, but in suitable geology could produce very large amounts of energy for a very long time.

More generally, in hot dry rock, it will be necessary to produce new heat-transfer surface as well as to introduce, circulate, and recover a heat-transfer fluid. A method frequently suggested for accomplishing this is to detonate high explosive in the borehole in order to fragment the rock around it, and then to introduce cool water either through the borehole or through an insulated pipe down its center. The water would circulate convectively through the rubble, and would be recovered as either hot water or steam through a second hole or through the annulus around the insulated pipe. Unfortunately, to produce a usefully large volume of broken rock would require an intolerably and uneconomically large amount of explosive. Accordingly, the use of nuclear rather than conventional explosives has frequently been suggested.

An excellent Plowshare study (PNE-1550 Chapter 4) concludes that the use of nuclear explosives for this purpose would become economical only if a rather large number of very high-yield nuclear devices were detonated sequentially in a systematic, three-dimensional array. In addition to the difficulty of permanently containing the radioactive debris from these explosions, the ground-shock produced would be so intense that it could not be tolerated in most parts of the world. However, there may be solutions to these problems, and the Plowshare geothermal concept has not been abandoned.

An alternative method now being investigated is the use of hydraulic fracturing to produce the connected permeability and heat-transfer surface required to extract energy at a high rate from hot, dry rock (See Chapter 3). This is a standard technique of well-completion in oil and gas fields, used to increase the permeability of the producing formations around the well. It involves the use of a high-pressure pump at the surface to develop fluid pressure in the borehole sufficient to crack the rock and then to open and extend the crack. As is illustrated schematically in Figure 1.3, the proposed method would involve drilling a hole to the depth required to penetrate usefully hot rock, and then producing a very large, pancake-shaped hydraulic fracture extending outward from this hole. The underground circulation system would be completed either by inserting an insulated pipe in the original hole or, as shown in the sketch, by drilling a second hole that intersects the crack system.

Cool water injected through the deeper hole would be heated in its passage through the fracture and, as a result of its reduced density, would rise buoyantly through the second hole (or through the annulus around the insulated pipe)—so that circulation could probably be maintained continuously without pumping. The possibility exists that cooling of the rock near the bottom of the injection hole will develop tensile stresses sufficient to cause new cracks to form, thus extending the circulation system outward and downward in three dimensions. If this should occur, new surface may be developed in the hot rock at a rate which is sufficient to compensate for cooling of the original crack surface, in which case the energy-extraction system will be self-perpetuating.

There are obvious problems associated with the creation and operation of such a system. Very large hydraulic fracturing has not been demonstrated in hot crystalline rock, and represents a significant extrapolation of the state of the art. Creating and maintaining interconnected circulation channels of sufficiently low impedance to permit bouyant circulation of a large volume of fluid may be difficult, and the possibility exists of short-circuiting of cool water from the injection hole directly to the recovery well. If the formation containing the fractures has an appreciable permeability, the rate of fluid loss from the system may be too high to permit operation of the pressurized-water loop described above.

Presumably this could be avoided by permitting the water to flash to steam at any desired level in the system. This, however, introduces the danger of plugging the crack system with minerals dissolved at some lower level, and would also greatly reduce the rate at which heat was carried upward through the recovery well. Alternatively, a down-hole pump might be used in the recovery well, to maintain pressure in the upper part of the hole sufficient to prevent boiling while the rest of the system was kept at a reduced pressure. Unfortunately, the high-temperature down-hole pumps required by such a system have not yet been developed.

FIGURE 1.3: A DRY-ROCK GEOTHERMAL ENERGY SYSTEM DEVELOPED BY HYDRAULIC FRACTURING

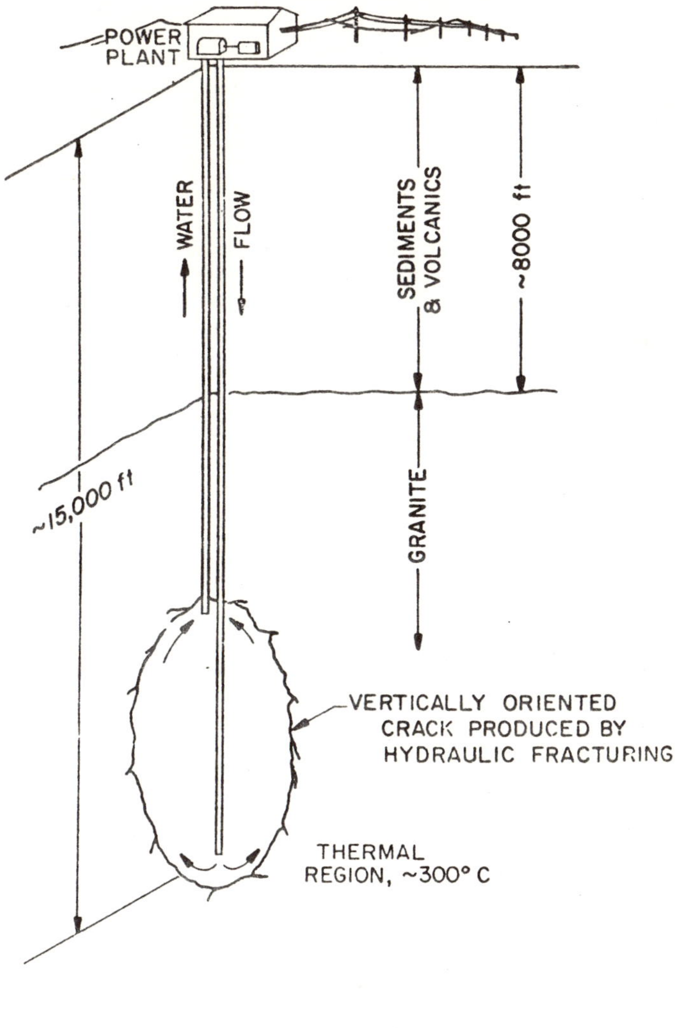

Source: LA-5289-MS

Other circulation systems and operating modes are evidently possible to inject cool fluid into dry geothermal reservoirs and recover it after it has been sufficiently heated by the rock. In general such systems will be completely contained, and therefore essentially nonpolluting. To some degree, however, they will all be faced with the same chemical and corrosion problems as will the natural hot-water and steam systems discussed above.

Perhaps the greatest problems facing the large-scale development of geothermal energy in the United States are not the scientific and engineering ones so far considered, which are straightforward, but rather are traditional, organizational, financial, and legal problems, which are very complex. For example, where surface, mineral, and water rights exist separately, it is not clear who owns the geothermal energy or how this ownership can be established or transferred. A large fraction of the known geothermal resource of the United States is on public land, where prospecting in private is virtually impossible and for which a geothermal leasing system is still being developed.

Local, state, and federal laws, tax structures, and licensing and regulatory systems are incomplete, overlapping, occasionally contradictory, and generally uncorrelated and incompletely interpreted. Legal requirements on public utilities require that a reliable, long-term fuel supply be certified before a new power plant is built. In the case of geothermal energy, this now requires that enough production wells to support the plant be drilled and demonstrated by venting to the atmosphere, for months or years—before construction of the power plant can begin. This long delay in securing a financial return makes the financing of a geothermal development very difficult. There are other institutional problems facing the geothermal industry of the United States, and in sum they are formidable.

COSTS

The major cost of exploring for, developing, and producing a geothermal energy supply is that of drilling and casing large, deep holes in hot formations, which is very expensive. However, to compensate for this, the surface plant required to use the thermal energy taken from the earth is in general relatively simple and inexpensive. An electrical generating plant powered by geothermal heat does not require fuel-handling space or equipment, boilers, smokestacks, or a smoke-abatement system, so that the plant cost per kilowatt of installed generating capacity is low. If the system is to be used simply as a heat-source, for example for desalination, the surface plant is principally plumbing and heat-exchangers and the plant cost per thermal kilowatt again is low.

In either case the subsequent "fuel" cost is the capital charge against the wells and collection system, which is generally low relative to the costs of fossil fuels or of producing and reprocessing fuel elements. Operating and maintenance costs again are typically low because of plant simplicity and relatively low operating temperatures and pressures. In general, then, for conventional geothermal energy systems both plant and generating or heat costs are low.

In the United States, the only firm cost experience with geothermal power systems is at The Geysers, in California. Based principally on that experience, Bowen and Groh have produced the cost estimates listed in Table 1.2 for developing, building, and operating a 110 Mw natural-steam power plant. An exploration cost of $2 million is assumed to find the natural-steam field. In Table 1.3 these costs are compared with those similarly estimated for other power systems. (R.G.

Bowen and E.A. Groh, "Geothermal—Earth's Primordial Energy," *Technology Review,* Oct/Nov 1971, pp 42-48). Both this comparison and the commercial success of The Geysers operation indicate that a geothermal power plant using natural steam is at least competitive with plants of other types. This is confirmed by the long and successful history of the natural-steam system at Larderello, Italy.

TABLE 1.2: COSTS ASSOCIATED WITH A 110 Mw NATURAL-STEAM POWER PLANT

Steam winning:	
Wells necessary to produce steam for 110,000-kw. plant (16 wells at $150,000)	$2,400,000
Steam transmission lines at $10/kw.	1,100,000
Total	3,500,000
Overhead charges:	
Annual fixed charges at 14 %	$490,000
Royalty payment to landowners	175,600
Total annual	665,600
Heat cost:	
Steam winning cost in mills/kwh.	0.80
Exploration, mills/kwh.	0.33
Disposal charge, mills/kwh.	0.02
Total energy cost, mills/kwh.	1.15
Conversion cost :	
Fixed charges, mills/kwh.	1.95
Operating cost, mills/kwh.	0.25
Total power cost, mills/kwh.	3.35

Source: LA-5289-MS

No comparable cost information is available for wet-steam systems. Undoubtedly both plant and generating costs are higher than for dry-steam systems because more production wells are required per unit of generating capacity, plant equipment is larger, chemical problems are more serious, and a very large volume of hot water—produced with the steam—must be separated, handled, and disposed of by reinjection or some other means. However the cost increment is generally estimated to be small, and wet-steam power plants in New Zealand and elsewhere have been very successful.

American industry will be watching with particular interest the operating experience of a wet-steam plant just being completed at Cerro Prieto in northwestern

Mexico, where the superheated geothermal water is a relatively concentrated brine very similar to that in some parts of the Imperial Valley of California

TABLE 1.3: PLANT AND GENERATING COSTS FOR SEVERAL TYPES OF POWER PLANTS IN THE UNITED STATES IN 1970

	Electric Power Costs			
	Geothermal	Nuclear	Hydropower	Coal
Plant investment, $/kw.	$110.00	$225.00	$250.00	$150.00
Fixed charges, 14%/year/kw.	15.40	31.50	35.00	21.00
Fixed charges, mills/kwh.	1.95	4.00	6.10	4.36
Operating costs, mills/kwh.	0.25	0.50	0.10	0.25
Energy costs, mills/kwh.	2.66	2.00	----	3.00
Total costs:				
Variable load factor, mills/kwh.	4.86	6.50	6.20	7.61
90% load factor, mills/kwh.	4.86	6.50	4.55	5.92

Source: LA-5289-MS

Very little information is available concerning the USSR plant on Kamchatka, and there is no US operating or cost experience with contained hot-water geothermal systems, use of relatively low-temperature geothermal water for generating electricity, or two-fluid power plants in this type of service. However a 19 Mw isobutane-turbine demonstration generating plant is now being built in the Imperial Valley. If it is successful, it will undoubtedly lead to a rapid development of low-temperature geothermal power systems and of contained two-fluid systems in general.

Cost projections for dry-rock systems are of course still more uncertain than those for natural hot-water systems, since none has ever been built or operated. However, reliable estimates are available for the costs of drilling, fracturing, and constructing the surface plant. Mineral and gas contents of the circulating fluid and the accompanying scaling, plugging, corrosion, and pollution problems should be less than with most superheated natural brines. Accordingly, if such a system can be made to operate at all, its plant and operating costs should be no greater than those of natural wet-steam and hot-water geothermal systems, which are expected to be at least competitive with fuel-fired and nuclear power plants.

ENVIRONMENTAL IMPACT

No power system will have zero impact on its environment. However, of those so far in existence or proposed, geothermal systems appear to be the least objectionable in this respect.

With regard to land usage and effect on the landscape, a geothermal power plant will have the disadvantages of any power plant, mitigated by a typically small plant size and the absence of fuel-storage and handling yards, combustion equipment, and smokestacks. A steam- or water-collection system will be required and, like that for a municipal water system or an oil or gas field, it can be either unobtrusive or very unsightly according to local environmental regulations and the conscience of the operator.

Noise and odors from geothermal systems can be dealth with, although some technological development will be required before this can be done effectively in natural steam fields at a cost that the industry can afford. Here completely contained hot-water and dry-rock systems will have an advantage.

Concentrations of undesirable minerals in steam condensates and in unflashed water from wet-steam fields have traditionally been disposed of by draining the mineralized water into the nearest body of surface water. This has usually resulted in a high degree of both chemical and thermal pollution, and in most developed countries it is no longer acceptable. The emerging solution is to reinject this water into the producing formation through unproductive wells or injection wells drilled especially for that purpose. By replacing part of the fluid removed from the reservoir, this should also reduce the danger of significant surface subsidence.

Subsidence is a common problem wherever a large amount of any material, including water or steam, is removed from beneath the earth's surface. Dry-rock systems, unless developed by water-flooding of a large area, are expected to affect only a deeply buried region of limited lateral extent, and are unlikely to produce a detectable subsidence of the ground surface above them. Wet systems, however, in general involve larger areas, shallower formations, porous rock, and withdrawal of large volumes of fluid, so that for them the probability of subsidence is high. Fortunately, oil-field experience indicates that reinjection of a volume of fluid sufficient to maintain the reservoir pressure can prevent significant subsidence. This may not always be possible in geothermal fields, but it will certainly be attempted.

Microearthquake activity seems commonly to be concentrated in potentially productive geothermal regions and may, in fact, be a useful guide to their locations. Earthquakes intense enough to cause even local damage at the surface seem if anything to be less common in geothermal reservoirs than in geologically similar regions elsewhere, perhaps because the small energy releases associated with microearthquakes prevent accumulation of the large amount of stored energy required to produce a major shock.

The only real danger of triggering earthquakes appears to be that associated with reinjection of fluids into natural geothermal reservoirs and of fluid pressurization in the creation and operation of man-made systems. Studies by the US Geological Survey and others of the Denver earthquake experience and of controlled earthquake experiments at Rangely, Colorado, suggest that significant earthquake activity can be induced only if very large volumes of fluid are injected under relatively high pressures into active fault systems. In general this combination of conditions can be avoided, and the possibility of initiating a damaging earthquake by developing or operating a geothermal-energy system appears to be very remote.

ADVANTAGES AND DISADVANTAGES

Geothermal energy, the natural heat of the earth's interior, is the largest energy reservoir that is directly accessible to man. Where nature delivers energy from this reservoir to the earth's surface in the form of steam or superheated water capable of producing steam, its direct conversion to electricity has proven to be economical and to produce minimal environmental damage. Active prospecting for new geothermal reservoirs of this type is in progress in many places, and new ones will undoubtedly be found—although probably never in large numbers. Where they are found, industry will be anxious to develop and produce them, and the technology for doing so exists—although incremental improvements in it are still badly needed.

Natural hot-water reservoirs at lower temperatures are both larger and more common than dry- or wet-steam systems, and the attention of the geothermal community is now being refocussed on them. Wherever a hot spring exists or hot water flows into a drilled hole, the possibility of extracting useful geothermal energy is demonstrated to exist. To this point, such low-temperature geothermal energy is being used almost exclusively for space-heating, and in the United States even this usage is small. Nevertheless this type of use is important and should be encouraged, particularly because it saves higher-grade heat sources for higher-temperature uses, and can significantly reduce the balance-of-payments and national-security problems associated with importation of fossil fuels from abroad.

Typically, hot-water systems are both lower in temperature and higher in dissolved mineral content than are natural steam systems, which presents problems with regard to their use for generating electricity. However, the engineering and geochemical developments required to solve these problems are easily defined and straightforward, and are now being pursued vigorously. Growth of this type of system into a major producer of electrical energy can be predicted with confidence, and the time scale for it can be short.

The individual units of wet geothermal energy systems are naturally small, which has both advantages and disadvantages. Because steam cannot economically be transported over long distances, a geothermal steam plant is necessarily located at the steam field—which is likely to be remote from major markets for heat and electricity. Hot water at moderate pressure can be piped efficiently over much longer distances, up to at least tens of miles, but to a lesser degree this same limitation on plant location also applies to hot-water systems.

The possibility of extracting energy from dry geothermal reservoirs is therefore attractive for more than one reason. Dry hot rock represents not only our largest but also our most broadly distributed energy supply. If economical methods can be developed to extract energy from it, it may become possible to create a geothermal-energy system almost anywhere that heat or electricity is needed, which would have the incidental advantage of greatly reducing the need for new transmission lines, pipelines, and other transportation systems.

Investigations of the possibilities, problems, and economics of such systems are just beginning, but the results of initial studies make it appear probable that dry-rock geothermal systems will eventually contribute significantly to the long-term solution of the world's energy and pollution problems.

RECENT EXPLORATION TECHNIQUES

A summary of recent geothermal exploration techniques is given in PB 218,830 *Exploration and Exploitation of Geothermal Resources in Arid and Semiarid Lands* prepared by the staff of the Office of Arid Land Studies, University of Arizona, 1973 from which this section is derived.

Until recently, geothermal fields had been discovered only by drilling in areas of thermal manifestations such as geysers, fumaroles, and hot springs. Unfortunately, surface phenomena do not readily depict subsurface conditions (McNitt, J.R. 1965, *Exploration and Development of Geothermal Power in California* California, Division of Mines and Geology, Special Report 75). Many regions in which there are thermal manifestations are not amenable to geothermal development. Other regions lacking such thermal manifestations have been found to have subsurface conditions suitable for the development and exploitation of geothermal resources. Thus it has become necessary to utilize more sophisticated exploration techniques than simply surface expression.

The initial stage of any exploration program is a reconnaissance of the field geology and geologic mapping. Structural geology is of primary importance in determining geothermal anomalies. Steeply-dipping, normal, or strike-slip faults are common in thermal areas and probably aid in providing deep circulation of meteoric waters (Summers, W.K. and Ross, S.H. 1971, *Geothermics in North America: Present and Future,* Earth Science Bulletin 4(1):7-22). Structural troughs, such as the Salton Sea area of California, or the Rio Grande trough in New Mexico, are favorable large-scale structures for the production of geothermal energy.

Photogeologic mapping, which facilitates rapid examination of a large area, is used to augment field mapping. Structural and lithologic information can be obtained readily from air photos, including the delineation of surface thermal phenomena (Miller, L.D. 1968, "Steaming and Warm Ground in Yellowstone National Park: Their Location, Geophysics, Vegetation and Mapping with Aerial Multispectral Imagery." University of Michigan (Ph.D. Dissertation). Moxham, R.M. 1969, *Aerial Infrared Surveys at The Geysers Geothermal Steam Field,* US Geological Survey, Professional Paper 650-C:106-122. Moxham, R.M. 1970, "Thermal Features at Volcanoes in the Cascade Range, as Observed by Aerial Infrared Studies," *Bulletin Volcanologique* 34:77-106.)

Closely associated with photogeologic mapping is infrared aerial reconnaissance (Moxham, 1969, 1970). Because infrared film is sensitive to heat, it is capable of detecting surface and near-surface areas of anomalous heat flow (Hase, H. 1971, "Surface Heat-Flow Studies for Remote Sensing of Geothermal Resources." *Japanese Society of Photogrammetry, Journal* 10(3):9-17).

Recently, vegetative changes have been found to be useful criteria for locating geothermal phenomena (United Nations 1964, *United Nations Conference on*

New Sources of Energy: Solar Energy, Wind Power and Geothermal Energy, Rome, 21-31 August 1961, Vol. 2-3: *Geothermal Energy,* Vol. 1-2). Certain types of vegetation are modified in a characteristic way by variations in temperatures. A sudden change in vegetation may denote a change in the groundwater regime.

After the surface reconnaissance, the next step in an exploration program is a geochemical analysis of the groundwater, including sampling and testing of the steam emanations, as well as the condensed steam from the hot waters. Such a geochemical analysis should provide three important facts: temperature, age, and chloride content of the water. First, the maximum temperature the water has reached provides information concerning the potential of the location to produce geothermal energy. Second, the period of time the water has been in the system: "If the water is relatively old it may indicate that the reservoir is recharged slowly and therefore would be subject to depletion if developed unwisely" (Summers and Ross, 1971).

Finally, White, Muffler, and Truesdell have proposed a method for distinguishing between dry steam areas and wet steam areas. Dry steam areas are characterized by acid-sulfate springs, with insignificant chloride content, but springs from wet steam fields have a high chloride content [White, D.E., Muffler, L.J.P. and Truesdell, A.H. 1971, "Vapor-Dominated Hydrothermal Systems Compared with Hot-Water Systems." *Economic Geology* 66(1):75-97].

A new exploration approach presently being developed is the application of fluid dynamics to natural steam systems (Summers and Ross, 1971). This method describes quantitatively the movement of groundwater on a regional scale. A study of fluid dynamics also facilitates the determination of the life span of a particular geothermal field after it is developed. Other regional studies, such as temperature and heat flow analysis permit localization of favorable zones for the presence of super-heated fluids.

Electrical prospecting methods facilitate relatively rapid and complete reconnaissance programs. The resistivity of a formation is partially dependent on its temperature; an increase in groundwater temperature would appear as a decrease in resistivity. Variations in resistivity also denote changes in groundwater salinity or changes in geologic formation.

In New Zealand seismic methods have been incorporated into geothermal explorations. When groundwater comes in contact with a heat source, a seismic noise pattern is produced that can be used to determine the extent of a favorable geothermal aquifer (Summers and Ross, 1971). Refraction and reflection seismic methods have also been used to investigate the subsurface geologic structure of a region.

Gravity and magnetic surveys are two additional techniques used in the exploration for geothermal resources (Crosby, J.W., 1971, "Geothermal Exploration," *Northwest Conference on Geothermal Power, 1st,* Olympia, Washington. Griscom, A. and Muffler, L.J.P., 1971, *Aeromagnetic Map and Interpretation of the Salton Sea Geothermal Area, California,* US Geological Survey, Geophysical Investigations Map GP 754). Because hydrogen sulfide near heat sources causes magnetite to change to nonmagnetic pyrite, magnetic lows are often associated with hydrothermal activity. Gravity surveys can be utilized to study deep subsurface structure and help delineate feeder channels or reservoir characteristics.

The final and most important phase of the exploration is the actual drilling. No conclusion as to the productive capacity of a region can be formulated without drilling. Information derived from the drilling program should include the porosity and permeability of the rock strata and the nature of the permeating fluid.

In exploring for geothermal resources, it should always be kept in mind that different geologic settings require different exploration techniques. Because the expense involved in drilling can be prohibitive, a very diverse and intensive yet low-cost exploration program should be carried out prior to actual drilling.

RESEARCH AND DEVELOPMENT IN THE UNITED STATES

Research and development programs in geothermal resources have been carried out in the United States by several Federal agencies, state agencies in most of the western states, and private industry. The following descriptions of these programs and the ultimate potential of geothermal energy resources in the United States are from PB 224,882 *Assessment of Geothermal Energy Resources* D.L. Peck, Coordinator, a report prepared by the Panel on Geothermal Energy Resources for the Committee on Energy Research and Development Goals, Federal Council of Science and Technology, June 1972.

The summaries of the Federal agency programs were prepared from material supplied by the agencies. As of June 1972, fiscal year 1973 programs had been submitted to Congress by the agencies but funds had not yet been appropriated and allotted. Fiscal year 1974 programs for all agencies were still in the planning stage.

US Geological Survey

Prior to 1971, the US Geological Survey had no specially funded geothermal resources program, although modest investigations of hot springs and geothermal phenomena were carried out as part of the Geologic Division investigations of energy and mineral resources. Modern Geological Survey hot-spring studies were begun by Donald E. White in 1945 at Steamboat Springs; the initial emphasis was on mechanisms of ore transportation and deposition, but the results of a decade of study proved to have important application to the understanding of geothermal resources.

Beginning in 1963, the focus of Geological Survey geothermal research shifted to the Salton Sea geothermal field, in particular, to cooperative studies with private industry focused on the geochemistry and petrology of the hydrothermal system. In 1966, the Geological Survey began a major investigation of all aspects of the geology of Yellowstone National Park, including intensive studies of the hot-spring areas. Studies included geologic mapping, geophysical surveys, sampling and analysis of hot springs, and the drilling of 13 research holes to depths as great as 1,100 feet.

Several other hot-spring systems in the western United States have been investigated by the Geological Survey, including Sulphur Bank or Clear Lake, California, and the Lake City area. A very useful source of information for geothermal exploration is the Geological Survey compilation of temperature and other data on thermal springs (Waring, G.A., 1965, *Thermal springs of the United States and other countries of the world - a summary:* US Geol. Survey Prof. Paper 492). In addition, other studies by the Geological Survey have proved important to geothermal resources, including heat-flow studies, determination of silica solubility, development of the silica, Na/K, and Na-K-Ca geothermometers, and experimental studies of rock alteration.

In August of 1971 the 92nd Congress appropriated an increase of $500,000 to the Geological Survey for an expanded geothermal program. As a result:

(1) The staff for classifying Federal lands for geothermal resources was increased and preparations were made for lease management.

(2) Projects were started at The Geysers/Clear Lake and Long Valley, California, using detailed geologic, geophysical, geochemical, and hydrologic techniques.

(3) Reconnaissance geologic, geochemical, and hydrologic surveys were started in northwestern Nevada and southeastern Oregon.

(4) A program was started for developing and testing new geophysical techniques (electrical, electromagnetic, passive seismic, and thermal infrared); and

(5) Laboratory studies were started on hydrologic modeling, solution chemistry, and geochemical indicators of reservoir temperatures.

A $1,800,000 increase in the Geological Survey geothermal program was proposed to Congress for FY 1973 to support an augmented program including a number of new research projects, supervision of operations on geothermal leases, and the classification of additional geothermal resource areas. New hydrologic studies will be undertaken on the Island of Hawaii and in the Jemez Mountains, New Mexico. Geophysical studies of these and other areas will be broadened to include microseismic and telluric methods as well as heat flow at shallow to moderate depths. The physics, temperatures, and hydrology of several geothermal areas will be modeled using computer technology. Subsurface temperatures and hydrologic parameters in the Gulf Coast area will be studied.

Reconnaissance surveys will be started in Alaska and expanded in the western states, and chemical analyses of water and rock samples will be supplemented with isotopic analyses. Environmental effects of geothermal developments, such as ground subsidence and earthquake activity—will be monitored with initial emphasis on the Imperial Valley, California.

The Geological Survey has planned an expanded 5-year program starting in FY 1974 to be carried out if funds are made available. The program would include (1) classification and lease supervision, (2) environmental monitoring and evaluation, (3) pre-lease evaluation of geothermal areas and development of better exploration techniques through detailed field studies including research drilling, (4) regional reconnaissance studies, and (5) brine disposal studies.

Bureau of Reclamation

The Colorado River Basin Act (Public Law 90-537) requires the Secretary of the Interior to make reconnaissance studies for the purpose of developing a general plan to meet future water needs of the Western United States and to determine the most economical method of augmenting the flow of the Colorado River by 2,500,000 acre-feet per year. This work has been delegated to the Bureau of Reclamation. A promising source to meet both of the above requirements is the geothermal resources in the Imperial Valley of California, by desalting the hot geothermal brines underlying the Reclamation withdrawn lands. Reclamation and the University of California, Riverside, began a joint exploratory program to evaluate the goethermal potential of these lands in 1968.

The Bureau of Reclamation and the Office of Saline Water have recently undertaken a joint 7-year research and development program with an estimated cost of $16 million ($9.7 million for the Bureau of Reclamation) to determine the feasibility of multipurpose development of electric power and desalted water from the geothermal resources of Imperial Valley. This program was begun in FY 1972 with an appropriation of $800,000.

An increase of $400,000 in the Reclamation geothermal program was proposed to Congress for FY 1973 to accelerate the program, which is designed to determine the quantity and quality of hot geothermal brines underlying Reclamation withdrawn lands, to develop the technology for total development of the geothermal resources, to demonstrate the feasibility of developing large quantities of desalted water, and to determine on the best plan for delivery of desalted water to the Colorado River. This effort is and will continue to be coordinated with other investigations aimed at determining the potential of the resource for power generation and recovery of chemicals and gases.

An additional $5.7 million effort is needed by the Bureau to explore in areas of critical water shortages, other promising geothermal areas having a potential for producing low-cost water supplies in the western states. This exploration is necessary to assure that all potential alternatives for water supplies have been evaluated. In the early stages, this program would concentrate on the exploration and the evaluation of the resource for water production rather than on research and development similar to that presently being carried out in the Imperial Valley. To meet the requirements of Public Law 90-537, this evaluation must be completed no later than 1977. The components of the Bureau of Reclamation geothermal program are listed below.

Exploration: The exploration program is designed to secure a reasonable appraisal of the quantity and quality of geothermal resources on Reclamation withdraw lands in the Imperial Valley in the shortest practical time. Methods employed will be geologic and geophysical investigations combined with relatively shallow (1,500 feet or less) temperature gradient drill holes and a limited number of production wells to production depths of 5,000 to 7,000 feet. Verification of water quality will be necessary to assure that desalting and electric power generation techniques will be applicable to the entire production area.

A reasonably reliable estimate of the quantity of the geothermal resources on Federal lands in the Imperial Valley (almost all of which are Reclamation withdrawn lands) will be needed at an early date for a decision on the source of replacement water to prevent subsidence due to fluid withdrawal. If the resource proves to be large, the most economical source is the ocean. If the resource is moderate or small, local replacement water from such sources as near surface ground water, the Salton Sea, and the Wellton-Mohawk Drain would be used.

If the total quantity of the resource in the Imperial Valley is of sufficient size to permit large-scale importation of replacement water from the ocean, then the entire resource could be developed as a joint venture by Federal and non-Federal interests to develop water and power at the lowest cost.

Production and Injection Wells: Sufficient production wells must be drilled in the research and development stage to verify the results of the exploration program. The production wells must also be operated in conjunction with injection wells for a sufficient period of time to determine the best well spacing and well diameter. Environmental effects such as seismic activity, subsidence, and air pollution will be monitored in conjunction with the operation.

Research and Development: Research and development activities will be required in various fields in order to support the desalting efforts. The best size and spacing of both production and injection wells will be analyzed. Injection techniques including disposal of byproducts will be studied. The integration of electric power production and desalting will be studied with the goal of optimizing benefits from both these products and possible mineral byproduct recovery.

Subsidence, seismic activity, and other environmental impacts that could result from geothermal resource development will be investigated and procedures developed to modify or eliminate adverse effects. The answer to these problems should be applicable in whole or in part to the nationwide development of geothermal resources.

Appraisal: The results of the research and development program must be carefully analyzed and appraised to provide answers to environmental problems and the character and magnitude of the resource and the economics of desalting appraised to assure that the total potential is realized in the national interest.

Office of Saline Water

The Office of Saline Water started a $6,300,000 7-year geothermal program in FY 1972 aimed at developing the technology necessary to the desalting of geothermal brines. The program objective is to develop economical and technically feasible processes for the production of high quality water. In addition, the desalting processes must be made compatible with other processes based on the geothermal resource, such as power production and mineral producing processes, to permit an optimal and comprehensive geothermal development.

The studies are centered on the geothermal brines in the Imperial Valley, Calif., and are carried out in close collaboration with the Bureau of Reclamation. A large part of the technology developed there should be applicable to the desalting of other geothermal brines.

In FY 1973, the program will be carried out at a level of $800,000, subject to congressional approval of appropriation, and will include research on brine chemistry, materials of construction, process analysis and testing, environmental effects of the desalting process, and effluent disposal techniques. Efforts in brine chemistry will include research on the composition of geothermal brine and its reactions within the desalting plants and the effluent disposal system, determination of the extent of scaling of process equipment and research on means to reduce the effects of scaling to a minimum, and research on the chemical nature of the deep-well disposal system and related chemical reactions.

Attention will be directed towards finding economical and reliable materials of construction for the potentially corrosive geothermal fluid—a hot, acidic brine containing sulfide and carbonic gases. Process parameters, including heat-transfer coefficients and flashing characteristics, will be determined, as well as the effective design of equipment components. Methods will be developed to permit the inclusion of the desalting plant into an integrated power-water-mineral geothermal utilization system. The 7-year program will include the design and construction of a 200,000 to 500,000 gpd pilot desalting plant in FY 74 and a 2 to 3 million gpd prototype plant in FY 76.

Bureau of Mines

The Bureau of Mines has for many years carried out programs that relate generally to geothermal resources research needs, such as extraction, processing, use, reuse, and disposal of minerals and mineral fuels, but has not carried out an identified geothermal program until the present. In FY 1973, the Bureau will undertake a $160,000 geothermal effort aimed at developing improved drilling muds and improved cements for use in drilling and casing geothermal wells at temperatures as high as 350°C. Starting in FY 1974, the Bureau of Mines proposes to begin a 5-year expanded program principally aimed toward the recovery of minerals and gases dissolved in geothermal fluids.

Reject brines from a water desalting plant would constitute the feed material for mineral recovery processes; noncondensible gases probably would be delivered from a separate power plant. Research investigations would concentrate on the geothermal fluids of the Imperial Valley in addition to other promising geothermal resource areas. If research indicates the presence of mineral and gas constituents of sufficient economic value in the brines, a pilot plant for their recovery could be built in close proximity to a desalination plant in order to receive and process the plant effluents.

The Bureau of Mines geothermal program has the following six goals: (1) to develop a system for the evaluation of brines, determining constituents of potential economic value; (2) to develop methods for recovering constituents so determined, (3) to develop methods for separating noncommercial constituents and disposing of them, (4) to determine criteria for selecting materials of engineering construction needed in the handling of hot commercial brines, (5) to find new metallurgical processing uses for residual brines, and (6) to improve drilling muds and cements so as to lower costs of tapping geothermal resources.

Bureau of Land Management

As manager of the Public Lands, the Bureau of Land Management is responsible for leasing of Federally-owned geothermal resources in compliance with the spirit and objectives of the Geothermal Steam Act of 1970, the National Environmental Policy Act of 1969, other legislation, and supporting Executive Orders and regulations. The Bureau's geothermal leasing program will be at a level of $200,000 in FY 1973, subject to Congressional approval of appropriation, and the Bureau has proposed an expanded 5-year program beginning in FY 1974.

Initially, the Bureau's geothermal leasing program will be directed primarily at adjudicating conversion rights established by Sec. 4 of the Act, and to leasing of lands of known or potential value for geothermal resources. Competitive lease offerings will result for lands confirmed to be of value for geothermal resources. Other lands will be available for noncompetitive leasing, which will proceed concurrently with the competitive leasing program. The expanded program would consist of the following components:

(1) Identification of potential geothermal leasing tracts based on Geological Survey, Bureau of Mines, and Bureau of Land Management data and evaluation.

(2) Development of a tentative leasing schedule.

(3) Consideration of environmental factors and land-use plannings for each lease tract.

(4) Evaluation of resource, cost, and market data including the refinement of Geological Survey estimates of bonus bids for each leasing tract.

(5) Post-sale evaluation of tracts.

(6) Development and maintenance of professional competence.

(7) Development and maintenance of a computer system including all available resource and land-use data on areas identified for geothermal leasing.

(8) Conduct of a study program to update and refine leasing regulations and procedures, pre- and post-sale evaluation procedures, and environmental considerations.

National Science Foundation

The Research Applications Directorate of the National Science Foundation began a careful consideration of the need for geothermal energy research at the end of Fiscal Year 1971 but did not sponsor any projects until FY 1972. Four projects have been sponsored in FY 1972, and several others are under serious consideration. The sponsored projects are described below.

"Investigation of the Thermal Regime of the Rio Grande Rift System of New Mexico" by Marshall Reiter of the New Mexico Institute of Mining and Technology ($24,900/24 months). A geophysical investigation of the thermal regime of the Rio Grande rift system of New Mexico is being conducted in order to help assess the rift's possible potential as a large scale source of geothermal power. The Rio Grande rift displays many geological and geophysical characteristics indicative of both the mid-oceanic ridges and presently producing geothermal sites. Heat flow data are being acquired over the extent of the rift in New Mexico and several profiles normal to the rift will be sought.

"Geothermal Resources Research Conference" by Walter J. Hickel of the University of Alaska ($56,500/8 months). The objective of the conference on geothermal resources research is to develop an assessment of the state-of-the-art and to recommend a research program to provide the requisite knowledge for establishing the proper role of geothermal resources in providing (1) additional energy to alleviate the Nation's impending shortage, (2) water to supplement present supplies, (3) mineral resources.

"Stimulation of Geothermal Aquifers" by Paul Kruger and Henry Ramey of Stanford University ($141,500/12 months). The research has three major objectives: (1) development of experimental and numerical data to evaluate the optimum performance of explosion-stimulated geothermal aquifers, (2) development of a geothermal steam reservoir model to evaluate the thermophysical, hydrodynamic, and chemical parameters involved, (3) development of a laboratory model of an explosion-produced chimney to obtain experimental data on the processes of in-place boiling, moving flash fronts, and two-phase flow in hot porous media, as well as chemical and radiochemical data for the fluids produced.

"Investigation of Hydrothermal Systems at Kilauea Volcano, Hawaii" by George Keller of the Colorado School of Mines ($462,500/12 months). Kilauea volcano, Hawaii will be used as a geothermal field research laboratory to evaluate groundwater movement in the vicinity of a magma chamber. Kilauea volcano can provide a means for testing some of the physical concepts that have developed about the characteristics of hydrothermal systems. The project goal is to provide positive evidence about groundwater dynamics in a hydrothermal system and to provide a model for other geothermal systems so that efficient exploration approaches can be developed.

The major part of the program will be the drilling of a test hole at a site selected to provide the best chance of intersecting a hydrothermal convection cell above the Kilauea magma-chamber, but the program would also include supplementary geophysical studies of the summit area of the volcano, physical tests in the drill hole, and measurements on recovered samples.

The NSF Research Applications Directorate estimates (Congress willing) in FY 1973 that it will provide approximately $1,500,000 in the Energy Conversion section of the Energy Research and Technology program and $500,000 in the Energy Resources section. The Energy Conversion portion will concentrate on the science and technology of utilizing geothermal resources whereas the Energy Resources portion will emphasize evaluation, exploration, etc., for the resource.

Atomic Energy Commission

Until the present time, the Atomic Energy Commission has not had a formal geothermal resources program, but several investigations carried out under AEC's auspices during the last few years relate directly to research and development needs of geothermal energy. In 1970 and 1971 a detailed study was conducted by a team from Battelle Northwest, American Oil Shale Co., Westinghouse, Inc., AEC and the Lawrence Livermore Laboratory of the technical and economic feasibility of extracting thermal energy from relatively dry geothermal reservoirs through the use of nuclear explosives to fracture the rocks and the subsequent injection of water to provide a carrier for the heat.

At the Los Alamos Scientific Laboratory another project involves the development of a thermal drill using heat generated electrically. A prototype of the drill has been successfully tested in the laboratory and in the field. Preliminary studies have been conducted at the same laboratory of the feasibility of the use of hydraulic and thermal fracture techniques to expose large heat transfer surfaces in a hot but relatively dry geothermal reservoir in order to recover the energy for power generation. In addition, Battelle Northwest has conducted laboratory studies of silica and radio-nuclide leaching from rock melted by underground nuclear explosions.

In response to inquiries, the AEC stated before the Senate Appropriations Committee that if funds were made available a program of research and development contributing to the wide-scale use of geothermal resources and their conversion into economic electric power could begin at about a level of $2,000,000 in FY 1973. Primary emphasis would be placed on initial examination of the feasibility of extracting energy from relatively dry geothermal reservoirs through the injection and recovery of water in a high-pressure system.

This effort would include study of the behavior of dry geothermal reservoirs as they are cooled and energy is extracted in order to determine the effect of a hydraulic fracture system and the possible expansion of this system by thermal-stress cracking to expose additional heat transfer surface. Laboratory and preliminary field preparation would be undertaken on the feasibility, methods, and economics of developing dry geothermal reservoirs by circulating cooling water through hot rock exposed by hydraulic fracturing techniques. The geothermal provinces of concern would be those where there is abundant hot rock (about 300°C or more) but where there is little or no mobile ground water because of the low permeability of the rocks.

Assessment and evaluation would also be undertaken of methods to stimulate production of geothermal fluids from known geothermal provinces which do not produce steam or hot water of sufficient temperature, quantity, or purity as to be economically usable in power generation. It is believed that stimulation techniques similar to those that have been successful in the oil and gas industry may be of value in these areas.

Studies would be undertaken of the mass transport of silica as well as the chemical, hydraulic and thermal characteristics of natural geothermal systems in order to develop methods for removal of impurities and to develop generating plant equipment compatible with geothermal fluids. These studies would provide information leading to the development of models for evaluation of the economic potential of geothermal systems and techniques for power generation as well as the cost-effectiveness of geothermal research and development programs.

Beginning with engineering analysis of existing technology, a program would be undertaken to develop efficient binary-fluid conversion systems for lower temperature or highly corrosive geothermal fluids in order to maximize the electrical generating capacity of the thermal energy and to enlarge the range of productive geothermal fluids.

Department of Defense

The Advanced Research Projects Agency has started a study to investigate the potential applications of geothermal energy sources to specific military needs. It is possible that a number of power requirements for the Services can be satisfied with geothermal energy sources much smaller than those necessary for commercial power production. Possible military applications of geothermal energy of interest to the Department of Defense include: (1) primary source of reliable power at remote locations, (2) an emergency or back-up power source in the event conventional power sources are disrupted, and (3) a self-contained, standby power source for hardened underground or underwater facilities capable of providing space heating and cooling, and usable water as well as electric power.

The ARPA effort is limited to the definition of Department of Defense needs, definition of the research and development required to apply geothermal energy technology to a variety of military requirements, and to a pilot study of a specific geothermal reservoir adjacent to an existing military facility to determine the research necessary to develop this type energy source for selected military needs.

National Aeronautics and Space Administration

The National Aeronautics and Space Administration has carried out for several years cooperative research with the US Geological Survey in the remote sensing of thermal areas. In this program, data from Geological Survey field and laboratory studies and aerial photographs were augmented by infrared scanner data and high altitude serial photographs provided by NASA. This cooperative program will continue into space through Geological Survey proposals selected by NASA for analysis of data from ERTS-A (July 1972 launch) and Skylab (April 1973 launch).

ERTS television and multispectral imagery will contribute indirectly to geothermal studies through analysis of snowmelt and vegetation anomalies. Addition of emitted infrared imagery with Skylab will permit direct observation of thermal anomalies, with anticipated ground resolution of approximately 80 M and 0.4°K temperature sensitivity. In addition to the above, NASA plans to keep abreast of research and development in geothermal resources in order to assist in basic technology wherever possible.

In carrying out its responsibilities NASA has developed during the last decade capabilities in many disciplines, technologies, and techniques. The particular problems in the field of geothermal resources to which NASA may be able to make contributions in cooperation with user agencies include material development, heat conversion cycles, silencers development, pipe flow and compressor design.

State Organizations

Natural resource agencies of many of the western States are carrying out geothermal studies. In California the Division of Mines and Geology as well as the Department of Water Resources and Division of Oil and Gas have conducted noteworthy geothermal programs for many years including appraisal of the State's resources and geologic investigations of the producing field at The Geysers and several promising areas in the State. The Oregon Department of Geology and Mineral Industries and the New Mexico Bureau of Mines and Mineral Resources have also conducted productive geothermal programs for many years, resulting in the cataloguing of hot spring data in both states and the geologic analysis of favorable areas.

Useful geothermal resources programs are also being carried out by the Nevada Bureau of Mines and Geology, the Idaho Bureau of Mines and Geology, and natural resource agencies of Alaska, Arizona, Colorado, Montana, Utah, and Washington. Altogether, geothermal research and development studies by state agencies, most of which are aimed at resource appraisal, involve several dozen scientists and engineers at an annual funding level of perhaps half a million dollars.

Private Industry

Significant exploration and development programs for geothermal resources in the United States have been conducted by private industry only for the last two decades. The first plant in the United States to generate electricity from natural steam was put into production in 1960 by Pacific Gas and Electric Company, based on steam produced by Magma Power Company and Thermal Power Company.

In the ensuing years these latter companies and Union Oil have carried out successful development of the Geysers field together with exploration of promising areas elsewhere in the western United States. A few other companies, largely petroleum companies, joined in these activities in the 60's, but interest on the part of private industry was not widespread until passage of the Geothermal Steam Act of 1970 opened the possibility of leasing Federal lands for geothermal development.

Private industry has supported geothermal research and development by industry staff, university consultants, and other companies in many areas. Because of the general lack of knowledge in the field, each exploration program is a research project. The development of The Geysers field has led to increased knowledge of the resource and the development of new techniques of reservoir development and production.

The attempted development of the Salton Sea geothermal field required research and development on power generation and mineral recovery technology using concentrated geothermal brines. Industry has supported engineering design studies of binary-fluid power generating technology, and plans to construct a pilot plant of this nature if sufficient private capital is found.

Some of the knowledge resulting from industry research and development has been made available to other scientists and engineers in the field of geothermal resources through the exchanges at scientific meetings, formal publications, etc., but much remains confidential because of its competitive value. The total current effort in geothermal research and development by private industry in the United States is not known, but it very probably amounts to several tens of millions of dollars per year apart from costs of land purchase, production drilling, and plant construction.

ULTIMATE POTENTIAL

Estimates of the geothermal resources of the United States and the impact of the utilization of these resources on the Nation's energy needs differ by more than a thousand times. For example, calculations based on D.E. White's estimate indicate that 5,000 to 10,000 Mw could be generated and maintained for at least 50 years under present economic conditions and technology (White, D.E., 1965, *Geothermal Energy:* U.S. Geol. Survey Circ. 519).

Kilkenny, in the course of a study by the "New Energy Forms Task Group" of the National Petroleum Council, estimates that generating capacity based on geothermal fluids in the United States, specifically in California and Nevada, will increase to between 7,000 and 19,000 Mw by 1958 (Kilkenny, J.E., 1972, Geothermal Energy, part II, General appraisal of the Earth's prospective energy, in *An Initial Appraisal by the New Energy Forms Task Group, 1971-1985;* National Petroleum Council p. 27-35).

R.W. Rex in testimony before the Senate Interior and Insular Affairs Committee on Fuels and Energy Policy (Congressional Record, June 15, 1972) stated that with a successful augmented Federal research and development program, a generating capacity of 400,000 Mw based on geothermal resources could be developed in the western United States within the next 20 years.

The wide variation in resource estimates reflects a number of factors including different assumptions about technological developments and cost factors, preparation of different estimates for different reasons (for example, legal requirements versus program justification), and different uses of the terms "resources" and "reserves", but the major cause of the broad spread in estimates is a lack of factual knowledge of the resource itself.

Conservative estimates of geothermal reserves and resources are shown in the diagram of Figure 1.4, which is based on the format of McKelvey (McKelvey, V.E., 1972. "Mineral Resource Estimates and Public Policy," *American Scientist* v. 60, no. 1, 1972). Reserves and resources should be distinguished from the "resource base," which is defined by Schurr and Netschert as all of a given material present in the earth's crust, whether its existence is known or unknown and regardless of cost considerations. (Schurr, S.H., and Netschert, 1960, *Energy in the American Economy 1850-1975: An Economic Study of Its History and Prospects:* Johns Hopkins Press, Baltimore).

For geothermal resources, the resource base is all of the heat in the earth's crust above mean surface temperature (about 15°C). White estimates the heat in the earth's crust to a depth of 10 km to be about 10^{24} Btu, equivalent to 2000 times the world's coal reserves. Grossling estimates that the heat stored in the

FIGURE 1.4: ESTIMATED RESERVES AND RESOURCES OF GEOTHERMAL ENERGY

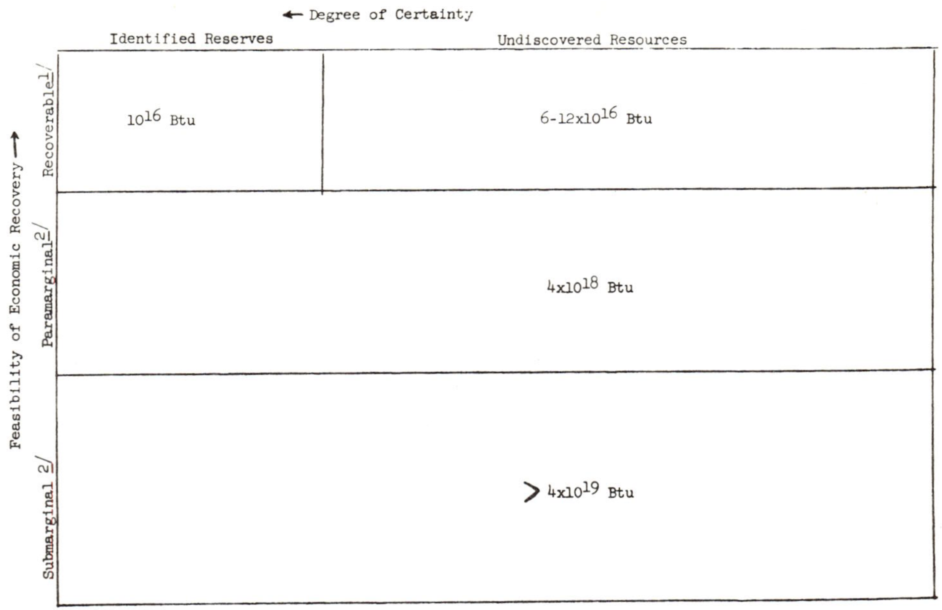

1/ figures shown are energy at the well head

2/ figures shown are energy in situ

Source: PB 224,882

crust of the United States in cratonic and platform rocks to a depth of 10 Km and in interstitial water in sedimentary basins to be 2×10^{22} Btu (Grossling, B.F., 1972, An appraisal of the prospects of geothermal energy in the United States, part 1, in *An Initial Appraisal by the New Energy Forms Task Group 1971-1985:* National Petroleum Council, p. 15-26). Only a small proportion of the geothermal resource base, however, can be considered a resource.

Extraction of all the heat in the crust not only would be uneconomic but also might cause problems such as ground movement and earthquakes caused by the resulting thermal stresses. Exactly what proportion of the geothermal resource base can be considered a resource depends on a number of factors, such as depth of extraction; assumed temperature distribution to that depth; effective porosity, specific yield, and permeability of the reservoir rocks; physical state of the fluid (water or steam); available technology; economics of various uses; and government policy with respect to research and development, leasing, environmental constraints, etc. Different assumptions as to technology, economics, and government policy have cuased much of the large variation in published estimates of geothermal resources.

The identified recoverable geothermal resources of the United States (excluding National Parks) are estimated in Figure 1.4 to be 10^{16} Btu, based on the estimate of Muffler and White (1972) of 6×10^{16} Btu in the ground, including measured, indicated, and inferred resources. This estimate includes the vapor-dominated reservoirs at The Geysers as well as several high-temperature hot-water reservoirs elsewhere in the western United States.

As only about 15 percent of this is deliverable to the well head with current technology, the total reserves (at the well head) are thus 10^{16} Btu or enough steam to generate 1000 Mw for 50 years at the current generating efficiency of 14 percent. In comparison, R.W. Rex estimates that reserves in the vapor-dominated reservoirs at The Geysers alone would support 1000 Mw for 100 years.

Undiscovered recoverable geothermal resources (defined as those resources recoverable at cost competitive with alternative forms of energy) are estimated in Figure 1.4 to be $6\text{-}12 \times 10^{16}$ Btu at well head, based on White's estimate of geothermal resources recoverable as electricity to a depth of 3 km in the United States (assuming that 1 percent of the in situ heat is recovered as electricity). This resource would support 3000-6000 Mw generating capacity for 50 years.

The estimate is for the generation of electricity using proven and conventional technology, thus the geothermal reservoirs included in the estimate must have temperatures of at least 180°C, because at lower temperatures the quantity of flash steam is inadequate, and would include vapor-dominated and high-temperature hot-water systems.

In comparison, Rex estimates that there are sufficient vapor-dominated reservoirs, primarily at The Geysers, to support 15,000 Mw generating capacity for 100 years. Kilkenny has not estimated resources explicitly, but his forecast of future geothermal power production implies at least minimum estimates of resources. He has, for example, estimated that there will be 7000 Mw installed capacity by 1985 with costs of 5.25 mills/kwh (i.e., production based on vapor-dominated systems).

Paramarginal resources are estimated in Figure 1.4 to be 4×10^{18} Btu in situ, sufficient to support 40,000 Mw generating capacity for at least 50 years. These resources would consist largely of high-temperature (>180°C) hot-water systems that might be developed economically at the present time or in the near future. Critical economic factors will be the cost of reinjecting waste fluids and the successful development of binary-fluid or other efficient generating systems and multipurpose (e.e., power-water-chemicals) utilization technology.

Much attention is presently being given to the generation of electricity from geothermal waters, using a system whereby the geothermal heat is used in a heat exchanger to boil a secondary fluid such as isobutane or Freon, which drives the turbine. A generating unit based on the heat-exchange principle and using intake water at 81°C is reported to be in pilot operation at Paratunka, Kamchatka, USSR. Industry interest in this generating mode in the United States is high, and engineering design for a pilot plant using isobutane has been completed, although construction of the pilot plant has not been undertaken.

Multipurpose developments yielding power, desalted water, and/or recovered chemicals improve the economics of power generation by sharing a significant part of the costs against other products. Several projects of this type are currently in the development stage, including the Imperial Valley, California, by the Bureau of Reclamation and the Office of Saline Water. Kilkenny estimates that successful application of these technologies at somewhat higher costs (5.75 - 6.25 mills/kwh) will result in the installation of an additional 6000 Mw generating capacity by 1980 and 12,000 Mw by 1985.

Rex includes in known reserves high-temperature hot-water reservoirs of this quality largely in the Imperial Valley, as well as deeper vapor-dominated reservoirs at The Geysers, which are sufficient to support a generating capacity of 30,000 Mw for 100 years. He estimates that additional resources of this type in the western conterminous United States are sufficient to support 2,400,000 Mw generating capacity.

Recoverable geothermal resources could be still further expanded by technological breakthroughs that would permit the economic development of resources that are clearly not economic at the present time. These submarginal resources are estimated to be more than 4×10^{19} Btu by Muffler and White and sufficient to support more than 400,000 Mw generating capacity for 50 years. Such possible developments include:

(1) Successful application of power generating technology to geothermal reservoirs of intermediate and low temperatures, that is, cooler than 180°C.

(2) Development of better methods for recharge of fluids in geothermal reservoirs, thus permitting fuller and longer utilization of the heat stored in both rocks and fluids of the reservoir.

(3) Development of techniques of artificial stimulation to increase the productivity of geothermal reservoirs, and in combination with the introduction of fluids, to tap the heat stored in hot dry rocks.

(4) Improvements in existing drilling techniques as well as the development

of new drilling techniques that would allow economical drilling to greater depths.

(5) New technology that would favor wider utilization of low-temperature geothermal resources for space heating, product processing, agriculture, etc.

As these techniques are successfully developed, the recoverable geothermal resource will expand within the limit set by the US geothermal resource base. R.W. Rex estimates that the resources present in intermediate-temperature hot-water reservoirs and in deep reservoirs of vapor-dominated and hot-water type in the western United States, the Gulf Coast area, Alaska, and Hawaii, could support 12,600,000 Mw for 100 years. In addition, he estimates that the successful development and application of technology for recovering the thermal energy from hot dry rocks could add resources sufficient to support more than 20,000,000 Mw generating capacity at depths of less than 6 km and an additional 40,000,000 Mw at depths between 6 and 10 km.

An estimate of the impact of geothermal resources on the Nation's energy needs over the next 30 years can be gained from the following estimates of the magnitude of installed power generation capacity based on geothermal resources:

	- Mw -
1972	192
1975	1,500
1980	10,500
1985	19,000
1990	35,000
2000	75,000

For the years 1975 through 1985 the estimates of John E. Kilkenny have been accepted because of his position as an official in one of the very few companies in the United States engaged in the exploration, development, and production of geothermal resources at a profit. The estimates are for power generating capacity at costs of 6.25 mills/kwh and less, and are based on the assumption that technologies will be developed for the economic generation of power from hot water reservoirs as well as for the multipurpose utilization of some geothermal resources such as those of the Imperial Valley.

A larger Federal research and development program than that assumed by Kilkenny could bring about an earlier application of the technology and result in greater installed capacity in 1985 than that shown. The estimated increase in power generating capacity beyond 1985 to 35,000 Mw in 1990 and 75,000 Mw in 2000 is based on the assumption that an expanded Federal research and development program will result in the identification of more geothermal reservoirs and the development of technologies for generating power from deeper and lower temperature geothermal fluids including the over-pressured geothermal brines of the Gulf Coast.

The successful development and application of technologies for the economic recovery of thermal energy from reservoirs with insufficient fluids or permeability might result in an even greater increase in geothermal-based generating capacity.

LEASING OF GEOTHERMAL LAND IN THE UNITED STATES

This section which is derived from PNE-1550, *A Feasibility Study of a Plowshare Geothermal Power Plant,* April 1971, discusses recent Congressional action that has removed a major obstacle to the development of geothermal resources by providing adequate statutory authority for the Secretary of the Interior to issue leases for developing geothermal resources on public land. The availability of both federal and private land is discussed. Site requirements and geologic descriptions of various heat sources are outlined and comments on the public acceptance of the Plowshare concept of geothermal-heat extraction are made.

Availability of Federal Geothermal Land

The department of the Interior has withdrawn 1,345,670 acres of federal land from all subsurface use because the acreage is either known to be valuable or is potentially valuable for geothermal-resource development. A state-by-state description of this land was published in the Federal Register on March 24, 1967, and amended on April 2, 1967. It breaks down as follows:

Land known to be valuable (1,052,140 acres)	
California	839,640 acres
Idaho	16,000
Montana	17,920
Nevada	38,400
New Mexico	140,180
Land potentially valuable (293,530 acres)	
Alaska	65,760 acres
California	83,610
Idaho	7,680
Nevada	81,280
Oregon	34,560
Utah	10,880
Washington	9,760

One reason for the lack of development of geothermal resources in the United States has been the absence until now of reliable statutory authority to permit such development on public lands.

On December 24, 1970, a Senate bill that had been introduced by Senator Bible of Nevada and seven other western Senators was enacted into Public Law 91-581. This law provides statutory authority for the Secretary of the Interior to issue leases for the development of geothermal resources on public land in much the same manner as he is authorized to issue leases for the development of oil and gas deposits.

The resources subject to disposition under Public Law 91-581 are "geothermal steam and associated geothermal resources," which are defined in Section 2(c) of the law. The law clearly covers energy from dry geothermal anomalies, and each lease includes any by-products that are associated with the goethermal resources except for oil, hydrocarbons, gas, and helium.

Leases cannot be issued for land in national parks, national monuments, national recreation areas, fish hatcheries administered by the Secretary of the Interior, wildlife refuges, wildlife ranges, game ranges, wildlife-management areas, waterfowl-production areas, land acquired or reserved for the protection and conservation of fish or wildlife threatened with extinction, or Indian-owned and certain other land. Thus, under no circumstances could the "Old Faithful" geyser in Yellowstone National Park or any other geysers or hot springs in parks or outdoor recreation areas be exploited.

As provided under Section 7 of the law, each lease is to embrace a reasonably compact area of not more than 2,560 acres, and the total acreage that a lessee may hold in any one state is limited to 20,480 acres. After 15 years of experience with the law, the Secretary of the Interior may increase the maximum holding in any one state to not more than 51,200 acres. In contrast, existing law (US Code, Title 30, Section 184) establishes limitations of 46,080 acres for coal and 246,080 acres for oil or gas.

If the land in a proposed lease lies in a "known geothermal resources area," Section 4 provides that the lease must be issued to the highest responsible qualified bidder by competitive bidding under regulations formulated by the Secretary of the Interior. If the land does not lie in a known geothermal-resources area (otherwise known as "wildcat land") it can be leased to the first qualified applicant without competitive bidding. This feature is patterned after the law governing the leasing of oil and gas deposits.

Under that law [US Code, Title 30, Section 266(b) and (c)], leases in a known oil or gas field must be issued competitively, and land not in such a field can be leased to the first qualified applicant without competitive bidding.

Section 4 also provides that holders of lands which were on September 7, 1965, subject to valid leases or permits issued under the Mineral Leasing Acts, or mining claims prior to September 7, 1965, and lands which were on that date the subject of applications for leases or permits under such acts, and who are otherwise qualified to hold geothermal leases, had the right within 180 days after December 24, 1970, to convert their leases, permits, claims, or applications into geothermal leases or applications for leases covering the same lands.

Section 5 of the law requires all geothermal leases to provide a royalty of not less than 10% nor more than 15% of the value of the steam or other form of heat or energy that is derived from production under the lease. In the case of by-products, a maximum royalty of 5% is provided except for leasable minerals, which are controlled by the mineral-leasing law. Also, each lease is subject to both an annual rental of not less than $1.00 per acre and a minimum royalty of $2.00 per acre after commercial production begins.

Section 6 provides that each geothermal lease shall last for a primary term of 10 years. If steam is produced or used in commercial quantities during this period, the lease can continue for up to 40 more years as long as such production or use continues. If steam is still being produced in commercial quantities at the end of this 40-year period, and if the land is not needed for other purposes, the lessee has a preferential right to renew the lease for a second 40-year period in accordance with such terms and conditions as the Secretary of the Interior deems appropriate.

A comparable provision exists for land leased under approved cooperative and unit plans. Leases that are extended but are later found to be incapable of the commercially valuable production of by-products can be converted to mineral leases under the mineral-leasing law. In the case of non-leasable minerals, a lessee is entitled to locate such minerals under the mining laws within 90 days after his geothermal lease ends.

Section 8 permits the Secretary of the Interior to readjust the terms and conditions, other than the rentals and royalties, of any geothermal lease at not less than 10-year intervals, beginning 10 years after the date on which geothermal steam is first produced. Rents and royalties, on the other hand, can be readjusted by the Secretary at not less than 20-year intervals, beginning 35 years after the date on which geothermal steam is first produced. In a readjustment, neither the rental nor the royalty can be increased by more than 50% over the rental or royalty paid during the previous period, and in no case can the royalty exceed 22.5%. Provision is made for notifying a lessee of any proposed readjustment in the terms of his lease.

In recognition of the possibility that substantial amounts of usable water, including commercial demineralized water, may be associated with geothermal steam in certain areas, Section 9 permits the Secretary of the Interior to require its production or use in circumstances where warranted. Such production or use would be in accordance with applicable state water laws.

In order to obtain an authoritative judicial determination of the ownership of geothermal resources on land whose surface (but not mineral rights) have passed from Federal ownership, Section 21(b) directs the Attorney General to initiate proceedings to acquire the title to such resources when their development either occurs or is imminent. Congress was aware that the Department of the Interior had expressed the view that geothermal steam is not subject to the mineral reservation in the Stockraising Homestead Act of December 29, 1961, but since the opinion of the Department is not a conclusive determination, it was believed that an early judicial determination of this question (upon which Congress took no position) was necessary.

At issue is the ownership of geothermal steam on more than 35 million acres of land. The bulk of this acreage was bought under the provisions of the Stockraising Homestead Act, but the mineral rights are subject to disposition under the appropriate mineral laws. The development of geothermal resources on this land will be retarded until the question of ownership is resolved.

Availability of Private Geothermal Land

The eleven western states of Arizona, California, Colorado, Idaho, Montana, Nevada, New Mexico, Oregon, Utah, Washington, and Wyoming cover about 753,000,000 acres, of which nearly half (360,000,000 acres) is classified as federally owned by the General Services Administration. The nonfederally owned land (about 393,000,000 acres) falls under two other kinds of ownership—state-owned land

and patented (also called fee or private) land. The location and possible geothermal development of such land is discussed below in general terms.

The hisotrical pattern of settlement and development in the western states provides a guide to the location of both patented and state land. Most of the patented land that was acquired under the Homestead Act of 1862 and similar laws is concentrated along rivers and other water sources, in valleys, and in areas that are generally suitable for ranching, dry-land farming, and (to some extent) grazing. This leaves much of the nonagricultural and mountainous areas, with the notable exception of developed mining regions, to federal ownership.

Additionally, railroads (such as the Union Pacific in Colorado, Wyoming, and Utah) still hold mineral rights to alternate sections granted along their rights-of-way. In the southern part of the western states, much of the patented land derives from Spanish land grants that predate statehood. In some states, notably Arizona, there is a hybrid category of land—Indian reservations—that can be leased from the various tribes as fee owners, but the Indians normally act on the advice and consent of the government.

State-owned land includes the so-called school sections or substitute selections granted at the time of statehood. However, on the basis of an average of two to four such sections to a surveyed township of 36 sections, it is obvious that the percentage of state land is low. For instance, Utah controls only 9% or 4.7 million acres of its area, and Colorado controls about 6% or 4 million acres of its area. Furthermore, because of its pattern of acquisition, state land is usually widely scattered.

A state-by-state outline of the general location of nonfederally owned land is as follows: Arizona—in the southeastern, east-central, and west-central parts; California—in the coastal and other scattered areas, in the interior valleys; Colorado—in the eastern half, in mining districts, and in river valleys; Idaho—in the eastern and southeastern parts, in the Snake River valley, and in the western side of the panhandle; Montana—Primarily east of the Rocky Mountains, and in western river valleys; Nevada—in river valleys and in the Reno area (both negligible); New Mexico—in the eastern half and the central part, in the Rio Grande valley, and in land grants reaching to the northern border; Oregon—in the Columbia and Willamette River valleys, and in the south-central and southeastern parts; Utah—in a belt extending from the Wasatch area in the north to the southwestern corner, in a railroad grant in the northern part; Washington—in the coastal areas, in the Columbia River valley, and in the central and southeastern parts; Wyoming—in the eastern third, in the western river valley, and in a railroad grant along the southern edge.

The concentration of both state and patented land in highly developed and densely populated areas eliminates such land from geothermal leasing. However, there are still areas in which private and state ownership will have considerable impact on geothermal leasing. The geothermal leasing of state and patented land is such a relatively new concept that few guidelines have been established. With a modified mineral-leasing form, it has been possible to lease patented land in Utah on a limited scale. This is done on the basis of a 10-year lease that calls for a relatively nominal bonus and an annual rental.

In such a lease, the landowner's royalty provision is based on a mining-like lease, 5% of the value of the initial production rather than the 12.5% royalty usually found in oil and gas leases. The heavy initial cost of geothermal development has been emphasized in order to obtain lower royalties. However, leasing terms and practices vary so widely that there is no practical way of predicting how much actual cash would be involved without having a specific case to evaluate. The leasing of state land is equally unpredictable, for as far as this study could determine, no state has a leasing form that covers either dry or wet geothermal areas.

In such cases, some states pattern their forms along the lines of federal leases (which in this case are nonexistent), while other states draft their own leasing forms. It is hoped that each state will adopt a modified mining-like lease with a corresponding royalty scale. As with patented land, the leasing of state-owned land would be a matter of negotiation when specific areas are located and the ownership of the land is determined. Also, bonuses and annual rentals on state leases are so variable that actual costs are not predictable. In summary, the pattern of land ownership in the western states is such that in most areas a leasing program will probably involve all three kinds of land—federal, state, and patented. The acquisition of leases on any one of these could unlock the door to development, but leasing on all three will eventually be necessary.

SURVEY OF MAJOR GEOTHERMAL INSTALLATIONS

J.B. Burnham and D.H. Stewart of Batelle, Pacific Northwest Laboratories made a world survey of the major geothermal areas during May 1971. The trip was sponsored by the USAEC. The purpose of the trip was to obtain the first hand experience of geothermal plant geologists, drillers, and plant operators. The material in this Section is derived from their report, BNWL-B-110, July 1971, *Foreign and Domestic Discussions on Natural Geothermal Power and Potential Use of Plowshare to Stimulate These Natural Systems.*

Iceland

Iceland has widely dispersed geothermal resources. These are characterized as low and high temperature fields. The latter are concentrated near the Quaternary volcanic activity associated with the crest of the Atlantic Ridge which crosses the country from SW to NE. The country has abundant energy resources. Most of the geothermal development has been concentrated in the heating of homes and hothouses. One experimental generating plant of 2.5 Mw capacity is operating at Lake Myvatn.

Location of Geothermal Areas: While Iceland has widespread areas of geothermal activity, the most active fields are located in a zone which runs through the country from southwest to northeast. This area corresponds roughly to the crest of the Mid-Atlantic Ridge. Geothermal areas of Iceland are characterized as low-temperature—below 150°C, or high temperature. The development of low temperature fields has been concentrated around Reykjavik.

High temperature areas which have been studied in some detail are the Reykjanes areas on the southwest tip of Iceland and the Namafjall area in the north-central part of the island. The former has been studied as a possible source of chemicals produced from geothermal brine. The latter has been developed as a source of process heat for the production of diatomaceous earth from Lake Myvatn. An experimental generating station of 2.5 Mw is connected with the Myvatn installation.

Geology: The predominate rocks of Iceland are volcanic—basalts, andesites, and rhyolites. The basaltic lava flows of the Tertiary age are found in the western, eastern, and the north-central regions and the NW Penisula of Iceland. Separating these regions are the younger volanic rocks of the Quaternary age—basaltic tuffs, palagonite breccias, lava flows, and pillow lavas.

Iceland is in the unique position of being energy-rich. They have developed only 5% of their available hydro power. As a result, they generally use their geothermal resources (which are extensive) for space heating. The exception to this is the 2.5 Mw experimental station at Myvatn.

Production: About 75,000 inhabitants of Reykjavik live in homes heated by geothermal gnergy. The discovery of a new field (Ellidaar) near Reykjavik has caused postponement of a project to pipe high-temperature steam from the more distant Hengill field. It has been found economical to use supplemental oil firing of the low-temperature steam in periods of high demand.

Wells are rotary drilled and cased only for the first 50-100 m. The operation of the production, collection, and distribution system is well automated. The cost of this energy to the homeowner is about 96¢/M Btu of which 66% is distribution costs. This is only a little over half the cost of oil heating in that area.

The Hveragerdi area was visited. In this region is concentrated most of the country's 105,000 square meters of greenhouses. Iceland produces most of its tomatoes, cucumbers, and lettuce in these hothouses heated by geothermal water. About a third of the area is devoted to the raising of flowers.

The government-owned Kisilidjan Company developed the Myvatn experimental power plant in cooperation with the Johns Manville Corp. in 1966. The plant producing 12,000 tons per year of diatomite was started up in 1967. Production was doubled in 1970.

Power Plants: In 1959 the Laxa Power Works built a 2.5 Mw generating station at Lake Myvatn. This is an atmospheric exhaust installation. As such, it utilizes only a small part of the energy available in the wet steam. A condensing system can be added later, though, if demand warrants the expansion. The turbine operates at 147 psia and a back pressure of 15 psia. Under these conditions it has a heat rate of 42,000 Btu/kwh. The turbine blading is a special 10% chromium-35% nickel alloy.

Steam is produced from four production wells about 1000 m deep. The wells are rotary-drilled. They are cased as follows: 16 inches to 30 m, 9-5/8 inches to 200 m, and 7-5/8 inches to 600 m.

Future Plans: A company town is under construction at Namafjall (Lake Myvatn).

It is planned to pipe by gravity flow the hot (183°C) water from the well-head separators to this town some 3 km away. To avoid the corrosion problems associated with using the high-sulfur water in home heating systems, a fresh water heat exchanger will be used. It has recently been reported that Iceland is planning one goethermal power plant in the north that would generate 55 million watts.

Italy

The Larderello field produces around 400 Mw from dry superheated steam. A new field, Monte Amiata, also shows great promise. These two fields lie in a band which runs north and south along the West Coast of Italy.

Locations: Two proved geothermal areas lie about 60 kms south of Pisa and Florence. One area, Larderello covers roughly 90 km^2; the other known as Monte Amiata, perhaps 40 km^2.

Geology: The Larderello field areas lie under a surface of neogenic deposits, flysch-facies formations which constitutes an impervious layer. The porous reserve consists of limestone and dolomites and the area is well faulted. Larderello can be considered a hoist—a tectonic high. The underlying boraciferous region corresponds to a vast gravity deficit due to presence of a magmatic acid mass at a depth of 6 - 8 km. The water derives its heat from the magmatic mass through the country rocks of the basement. The Monte Amiata area is somewhat similar. Thermal gradients as high as 50°C/100 meters have been measured. Electrical resistivity has proved a valuable exploration technique to 1,000 meters with values $\geqslant$100 ohm meters.

Monte Amiata may be the only geothermal field discovered by geologic evaluation alone—no surface indications were present. This field also appears to be extensive and further wells are being drilled.

Production: Most of the bores are less than 1,000 meters deep and produce steam containing gas at an average of 5% by wt. Temperatures run from 150°C to 260°C at the well head. Temperatures tend to increase with time. Flow rates average between 50 and 100 tonnes/hr. Shutdown pressures range from 88 psi to 617 psi. The Monte Amiata fields of Italy, Geysers in California and Matsukawa (Japan) are the only geothermal areas now producing dry steam.

Bore diameters are larger than those in New Zealand and are generally 13-3/8". Heavy mud is used when drilling and blowout preventers are always installed. Loss of drilling fluid is generally an indication that the producing layer has been reached. Casings are carefully cemented.

When a well is first produced the flow rate drops rapidly with time until a steady state level is approached asymptotically. Larderello is apparently in equilibrium with water inflow and no substantial increase in power from this field can be expected. It would be interesting to try reinjection in the production area. The geologists have estimated the natural water inflow takes some 40 years for fresh surface water to be removed as steam. Steam temperatures run from 140°C to 190°C at the power plant.

Power Plant: On remote isolated wells, noncondensing turbines are used. Although simple and reliable, about 20 kg of steam at 4.8 atmosphere and 185°C is required for one kwh net. It is also a useful cycle when gas content is high (i.e., as high as 30% by wt) as the gas is troublesome to remove in a condensing cycle. A new—15,000 kw—noncondensing cycle achieved a steam rate of 13.8 kg of steam per kwh.

In the condensing cycle, 10 kg of steam per kwh is required. The steam is condensed, the water cooled in large natural draft cooling towers, and reused to condense more steam. Total capacity installed is 470 Mw and production is running around 400 Mw. It is believed that further production at Larderello cannot be easily achieved as the field is about in equilibrium with water recharge. In a major power station, gross electrical output was much greater than the net. This was mainly due to the large amount (>20%) of power which was required to operate the noncondensable gas pumps.

New Zealand

Wairakei is a large hot water-steam field. The hot water is separated from the steam and reflashed at low pressure. About 180 Mw are being generated. A new field, Broadland, shows similar production capability.

Location: Wairakei is part of a large geothermal area extending north from Lake Taupo to Rotarura. It is approximately 20 miles wide and 150 miles long. Wairakei appears to be in a graben. As steam leaks and boiling pools of mud are quite prevalent in the area, its location has been known for centuries.

Geology: The area is covered with about 300 ft of light pumice. Below this is 100 to 200 ft of impermeable, dense, fine sandstone and mudstone rock. The next layer consists of a porous mass of cemented and fissured pumice and rhyolite. 2000 ft below this lies a thick layer of massive ignimbrite. The ignimbrites are fragmented rocks which were erupted explosively at high temperature and the dense clouds of incandescent particles became welded and compacted into hard dense rock as they settled. The area is approximately 500,000 years old. Extensive faulting of the earth has occurred in this area. This is believed to provide steep permeable zones along which the hot fluids ascend from depth.

Production: Boreholes in the area produce steam and water mixtures. Approximately 1/2 of the wells produce at 80 psi and the other 1/2 at 180 psi. The deepest borehole in use is 4,000 feet deep and the shallowest 563 ft. Average production depth is 2,200 ft and the pipe diameter is 8 inches. Average steam production is 48,000 lb/hr and maximum is 1,400,000 lb/hr. Steam lines up to 12,000 ft run to the power house. The highest steam temperature is 500°F. Bores only 90 ft apart do not appear to interact. There has been a general decline in both temperature and pressure although it is believed the present power output can be maintained by utilizing more of the waste hot water with additional flashing to steam. Only about 7% of the total heat is actually utilized.

The site must be consolidated by pumping an average of 65 tons of concrete in the ground before drilling. Heavy drilling muds are used to avoid blowout. Loss of circulation is common when the steam producing formations are located. Blowout preventers are always used.

The bores produce about 80% by wt water and 20% steam. The water, after separating the steam, is flashed to produce low pressure steam. Three piping systems are used, one for high pressure steam (160 psi), one for intermediate pressure steam (52 psi), and one for low pressure steam (2 psi).

Power Plant: Installed maximum capacity is 240,000 kw and approximately 180,000 kw is being generated—limited by steam production. They believe water inflow limits steam production at this level. No reinjection has been tried. The three steam pressures are fed to separate turbines and condensed by a river water cooling system. Most (80% by wt) of the hot water is run through concrete trenches into a river. Silica deposits in these trenches as the water cools and these must be removed annually. No silica deposition on the turbines has been noted and one turbine in operation 12 years appeared in good condition. Noncondensables are low—0.26% by volume—and are removed by a steam jet ejector. Carbon steel used extensively has good corrosion resistance providing oxide is kept from unused equipment.

Plans: Although geothermal costs are low and a large new field (Broadlands) has been proved nearby, it is unlikely new geothermal plants will be constructed soon. A natural gas field has been discovered off the shore of New Zealand, and its development will cost on the order of $150 million. This in turn requires a market so that new power plants over the next few decades are likely to run on natural gas.

Japan

The Japanese Archipelago is widely covered with active geothermal regions. Two of these—Otake, Kyushu (10 Mw) and Matsukawa, Honshu (20 Mw)—are producing electricity. The operating history of these plants has been excellent with a 96-97% capacity factor and a minimum of maintenance problems. The Matsukawa plant has acidic conditions which cause some corrosion problems but a minimum of scale deposition. The Otake plant with less acidic conditions has no corrosion problems but must hold up the hot water effluent for one hour to allow settling of silica.

Location of Geothermal Areas: The major geothermal areas of Japan are listed in Table 1.4 below. While there are some 1300 hot springs and fumaroles in Japan, only those in which geothermal development has been (or will be) emphasized are listed here:

TABLE 1.4: MAJOR GEOTHERMAL FIELDS OF JAPAN

Geothermal Area	Island	Area (km^2)	Total Heat Flow (10^6 cal/sec)
Atosanupuri	Hokkaido	0.9	7.8
Sounkyo	Hokkaido	0.8	1.9
Noboribetsu-Yunuma	Hokkaido	0.07	19.3
Noboribetsu-Zigokudani	Hokkaido	0.1	11.2
Showa-shinzan	Hokkaido	(0.1)	(3)

(continued)

TABLE 1.4: (continued)

Geothermal Area	Island	Area (km^2)	Total Heat Flow (10^6 cal/sec)
Nigorikawa	Hokkaido	3.6	3.0
Shikabe	Hokkaido	0.3	1.0
Tamagawa	Honshu	0.1	15.2
Matsukawa	Honshu	0.2	(33)
Kusatsu	Honshu		24.4
Hakone-Owakudani	Honshu	0.06	4.8
Beppu	Kyushu	7	37.3
Otake	Kyushu	0.1	44.6
Obama	Kyushu	1.5	52.4

Source: BNWL-B-110

The geothermal areas are fairly well distributed over the islands of Hokkaido, Honshu, and Kyushu. Shikoku is the only one of the four major islands of Japan which has few hot springs. Most of the hot springs are located along Quaternary volcanic zones and derive their energy from volcanic activity of that period.

Geology: Japan lies in the Pacific orogenic zone—the boundary between the stable continent and the Pacific Ocean. Very extensive faulting occurred through most of the Japanese Archipelago during the Tertiary Period. This faulting was followed by a period of igneous intrusion which deposited lava and tuff in a large geosynclinal basin. This tuff is a distinctive green-spotted variety called "green tuff" and the geosynclinal basin is called the "green tuff" region.

Over this tuff thick deposits of marine sediments were laid down. Later intrusions of granite (and metamorphic rock) occurred in this green tuff region along with some additional faulting. Toward the end of the Tertiary Period, most of the green tuff regions were uplifted. During the Quaternary Period more deep fracturing occurred and the new volcanoes of the Quaternary arose.

Some 265 volcanoes came into existence during the Quaternary Period of which almost 50 are still active. These fall into two chains—the eastern one which runs from Hokkaido through northeast Japan to the Marianas and the western chain which runs from Kyushu through the Ryukyu Islands to Taiwan.

The geothermal areas derive their heat from the Quaternary volcanic activity. Most of the hot areas are closely associated with the volcanic zones. The hot springs generally have basements consisting of green tuff, although some occur in areas of Palezoic and Mesozoic formations and granite.

Production Details: The two areas of Japan which have been developed to date for power production are the Matsukawa of northeastern Honshu and the Otake of north-central Kyushu.

The Otake plant has a design capacity of 10 Mw. Five production wells were drilled, but calcium carbonate deposits caused a shut-off of one of these after three years of operation.

The Otake area was developed by the Kyushu Electric Power Co. starting in 1950. The five production wells were drilled in the time period 1963-1966. Power production began in August of 1967.

The basement rock in this geothermal area consists of two-pyroxene andesite, hornblende-bearing two-pyroxene andesite lavas, and associated tuff breccia. This basement is overlain by the Kujyu volcanic complex generally composed of hornblende andesite.

Geophysical prospecting in the area has been carried out with magnetic, electrical, and gravitational techniques. These surveys indicate that an area approximately 1.5 km by 3.0 km has been hydrothermally transformed to a depth of 800-1,000 m.

The Otake field produces a steam-water mixture of about 25 wt% steam from an average well depth of about 450 m. The wells are rotary-drilled and are typically cased to about 250 m depth with 8-5/8 inch casing. Steam is separated by a rotary separator at the well head, piped to a collection tank, passed through the turbine, a vacuum ejector (for noncondensible gas removal), a barometric condenser, and cooled by three forced-draft cooling towers.

There were some corrosion problems due to the SO_2 in the steam until the vacuum ejector was installed. SiO_2 was a problem in the hot water lines until a baffled holdup tank was added. The one-hour holdup time allows the SiO_2 to settle out in the bottom of this tank, the outlet of which runs to the Kusu River in an open channel. The turbine has been shut down once a year for inspection and water jet washing for SiO_2 removal. This schedule has been changed to a two-year one because of the excellent performance characteristics of the system which has run at a 96% capacity factor. The turbine throttle steam conditions are 127°C, 2.5 atm (37 psi), 0.1 atm (2.9" Hg) back pressure at a steam consumption rate of 110 t/hr.

The Matsukawa geothermal field was developed through efforts of Japan Metals and Chemical Company and the Geological Survey of Japan. Test wells were sunk in 1952 and an extensive survey program was begun in 1956. Four production wells were sunk in 1964 and 1965. The area started power production of 20 Mw in 1966.

Basement rock in the area is Paleozoic slate and chert. This is overlain by Tertiary green tuff, the Tertiary Yamatsuda formation (sandstone and shale), then dacite tuff and dacite lava of the Tertiary period. The upper beds are Quaternary volcanic rocks, primarily andesite.

The four production wells were rotary drilled to a depth of about 1,000 m. A standard casing program consisted of a 16-inch casing to a depth of 30 m, 11-3/4 inch casing to 250 m, and 8-5/8 inch casing to a depth of 500 m. Several wells produced superheated steam.

Special features of the turbine include stellite facing on the steam inlets of the last two stages for corrosion resistance, moisture extracting buckets, and grooves to remove condensate from the inlet nozzle.

Plans: A new production area is currently under development several kilometers to the south of the Otake installation on Kyushu. Four wells have been completed at Hachabara and two more are being sunk. Ten more wells are planned for this phase of the plant construction. The output will be 50 Mw. Eventually the production from this geothermal area will be increased to 150 Mw. The wells are about 1,000 m deep in this area and produce a steam-water mixture which is about 33 wt% steam.

There are currently under active development several geothermal areas on Hokkaido. The most promising of these are: the Shikabe area on the Nasu volcanic zone, the Showa-shinzan, the Tokachi-Daisetsu volcanic region, and the Kurile volcanic region.

On Honshu, the Onikobe and the Oshiraka regions show the most promise. Ten wells have been sunk at the former. One of them is producing 20-30 t/hr of superheated steam from 500 m depth.

According to E/MJ, Feb. 1974 The Mitsubishi Metal Corp. is starting up a geothermal steam power generating plant at its 8,100-tpm electrolytic zinc plant at Akita. The Mitsubishi geothermal generating plant is the third geothermal station to be built in Japan but the first to be used in nonferrous production.

In late 1973, Mitsubishi was testing three geothermal steam production wells capable of generating 6,500 kw in the mountains of Akita Prefecture on the Sea of Japan coast. Barring unforeseen developments, these three wells will be supplying power to the Akita zinc plant by the second quarter of 1974. A fourth production well was being drilled, and if it proves out, it will be tied into the zinc plant power supply in the fall. The combined output of the four wells will provide about one-fifth of the zinc plant's total electric power requirement of 40 million kilowatts.

The Geysers, California

The Geysers field in California is the largest known geothermal field in the world, covering at least 20 square miles. It is one of three fields producing dry, superheated steam. Current production of 82 Mw is being expanded by additions of 110 Mw power plants and ultimate production could run from 1,000 to 4,000 Mw.

It has been recently reported that the total capacity at The Geysers project is 400,000 kilowatts. Additions of 100,000 kilowatts each year through 1983 are planned. This will bring total capacity to 1.4 million kilowatts.

It would take 45,000 barrels of oil today to develop as much power. Steam reservoir potential in the area is now 1.6 million kilowatts.

Location: The Geysers is located in the coast range about 90 miles north of San Francisco, California. It covers an area of at least 20 square miles, and as numerous step-out wells have been successful, the full extent of the area is still unknown.

Geology: The reservoir at The Geysers consists of Franciscan graywacke. The graywacke itself has a very low porosity, but is highly fractured and the bulk permeability is very high.

It is likely that the heat source is volcanic as widespread young volcanic activity is a feature of this portion of the coastal range. G. Facca's theory is that The Geysers was a self-sealing reservoir, forming its own cap rock through deposition of minerals and thermal alteration of the upper layers of the graywacke. A major fault has been located by drilling, and it is likely that this fault provides a passage for the water to the volcanic heat at depth. There is agreement that at least 90% of the steam is of meteoric origin.

Production: Air drilling is generally used and the bores run from 1000 to 7000 feet. The drilling record is extraordinary with 70 producing wells against 5 dry holes. Present proven capacity is 400 Mw and the capacity of the drilled area is 1300 Mw. Very little ground water is encountered as on the whole the cap rock is impervious.

The wells are cased to from 700 ft to 1200 ft. Casing from 9-5/8 to 13-3/8 is used. On many wells no further casing is used and the well is produced as a barefoot completion. If objectionable quantities of water are encountered, further casing is required.

Union Oil supplies the steam to the Power Plant where it is purchased by PG&E for about 2.9 mills/kwh of electrical power generated. This results in an average income of $240/day per average well. Shut off well pressures range from 210 psig to 479 psig. Maximum steam flow from a single well is 220,000 lb/hr at 110 psig.

Power Plants: The original two units 1 and 2 provided 20 Mw and produced power to the system for 5.65 mills. Units 3 and 4 total 56 Mw and power costs dropped to 4.71 mills/kwh. The two units presently being installed are 55 Mw each and will show a further reduction in power costs. The original units used forced draft cooling towers and barometric condensers. The water surplus, formerly added to the creek is now being reinjected into the producing formation. No loss in production from adjacent wells has been noted as a result of reinjection.

The steam supplier is required to remove the carryover rock and particulate matter from the steam before delivery to PG&E. Noncondensables run quite low, 1 to 2% by wt for new wells, but this drops off as the well is produced.

Mexico

According to PB 212,753, 1972 (see Chapter 6) about 20 miles south of the international border in the Mexicali Valley, near the volcano of Cerro Prieto, is a steam field currently being prepared by the Mexican government for the production of electricity. Surface manifestations of geothermal activity had been noted by Mexican geologists, and drilling to the area was begun in the 1960's. Commercially exploitable steam and water at high temperature was found at depths of less than 2000 feet. From 1964 onward, producing wells were drilled and by 1971 more than 16 producing wells had been installed. Production depths vary, but average about 4500 feet.

A typical well will produce 450,000 lb/hr of steam and water mixture. Approximately 2.75 pounds of water for each pound of steam are produced and the enthalpy of the mixture varies between 400 and 600 Btu/lb. The maximum temperature at depth is 700°F and pressures range up to 1700 psi gauge. At this time the Mexican government is installing a power plant of 75 megawatts capacity to use the steam that will be produced from this field. The size of the known field approximates 4 square miles. The Mexican wells are in clusters with spacing between adjacent wells that would correspond to 10 acres per producing well.

If the entire 4 square mile field were filled in, as is not presently the case, approximately 250 wells would be installed. If the present average steam production per well was maintained, 120,000 lb/hr, the capacity of the field is then calculated to be about 1500 megawatts of electrical production. The water that is found at the producing depths at Cerro Prieto is saline in character with approximately 2 1/2% total mineral content. While far too salty for any direct consumptive use, it is an order of magnitude less mineralized than the water found at the Salton Sea area in Imperial Valley.

SOVIET GEOTHERMAL RESEARCH AND DEVELOPMENT

The information in the following section is derived from AD-754 947, *Soviet Geothermal Electric Power Engineering-Report 2*, December 1972, by V.A. Stevovich of Informatics, Inc. This report was prepared for the Air Force Office of Scientific Research, Advanced Research Projects Agency.

SOVIET GEOTHERMAL RESEARCH AND ENGINEERING

Historic Background Information

Geothermal research in Russia started in the first half of the 18th century with several scientific expeditions engaged in measuring terrestrial temperatures in various mines in the Urals, Siberia, Altay, Volga region, and in the Far East. In 1906, scientist A.N. Ogil'vi developed a method for geothermal prospecting and exploitation of mineral waters. This method was applied also in prospecting for minerals, primarily oil and balneological waters. Information on terrestrial heat played an important role in the early evolution of geological science in Russia.

In 1910, the first Geothermal Commission of the Russian Geographic Society was organized and is credited with the collection of data on soil temperatures in permafrost zones and geothermal anomalies in the Caucasus.

After the revolution, geothermal research was given practical and scientific attention in national industrialization plans, though on a rather moderate scale and at a slow pace. In 1929, the first Geothermal Laboratory of the Central Scientific Research Institute for Geological Exploration was established to assume responsibility for geothermal research in the search for new natural resources. In the late 1930's, systematic thermometric observations were being conducted at several permafrost stations (operated by the Permafrost Institute of the USSR Academy of Sciences), with primary emphasis on the thermophysical constants of specific regions and large reservoirs.

Prior to World War II, intensive surveying and mapping of geothermal fields over a network of test wells was initiated, but was soon interrupted by the war.

In the early 1940's, by merging hydrological and geothermal doctrines, research was directed toward the development and exploitation of mineral and thermal springs. The most outstanding work of this period, *Geothermal Characteristics of the Caucasus*, provided a new interpretation of the origin and distribution of therapeutic waters, terrestrial temperature, conditions for the formation and evolution of geothermy, depth of thermal flow, variations of heat gradients, and formation of ground waters. The All-Union Conference on Geothermy, held in Moscow in March 1956, stressed the great need for broader basic and applied research.

In 1963, the Plenum of the Central Committee of the Communist Party enacted a resolution, in a drive for overall expansion of the chemical industry, to intensify research on and utilization of geothermal resources, primarily on Kamchatka, and subsequently in the Caucasus, Siberia, and other regions of the Soviet Union.

To coordinate these activities, the Earth Sciences Section of the USSR Academy of Sciences (established in 1961) was reorganized in January 1964 into the Scientific Council for Geothermal Research under the USSR Academy of Sciences. This Council was strictly an administrative body, as all geothermal research was conducted and financed by various institutes, laboratories, universities, and several branches of the Academy of Sciences. Many valuable studies and collected works on geothermy and related subjects were published by the All-Union Scientific Research Institute for Geology and Engineering Geology, the Ministry of Geology, the Geological Institute, the USSR Academy of Sciences and its various branches, the Laboratory for Hydrogeological Problems imeni F.P. Savarensky, the Institute of Volcanology, and many others.

Presently, geothermal resources are considered as a new branch of the national economy. The Ministry of the Gas Industry has established several organizations and prospecting expeditions for broader development of geothermal resources in general, and multipurpose exploitation of hot water and steam in particular.

In connection with these, considerable activity is underway in Dagestan, the Chechen-Ingush ASSR, the Georgian SSR, Kabardino-Balkar, northern Osetiya, Stavropol', Krasnodarskiy Kray, Kazakhstan, and Kamchatka Oblast, while activities in the Caucasus and Kurile and Sakhalin Islands will be stepped up. To meet increasing demands for hot water, heating, and hothouse-greenhouse facilities, the Ministry's goal for the current Five-Year Plan (1971-1975) is to drill over 10,000 meters of additional geothermal test wells.

Since 1967, the utilization of geothermal water in the USSR has increased about 16 times. Several towns and industries have been supplied with hot water and heating, and the largest hothouse-greenhouse complex, completed at Paratunka in 1971, provides year-around fresh vegetables (previously imported from Vladivostok) to Petropavlovsk-Kamchatskiy and the surrounding area.

It has been estimated that the geothermal resources of the Soviet Union have a daily capacity of 22 million cubic meters of hot water (one of the world's largest reservoirs of hot water, extending over an area of about three million square kilometers, is situated in western Siberia) and 430,000 tons of steam. If fully utilized, this would save about 40 million tons of conventional fuel annually. The annual output set for 1975 is to produce about 15 million cubic meters of hot water and 470,000 tons of steam.

However, the development of geothermal resources is under the control of the Main Gas Production Administration of the Ministry of the Gas Industry, where a special Geothermy Department was established in 1963. The above Administration is primarily concerned with the extraction of natural gas, while geothermal resource development appears to be a secondary issue. The existence of this situation would not appear to be conducive to the fulfillment of the projected goals.

There is evidence, corroborated by several Soviet geothermal research scientists, that the economic benefits to accrue from geothermal resources have been somewhat ignored and not revealed to the public as being comparatively cheaper than conventional resources, such as gas, oil, coal, or peat. On Kamchatka, for example, the geothermal heat is 6 to 15 times cheaper than heat produced by conventional thermal plants.

Many geothermal experts are of the opinion that geothermal development cannot thrive as a separate discipline as long as it is subordinate to the gas industry. They further believe that now is the time for the establishment of a special organization which would be responsible for the promotion and development of this growing branch of the national economy.

Trends in Geothermal Research

In spite of all the organizational changes and complexities of geothermal research, Soviet basic and applied research in geothermy has achieved considerable results for both the scientific world and the Soviet national economy. In general, this research is concerned with broad theoretical investigations, improvement of equipment and methods, and more intensive field exploration and prospecting.

In the last two decades, the hydrogeological deep drilling activities have contributed to the development of the new science of hydrogeothermy, which was applied to regional projects in the exploration and mapping of the geothermal characteristics of the USSR, ground water table, oil-bearing areas, and coal regions in the Donbas and other areas.

In 1967, the Aerial Surveying Laboratory of the USSR Ministry of Geology conducted the first exploratory survey on Kamchatka peninsula to determine the possibllity of utilizing aerial infrared mapping techniques in studying volcanos and thermal anomalies. This survey concentrated primarily on the thermal characteristics of geysers and fumaroles, hot and warm springs, mud volcanos, mud pots, thermal streams and lakes, and other areas with thermal anomalies. The data obtained were correlated with field data and were used for mapping geothermal areas under consideration for development of their geothermal resources.

The Soviets have published a number of studies covering diverse subjects relating to various geothermal systems. Included among several subjects in these studies are: the measurement of the Moon's heat flow; the practical utilization of terrestrial heat; and the possible utilization of ground waters with low enthalpy, found in great quantities in the Soviet Union.

Based on regional synthesis of geothermal data, it is concluded that continental drift, which has separated the European platform from the Siberian platform, is manifested in the formation of a wide tectonic depression. The Soviets postulate that in the continental interior, the thermal energy released from the deepest part of the earth (left uncovered by drift movements) is not dissipated by the cooling effect of oceans (which would come to occupy dislocated areas) and the energy thus maintains itself over a long period of time, creating regions with high geothermal anomalies.

The sources of water and the nature of geological formations constitute a basis for determining the location and type of geothermal systems, which can be categorized as either an "open system" or a "buried system". The open systems are fed from the surface and are open throughout their extent. To these types are related some manifestations observed at several volcanos in the more elevated areas of Kamchatka.

The buried systems are those which utilize fossil waters or waters released by sediments under geostatic pressure. A similar situation is verified in the Caucasus where attempts have been made to utilize depleted oil deposit sites by injecting water into certain wells and extracting hot water from adjacent wells. Presently, such water is being used for industrial and agricultural purposes.

A different situtation exists in the Georgian SSR, where there are more than fifty groups of thermal springs with temperatures ranging between 35° and 105°C, and with a total yield of more than 1,000 liters per second. Although quite complex, the hydrogeological conditions in the Georgian SSR come within the scheme of ground hydrology.

Water circulation in some cases is at relatively deep horizons where the heat flow varies from moderate values for the Georgian Block (0.68-1.22 microcal per cm^3 sec) to rather high values in the Adzhar region (1.49-2.03 microcal per cm^3 sec). However, considering the situation in Kamchatka as a whole, and the Pauzhetka system in particular, there are indications that the cold, thermal, and hyperthermal waters are hydraulically interconnected.

In consequence, the piezometric levels of the Kamchatka thermal waters, measured in wells of the region's geothermal system, agree with the static levels of the local cold waters. This interference phenomenon between the cold descending flow and the hot portion of the system, results in a general decrease in the temperature of the extracted fluids, and in an increase of water quantity drawn to the surface from the wells. Flashing occurs only near the well bottom or actually inside the well thus creating scales which obstruct the fracture and reduce the diameter of the well, causing considerable reductions in yield and in some cases, an actual stoppage.

In specific cases, there is the possibility that a system has an almost direct connection between the recharge and discharge areas. This opinion has been supported by a recent Soviet study which has shed new light on the origins of geothermal water. Field data, collected over various areas, on the chemical components, isotopes, and organic material in geothermal waters, including the hydrothermy of recent volcanism, indicate regional characteristics and continuity of such waters in the upper hydrogeosphere. The interpretation of the data obtained is based on geotectonic, lithologic-facial, geothermal, and geochemical characteristics of the system under consideration. It is concluded that the basic sources of water, mineral salts, gases, and organic material in the thermal waters are the result of atmospheric precipitation and filtered sea water.

Considerable attention is given to the condensation of steam in contact with the coldest levels of the reservoir top, particularly with those levels which correspond to the peripheral parts of the production areas. Regular observations conducted over many years on Kamchatka, indicate that hydrothermal activity was unaffected by exploitation.

However, a different picture is seen for the Pauzhetka wells where, through exploitation, a considerable decrease in the piezometric levels (and hence in hydrostatic pressure) has been observed. It is noteworthy that, during test drilling of the Pauzhetka deposit, a sharp decrease in piezometric level was accompanied by the transformation of a constantly boiling spring into a geyser with reduced output from 10 liters/sec to 0.6 liter/sec. With the termination of test drilling and the restoration of piezometric level, the former constant discharge was resumed. This transformation, involving hydrodynamic variations in the hydrothermal water system, shows conclusively that geysers and boiling springs share the same nature, and also that geysers are a special type of boiling spring.

A decrease in hydrostatic pressure in areas with low piezometric level gives more intensive steam separation and thus, more surface steam discharge. This correlates with the fact that, at the Pauzhetka site, the surface thermal manifestations are not the result of changes in the temperature of the water-bearing complex. The above described examples demonstrate that geothermal systems must be classified and evaluated according to their types and potentials in order to predict the characteristics which they will display after a certain period of use. It is therefore important to broaden the base of experience and knowledge in forecasting the various modifications of a producing geothermal system and to predict the production time under varying conditions.

Soviet scientists stress the great importance of models based on numerical, analog, or experimental methods, in the study of geothermal systems. Heat and mass transfer, important components of a hydrothermal system, are the bases of a Soviet attempt to create a mathematical model of a selected system.

Presently, the classifications of geothermal fields are based on chemical or thermodynamic criteria. There are proposals by several scientists that classification should be based also on geological criteria. However, some Soviet scientists differ with the present classification system of geothermal fields and suggest a dual classification of occurrence of thermal waters based on tectonic setting and type of heat source. Other proposals suggest the use of geological age, as well as tectonic environment.

Based on the above classifications, Soviet scientists conclude that water having moderate to high thermal potential and in large quantities, can only be found in Cenozoic folded regions, in older regions reactivated in the Cenozoic age, and in the interiors of Paleozoic platforms. In addition, they distinguish between waters occurring in "fissure/veins" and those in permeable stratigraphic horizons called "stratal" deposits. The exploitable ratio is approximately 1 to 20, respectively. Based on the above geological differences, the thermal gradient, depending on the tectonic structure of the region, varies from formation to formation.

For example, in the western region of the Georgian SSR, the 50°C isotherm ranges in depth from 2,000 meters near the Black Sea coast, and only 800 meters inland averages 1,400 meters overall. This gives a maximum gradient of about 42°C/km, and an average gradient of only 21°C/km. In the eastern region of the Georgian SSR, the gradient ranges between 27° and 40°C/km. In general, the temperature gradients in these basins are from one and a half to two times the world average, making it possible to extract water at 80° to 100°C from relatively shallow depths over extensive areas.

Geothermal fields found in platforms and foredeep areas are generally characterized by a normal heat flow, and they have geothermal gradients close to the world average. The thermal gradients in the Ciscaucasus foredeep average about 30°C/km, and thermal waters ranging in temperature form 47° to 86°C are used extensively for balneological purposes, hothousing, and the dairy industry.

The occurrence of exploitable waters at a depth of 2 kilometers in the platform areas of the USSR, has been established in the West Siberian lowland (60° to 80°C) and in the Turana lowland, east of the Caspian Sea, (65° to 95°C).

In early 1972, during intensive field research on geothermal gradients, heat flow, and coefficient of conductivity in the area of the Koryak-Avachin and Paratunka depressions, the Soviets obtained conclusive data that the volume and distribution of anomalous heat flow rapidly attenuates near hydrothermal systems or magma chambers. This has been considered as a valuable indicator in prospecting for geothermal waters with the exploratory thermal corings.

Geochemistry has an important role in solving various problems connected with geothermal resource development, such as evaluation of reservoir energy reserve, reservoir depletion, power plant design, corrosion, scale formation, and effluent disposal. The following chemical indicators of subsurface temperatures in hot water systems is in broad use for evaluation reservoir potential:

Indicator	Comments
(1) SiO_2 content	Best of indicators; assumes quartz equilibrium at high temperature with no dilution or precipitation after cooling.
(2) Na/K	Generally significant for ratios between 20/1 to 8/1 and for some systems outside these limits.
(3) Ca and HCO_3 contents	Qualitatively useful for near-neutral

(continued)

Indicator	Comments
	waters; solubility of $CaCO_3$ inversely related to subsurface temperatures.
(4) Mg; Mg/Ca	Low values indicate high subsurface temperature, and vice versa.
(5) Cl dilution	Assumes dilution of lower-Cl springs by cold water, permitting calculation of subsurface temperatures from required mixing ratios with highest Cl waters.
(6) Na/Ca	High ratios may indicate high temperatures but not for high-Ca brines; less direct than 3?
(7) $Cl/HCO_3 + CO_3$	Highest ratios in related waters indicate highest subsurface temperatures and vice versa.
(8) Cl/F	High ratios may indicate high temperatures but Ca content (as controlled by pH and $CO_3^{=}$ contents) prevents quantitative application.
(9) H_2/other gases	High ratios qualitatively indicate high temperatures.
(10) Sinter deposits	Reliable indicator of subsurface temperatures (now or formerly) >180°C.
(11) Travertine deposits	Strong indicator of low subsurface temperatures unless bicarbonate waters have contacted limestone after cooling.

Based on the chemical analysis and measured subsurface temperatures from several drill holes, Table 2.1 shows data summarized for several Soviet geothermal areas presently under exploitation or in the final planning stage.

TABLE 2.1: INDICATED AND OBSERVED SUBSURFACE TEMPERATURES IN EXPLORED USSR GEOTHERMAL AREAS

Area	SiO_2 Content, ppm	Na/K Ratio Atomic	T_{SiO_2}, °C	$T_{Na/K}$ °C	T_{Max}, °C Observed
Pauzhetka,					
Spring P. 1 (~100°C)	166	23.2	164	~156	
Well 4	302	16.0	194	193	195
Uzon-Geyzernyy					
Geyser Velikan					
(~100°C)	294	16.9	200	187	
Bol'she-Bannaya					
Spring 4 (~100°C)	165	20.7	163	~167	
Well 35	223	22.0	177	~161	171
Paratunka					
Spring (42.5°C)	62	63.1	111	~83	
Well 2	25	35.0	70(?)	~123	106

Source: AD-754 947

The data in Table 2.1 are fundamental for designing geothermal electric power plant of respective capacity and for selecting corrosion resistive material.

Abundant textual and graphical data on geothermy have contributed extensively to present Soviet geothermal engineering activities in the Crimea, Caucasus, Soviet Central Asia, western and eastern Siberia, and the Far East, including Kamchatka peninsula and the Kurile Islands.

Geothermal Development

In the spring of 1941, the Soviet botanist Tatyana Ustinova made a major geographical discovery of powerful geysers in the southeastern part of the Kamchatka peninsula. Before that time, geysers were known only in Iceland, New Zealand, and the United States. Ustinova's discovery was to a certain extent a sensation, for in the middle of the 20th century, such a discovery is not frequent.

The first practical utilization of geothermal energy in Kamchatka started at the peninsula's southern tip near the large fish-processing plant at Ozernovskiy. Here, in the Pauzhetka river valley, between the Kambal'nyy and Koshelev volcanos, a vast territory of self-discharging hot water and steam vents became the site of the Soviet Union's first geothermal electric power station. Although its capacity was not great, this project began a new chapter in Soviet power engineering and the development of the Soviet Far East.

In general, a long delay between the prospecting phase and the ultimate generation of electric power is common to all geothermal plants. In many cases, a delay of five years or more noticeably increases the capital cost of geothermal energy and it becomes a very serious obstacle to procuring the large amount of capital required for the exploration.

During the last nine years, the installed capacity, of geothermal power in the world increased from 385.7 Mw in 1961 to 860.0 Mw, (projected for 1971), which is considered to be a very slow growth rate. Several projects scheduled for completion in the next few years will increase the world output to over 1,000 Mw. It seems a very large development, but it is a very small fraction of the power produced by various countries considering the enormous reserves of the global geothermal resources. Selection of power plant size is the most difficult decision for geothermal plant planners. The cost per kw decreases when the size of plant increases; however, a large plant requires more producing wells and a longer time for completion.

The Paratunka power plant (USSR), the first Freon-operated plant in the world, offers convincing evidence that a power plant using natural hot water below the boiling point may offer a convenient solution. This type of geothermal power plant can be designed and built within the framework of existing engineering knowledge, technology, and construction capabilities; no revolutionary new design or new methods are required. A Freon or isobutane hot-water plant may be designed and constructed without major innovations within existing production technology. In this regard, very large reservoirs of hot water above 80°C exist in many countries and can be easily harnessed by present technology.

Geothermal developments in the Soviet Union, for the most part, have followed an original pattern, differing significantly from the development patterns of other countries. In a recent study, it is estimated that 50 to 60% of the Soviet Union has deposits of thermal waters which are available for economically effective use. The use potential of these deposits are comparable to the coal, oil, gas, and peat resources of the Soviet Union taken together.

The thermal water reserves of the USSR at depths between 1,000 and 3,500 meters, with temperatures ranging between 50° and 130°C, have an estimated output of 7.9 million cubic meters daily. About 70% of these thermal waters are at depths between 1,000 and 1,500 meters.

Presently, there are eleven geothermal facilities in operation in the Soviet Union, with a total amount of consumed heat at 200 Gcal/hr, or about the equivalent of 125,000 tons of conventional fuel. An increase of at least 10 times this figure is forecast for the end of 1980.

The basic factors governing the technical feasibility of utilizing geothermal waters are: temperature potential, discharge rate, depth of occurrence, pressure, chemical composition (mineralization, hardness, content of free carbon dioxide and hydrogen sulfide), and the amount of organic material and suspensions.

The economic expediency in utilizing geothermal waters depends on the geographic location of the source, accessibility and the condition of roads, the overall economic development of the region, and the availability of other types of energy resources. Based on hydrogeological and geothermal characterisitcs, the Soviet Union is subdivided into nine regions having different priorities and prospects for geothermal resource development.

To meet the technical-economic criteria, the engineering requirements for practical exploitation of a geothermal site in various regions and for different industrial applications are given in Table 2.2 on the following page.

It is estimated that there are over 150 areas and groups of individual springs with temperatures above 40°C. Some 19 of these have near-boiling temperatures, and about 40 have temperatures ranging between 60° and 90°C. The total discharge from springs having temperatures above 40°C is estimated as being over 140,000 cubic meters per second, with a heat equivalent of 10 billion kilocalories daily. Recent research and prospecting activities in various Soviet republics foresee a complete inventory of geothermal resources as a contribution towards their more intensive utilization for heating, electric power, and other applications. The following are some of the major future developments planned for various republics and areas.

Ukrainian SSR: Utilization of all thermal waters with various temperatures (70° to 80°C) and different degrees of mineralization for experimental-industrial installations.

Moldavian SSR: Pilot projects contemplate the use of thermal waters with different degrees of mineralization for agriculture, with new wells to be drilled near the towns of Kishinev, Ungen, and Bendery.

TABLE 2.2: ENGINEERING REQUIREMENTS

Type of Utilization	Temperature, °C	Daily Output, m³	Depth of Water-Bearing Horizon, m	Mineralization, g/l
Production of electric power by direct steam cycle	100	10,000	3,100	2-4(50)*
Production of electric power by intermediate low-boiling agent (Freon, isobutane, etc.)	60-90	2,500	2,500	50
Heat supply for populated places	70-90	1,000	2,500	2(50)*
Refrigeration	70	500	1,500	50
Hot-water supply	40-60	1,000	1,500	1(50)*
Hothousing-greenhousing	40-70	500	1,500	10(50)*
Thermal irrigation	25-40	250	1,000	2
Heating of soil (for agriculture)**	25-50	500	1,500	50
Thawing of frozen soil***	25-50	250	3,000	50
Swimming pools & baths	25-40	250	1,000	50

*Figure in parentheses indicates permissible mineralization for geothermal plants with heat exchangers.
**During winger months or in permafrost regions.
***Mining in winter months, in permafrost, or in the Arctic region.

Source: AD-754 947

Crimea: Drilling is in progress at Dzhankoy, Feodosiya and Novoselovo, where surface manifestations of thermal water are present, but where additional surveying is required to determine daily discharge, chemical composition, and temperature.

Georgian SSR: Thermal waters will be used in growing citrus, tea, and grapes, as well as for resort centers in the areas of Pitsunda, Zugdidn, Alazan, and others. At Tbilisi, it is proposed to obtain thermal water for experimental-industrial installations.

Northern Caucasus: Plans call for tapping thermal waters having various degrees of mineralization (10 to 50 g/l) and temperatures (60° to 100°C) for agriculture purposes, as well as for heating the towns Cherkassk, Armavir, and Nal'chik.

Azerbaydzhan SSR: Exploratory drilling is planned for Lenkoran and Masallinsk areas in hopes of providing thermal waters at temperatures ranging between 50°C and 60°C for heating and hothousing complexes which will serve the European part of the Soviet Union. Presently, at Istisu, thermal water (54.5°C at the surface and 64°C at 50 m) is being used for balneological purposes.

Armenian SSR: Near the cities of Yerevan, Nalband, and Ankavana, prospecting work is scheduled to tap thermal waters with estimated temperatures ranging between 70° and 75°C.

Dagestan ASSR: Near the town of Makhachkala, drilling to a depth of 4,500 meters is in progress to reach steam-water layers for a planned experimental-industrial geothermal electric station. Near the town of Berekey there is a highly mineralized hot water and steam well with a capacity of 70,000 cubic meters daily and a temperature at the surface of 57°C. It is estimated that this well can produce about 94.54×10^{10} kcal of hot water annually, which is equivalent to 94,535 tons of fuel oil or 107,456 tons of coal.

Kazakh SSR: Extensive prospecting is being conducted to determine regional thermal water characteristics (pressure, discharge, chemical composition, temperature, depth and capacity of water-bearing horizons), especially in the regions of Chimkent, Turkestan, Turgay, and northern Pavlodar Oblast, for ultimate agricultural applications in arid and semiarid regions.

Uzbek SSR: Extensive exploratory works are in the areas of Tashkent and Fergana.

Kirgiz SSR: Prospecting is planned for the Przheval'sk area to develop thermal water resources for space heating. In the Issyk-Kul' lake region, additional drillings will be conducted to determine balneological capabilities and in particular, space heating potential for several resort centers under construction.

Tadzhik SSR: Near the city of Dushanbe, several drill sites for thermal waters are planned, to provide heating and airconditioning to the city. At the resort centers of Khodzha-Obi-Garm, Obi-Garm, and Obi-Garm, and Khabatag, the plans to increase output of thermal water and steam for balneological and heating purposes are in the final stage.

Western Siberia: To promote balneological resort centers, extensive drilling for thermal waters, ranging in temperature between 50° and 90°C with mineralization between 10 to 30 g/l, is in the planning stage.

Chukchi Peninsula: Exploratory drillings on Chukchi peninsula and in the coastal permafrost region bordering the Sea of Okhotsk will assist in studying the technical-economic indices of thermal waters for space heating, vegetable farming, and resort centers. Near the towns of Talaya and Motykleysk (about 125 km west of Magadan) exploratory drilling is in the preparatory stage, with the ultimate goal of obtaining fresh and low-salinity thermal waters for resort centers.

Under primary consideration for the production of electric power are the areas of Yevpatoriya, Nal'chik, Tbilisi, Tobol'sk, Tashkent, Bol'she-Bannaya, Pauzhetka and Sredne-Paratunka on the Kamchatka peninsula, Goryachiy Plyazh on Kunashir Island, Paramushir Island, and Chaplino on the Chukchi peninsula.

SOVIET GEOTHERMAL ELECTRIC POWER GENERATION

The enormous USSR reserves of geothermal waters having temperatures between 30° and 180°C could be utilized for purposes vital to the national economy, other than the generating of electric power. The most efficient way is to use a low boiling heat carrier as a second working fluid. In this line, the experimental Freon turbine power station represents a revolutionary innovation and appears to hold great promise for future Soviet geothermal power development and engineering. The following are some highlights of Soviet engineering efforts toward solving some problems in geothermal power generation.

The Soviets have conducted extensive studies on the corrosion effects of geothermal waters on metals and alloys having various degrees of corrosion resistance, such as steel, chrome-nickel steel, gray iron, galvanized steel, copper, and brass. It has been established that for geothermal water containing soluble hydrogen sulfide or carbon dioxide, at a temperature of 70°C, the corrosion rate of low-carbon steel increases by about 25 to 30 times; with the water aerated and at a temperature of about 55°C, corrosion increases only by 10 times. To capitalize on this, the geothermal water is not used directly from the well, but is collected in an open reservoir for aeration, and is then fed into the heating system.

The Electric Welding Institute imeni Ye. O. Paton of the Ukrainian Academy of Sciences has developed several types of cavitation and heat-resistant austenitic steels, characterized by high weldability and stability through work hardening.

Soviets use low-level turbines with external barometric condensers, while other countries use high-level turbines placed above the condensers. The extraction of noncondensable gases from the condenser is achieved by water ejectors, while some countries use steam ejectors, rotary exhausts, and reciprocating pumps.

Regarding the steam-control method, the steam is maintained at a constant pressure by blowing off variable quantities to waste, instead of allowing the steam to float to some extent by throttling at the turbine and thus reduce possible blow-offs. The turbine's stop valve pressures, at the first point of feed from the field (i.e., excluding turbines fed from the exhaust from other turbines or flash steam), are generally very low.

As the first phase in the utilization of geothermal resources, the Soviets are presently concerned with the construction of small capacity geothermal electric power stations and heat-supply systems.

An advanced idea, originated by the Institute of Thermal Power Engineering of the Ukrainian Academy of Sciences and supported by the Design Institute of the Thermal Electric Station Planning Enterprise, to design a very large geothermal power station for a highly superheated mixture of water and steam from a depth ranging between 7 and 10 kilometers, is under serious consideration. The basic concept of this plan is to pump the surface water into the well and then lift it heated to 300° to 400°C by the porous masses of the earth's interior. Such a method holds promise for building in the future a power plant of up to 10 million kilowatts capacity.

Paratunka Geothermal Electric Power Station

The Paratunka geothermal electric station is the first station in the world to operate on Freon. The Thermal Physics Institute of the Siberian Branch of the USSR Academy of Sciences, after extensive research designed a turbo-unit called Shatura HEPP-5. A prototype model of this Freon unit of 340 kw was constructed and tested. Another unit, the UEF90/65, designed on the same principles, was assembled and shipped to Sredne Paratunka thermal springs for subsequent testing. After construction of the pumping station and the pipeline system, this plant was completed in September 1967. A series of tests demonstrated that this power plant operates satisfactorily using thermal water with an 81.5°C temperature, instead of the design temperature of 90°C.

The operating principles of this plant, schematically shown in Figure 2.1, are as follows: the liquid Freon, preheated by the thermal water (temperature 83°C), boils and overheats to 55°C at an absolute pressure of 13.8 atmospheres in the main boiler **3** consisting of three preheaters, boilers, and superheaters. The Freon steam, preheated up to 65°C, enters the radial Freon-turbine **2** where it expands, reaching the required condensation pressure for the operation of the turbine.

The turbine is connected to a T2-75-2 generator **1**. From the turbine, the Freon steam enters two tube-type condensers **5**, where it is cooled by cold water (15°C), at an absolute pressure of 5 atmospheres. From the condensers, the liquid Freon flows into the line receiver and is returned again to the main boiler by the centrifugal pumps.

FIGURE 2.1: SCHEMATIC OF THE ENERGY CYCLE FOR THE PARATUNKA GEOTHERMAL ELECTRIC STATION

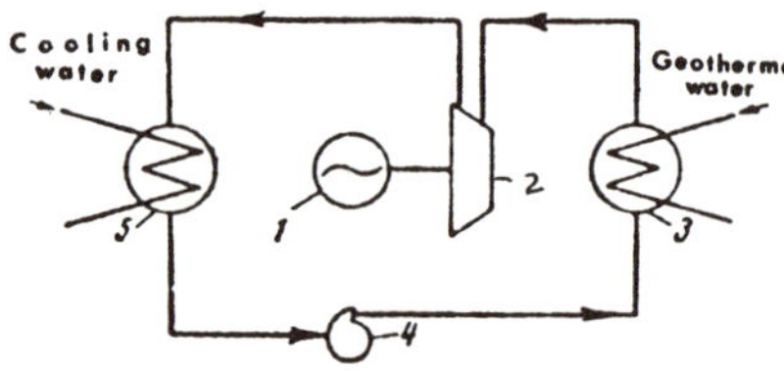

Source: AD-754 947

This plant has a receiving charge-discharge unit consisting of two receivers having a capacity of 6 cubic meters each, and two piston-type Freon compressor-condenser units designed for the removal of Freon from the system, the resupply of Freon, and the refilling of the system with Freon after installation or maintenance.

The maximum consumption of geothermal water (including water used to heat the station) is 289 cubic meters per hour while the consumption of water for cooling is 1,520 cubic meters per hour. In selecting the site for a power plant, besides the existing geothermal waters, the volume of water available for cooling cycles is also of considerable importance.

The distance between the geothermal wells and the station is about 1,000 meters. With the completion of the power station building in late 1964, eight geothermal wells had been finished, ranging in depth between 302 and 604 meters with the diameters ranging between 123 and 200 millimeters. At the wellhead, the thermal water attains a head of 30 meters.

The discharge of free-flow thermal water at each wellhead was registered at about 18 liters per second, and the average temperature was 83.7°C. To eliminate head decrease (about 0.55 liter per second per meter), booster-pumps were installed in order to maintain the required operating head.

To provide thermal water of the required discharge rate, six wells are used with one well in reserve. Water for cooling is brought from the Paratunka river into the pumping station by two suction pipes; from here it is pumped through pressure pipes into the power station.

The Paratunka station consists of two units. In the main unit is the machine hall, and in the annex is the control panel, machine shop, chemical laboratory, equipment and mercury rooms, ventilation chamber, and utility room. For the first two years, the annex unit was used as an experimental laboratory by the Thermal Physics Institute. The machine hall is 12 meters wide, 24 meters long, and 8 meters high, while the annex is 12 x 12 meters, with a height of 3.6 to 4.0 meters.

The electric station has the following technical-economic characteristics: installed capacity, 750 kw (1967), 1,000 kw (1972); consumption of electric energy by the power station, 35% (1967); and cost of 1 kw of installed capacity, 1,600 rubles (1967).

The high cost of the installed capacity (4 times higher than that of the Pauzhetka geothermal electric power station operating on a steam-water mixture) is explained by the comparatively small capacity of the station and the excessive construction of pipes and other installations for the supply of thermal waters to neighboring hothousing-greenhousing facilities. In addition, the final cost included all expenses in connection with planning, construction and testing of the Freon turbo-unit (Shatura HEPP-5).

Pauzhetka Geothermal Electric Power Station

The Pauzhetka geothermal electric station, located in the Pauzhetka river valley, was commissioned in July 1967. In the Pauzhetka river valley, during 1965, 22 wells were drilled to a maximum depth of 800 meters, the average depth being 300 meters. The highest temperature recorded at the surface was 187°C, with the average temperature at 170°C. Surface pressures ranged between 2.2 and 6.7 atmospheres, with mineralization between 3 and 3.5 grams per liter. Average discharge capacity of the hot water-steam mixture has been estimated at 170 kilograms per second.

According to original design estimates, this station was to operate on a steam-hot water mixture with a natural heat capacity of 170 kcal/kg at the surface obtainable from wells 100 to 400 meters deep, and was to have the following technical-economic indices. The 1963 estimate of electricity consumed by the station was 1.6%.

Initiating capacity	5 Mw
Total discharge of wells	500 tons per hour
Hours in operation	7,000 annually
Annual production of electric energy	35 million kw/hr
Annual loss of heat	25,000 Gcal
Cost of 1 kw of intalled capacity	400 rubles
Cost per kw/hr	0.65 kopeck
Electricity consumed by station	10%
Efficiency factor of the station (net)	9.2%
Operating personnel	21
Staff coefficient	4 men per 1,000 kw

In the first stage of construction (1966), an initiating capacity of 3 Mw was to be produced by two turbines operating with a pressure of 1.2 atmospheres and utilizing seven geothermal wells. An eighth well was used for research and as a reserve well.

In the second stage of the construction (1967), by drilling two additional wells, a considerable increase in discharge and pressure was to be obtained. Increased pressure from 1.2 to 2.0 atmospheres consequently was to increase the capacity of the station to 5 Mw, which is the current figure for the station. Below is a schematic outline of the Pauzhetka geothermal station (see Figure 2.2).

FIGURE 2.2: SCHEMATIC OF THE 5 Mw PAUZHETKA GEOTHERMAL ELECTRIC POWER STATION

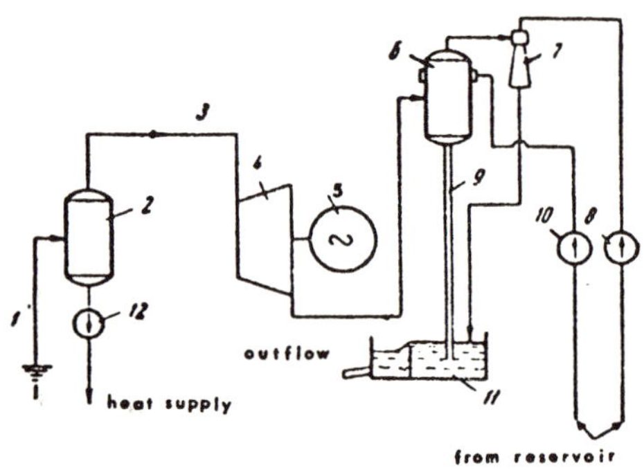

1 geothermal well
2 separator
3 steam pipe
4 turbine
5 generator
6 mixing condenser
7 water-jet ejector
8 ejector pump
9 barometric tube
10 cool water pump
11 outflow of hot water

Source: AD-754 947

The third stage of construction, planned 1968, to increase the capacity of the station to 12.5 Mw in 1968, did not materialize.

A schematic outline of the Pauzhetka geothermal electric power station with a planned capacity of 12.5 Mw is given below (see Figure 2.3). Besides the unfulfilled plans for expansion of this station up to 12.5 Mw, there reportedly have been other, additional plans to bring this station up to a final capacity of 20 Mw.

FIGURE 2.3: PLANNED PAUZHETKA GEOTHERMAL ELECTRIC STATION WITH 12.5 Mw CAPACITY

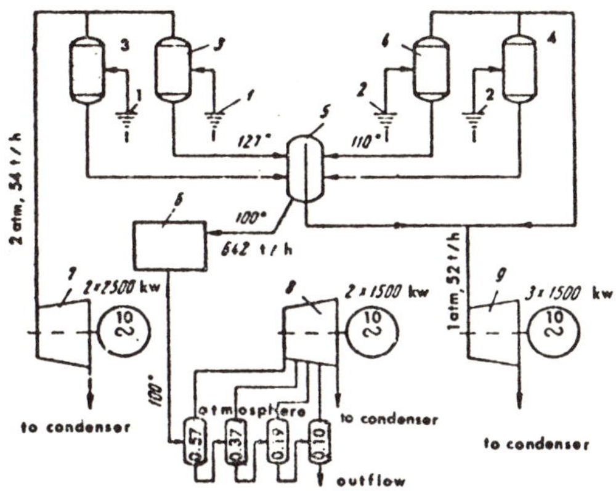

1 wells with 2.5 atm
2 wells with 1.5 atm
3 separators with 2.5 atm
4 separators with 1.5 atm
5 expander for 1 atm
6 hot water storage tank
7 2 turbines (each 2,500 kw at 2 atm)
8 2 vacuum turbines (each 1,500 kw)
9 3 turbines (each 1,500 kw at 1 atm)
10 generators

Source: AD-754 947

The main building is 4,400 cubic meters in size and is connected to the wells by pipes having a total length of 1.3 kilometers. The machine hall floor space is 33 by 9 meters, and the control room is housed in a 4-meter-wide auxiliary building. The Mk-2.5 turbines were manufactured by the Thermal Turbine Machine Construction Plant at Kaluga.

The Pauzhetka geothermal electric station supplies electricity over a 30-km-long, 35 kv power line from a 5,600 kv transformer to the Ozernovsk fishing combine, the "Krasnyy Truzhenik" collective farm, and the village of Pauzhetka.

The production cost of electricity at Pauzhetka has been estimated at 0.55 kopeck per kw/hr (1963), 0.65 kopeck per kw/hr (1965), and 7.2 US mills per kw/hr (1970). Considering the size of the plant and its remote location, the last figure given seems to be a very satisfactory one, and it will undoubtedly be much less when the plant is expanded. The Soviets claim that it is 30% cheaper than any alternative means of power supplied in the same area, and it has been estimated that this station will pay for itself within 3 to 4 years of operation.

FUTURE PROGRAMS

Besides the existing geothermal power stations, there are several prospective sites for the construction of additional geothermal power stations. For instance, in the Tersko-Kumsk artesian basin (towns of Makhachkala, Khasavyurt, Kizlyar, and others) geothermal waters have been found at a depth of 2,500 to 3,000 meters, with temperatures of about 120°C and discharge capacity ranging between 1,500 and 3,000 cubic meters daily. The pressures range between 5 and 10 atmospheres, with mineralization less than 2.5 grams per liter.

There are also several regions being considered for the future construction of geothermal power stations utilizing low-temperature geothermal waters. Additional potential sites for low-temperature geothermal power stations utilizing fissure waters are located in the northeastern USSR; i.e., the Mogoysk and Uakitsk hot springs (Buryat ASSR), and the Mechigmen, Sinyavina, and Chaplino hot springs in the Chukotskiy (Chukchi) Natsional'nyy Okrug. Below are available data on several geothermal power stations planned for construction in the near future.

Bol'she-Bannaya Geothermal Electric Power Station

The construction of this station, about 60 kilometers from the town of Petropavlovsk-Kamchatskiy, has been under consideration since 1965. At present, only 20 geothermal wells have been drilled, with only minor construction activities under way at the site.

During exploratory work in 1965, some wells produced about 73 kilograms per second of steam-hot water mixture with an average heat capacity of 150 kcal/kg at natural discharge. With additional wells, there is a possibility of increasing the volume up to 200 kilograms per second. It is estimated that this volume will warrant the operation of an electric station with an 8 Mw capacity. Below is a schematic layout of the first stage for the proposed power station (see Figure 2.4).

All wells will be directly connected to the separators **2**. Steam with pressure between 1 and 1.2 atmospheres will be channeled into the turbines **3**, while the hot water will be channeled into the expander **4** in order to reach the required pressure of one atmosphere. From the expander **4**, hot water (100°C) will enter the hot-water storage tank **5** and then the vacuum turbines **6**.

Based on Soviet estimates, the cost of the station will be about 2.5 million rubles. The initiating capacity of 8 Mw is rated as minimal, since the station will start operation with a steam-hot water mixture of only 200 kg/sec discharge. To increase capacity, the use of low-boiling agents (Freon, isobutane, and others) are under consideration. In this way it is expected to increase output by about 50%, and by operating with a steam-hot water mixture of over 200 kg/sec discharge rate, the station will have a capacity of 12 Mw. By fully utilizing the available resources (estimated at 400 kg/sec of steam-hot water mixture), the capacity eventually can be increased to 24 Mw.

The cost of electricity is estimated at 2 to 2.5 kopecks per kw/hr, representing $^1/_6$ to $^1/_7$ of the prevailing cost in the city of Petropavlovsk-Kamchatskiy.

However, the existing 20 wells represent only 50% of the geothermal resources required for the operation of a station with 24 Mw capacity, and additional prospecting drilling is needed to provide the necessary volume. The construction cost of this plant constitutes an additional problem, and the construction probably will not progress as planned.

FIGURE 2.4: SCHEMATIC OF THE FIRST STAGE OF CONSTRUCTION OF BOL'SHE-BANNAYA GEOTHERMAL POWER STATION FOR 8 Mw CAPACITY

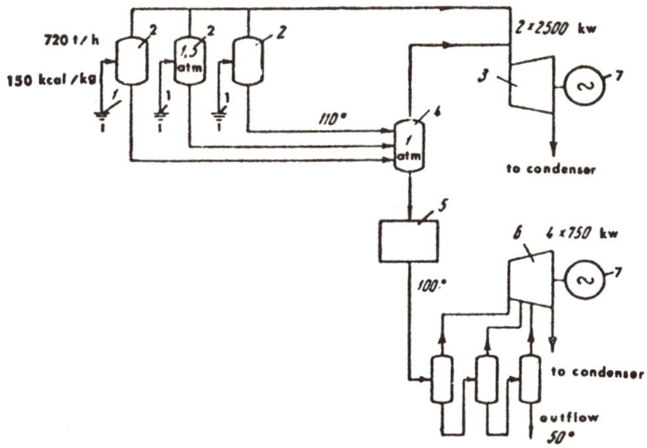

1 wells at 1.5 atm
2 separators at 1.5 atm
3 two turbines (each 2,500 kw at 1 atm)
4 expander at 1 atm
5 hot water storage tank
6 four vacuum turbines (each 750 kw)
7 generators

Source: AD-754 947

Makhachkala Geothermal Electric Power Station

This geothermal power station, designed by the Dagestan Branch of the USSR Academy of Sciences and the Design Institute of the Thermal Electric Station Planning Enterprise, in the first stage of development will have an initiating capacity of 12 Mw. There are indications that this station will be expanded to yield a much larger capacity, but for the present no definite data are available regarding the exact capacity and the date of completion.

In this region, at a depth between 2,500 and 3,000 meters, the geothermal waters attain a temperature of 120°C, with the discharge ranging between 1,500 and 3,000 cubic meters per day at a pressure of 5 to 10 atmospheres; mineralization is not more than 2.5 grams per liter. To reach geothermal waters with temperatures above 160°C, deep drilling to a depth of 4,000 to 4,500 meters has been in progress since 1965. These prospecting activities are related to Soviet plans for increasing the production capacity of this station. As part of this prospecting, at Berekey, near the town of Makhachkala in the Dagestan ASSR, drilling to a depth of 4,500 meters to reach the steam-hot water mixture is in progress. During the drilling, a highly mineralized steam-hot water mixture was obtained (from an undisclosed depth), with an estimated discharge capacity of 70,000 cubic meters daily, having a temperature of 57°C at the surface. The Soviets expect to tap about 94.54×10^{10} kcal hot water annually, which is equivalent to 94,535 tons of fuel oil, or 107,456 tons of coal.

The Makhachkala station is designed for geothermal water with a natural heat capacity of 160 kcal/kg at the surface, obtainable from depths ranging between 4,000 and 4,500 meters, to satisfy the electricity and heat requirements of the town of Makhachkala. The estimated technical-economic data for this station are summarized as follows:

Total discharge of wells	100 tons per hour
Initiating capacity	12 Mw
Number of hours in operation	8,000 annually
Annual production of electric energy	96 million kw/hr
Annual loss of heat	100,00 Gcal
Electricity consumed by the station	5%
Staff coefficient	2 men per 1,000 kw

Yuzhno-Kuril'sk Geothermal Electric Power Station

The Yuzhno-Kuril'sk geothermal power station on Kunashir Island (Kurile Island group), will have an initiating capacity of 5 to 6 Mw. The station, to be erected at the Goryachiy Plyazh site about 8 kilometers from the town of Yuzhno-Kuril'sk, will operate on geothermal water at a temperature of 130°C. There are also several steam-producing vents at this site, having temperatures ranging between 100° and 130°C.

The discharge rate of the steam-hot water mixture is not high, estimated at 1.5 kilograms per second. However, data presented by the Institute of Volcanology of the Siberian Branch of the USSR Academy of Sciences indicate that the average discharge volume is actually much larger (given as 77 kg/sec) with a total caloric value of 36 Gcal/hr. The minimum caloric value has been estimated at 130 kcal/kg. In January 1965, the Sakhalin Geological Administration started several drillings at the site, but at present no additional data are available regarding the process and the purpose of the acitivty.

The Goryachiy Plyazh area, abundant with geothermal surface manifestations, is being considered for exploitation for space heating, agriculture and other purposes, in addition to the generation of electric power. The Thermal Physics Institute of the Siberian Branch of the USSR Academy of Sciences presently is studying the technical-economic aspects of the construction of this station.

Additional Sites

Additional future developments of considerable importance have been planned for the following sites.

The Verkhne-Koshelevsk fumaroles are located about 22 kilometers southeast of the village of Ozernoye.

The Severo-Kambal'nyy fumaroles are located about 10 to 12 kilometers from the Pauzhetka station. By preliminary estimates, the discharge volume is put at 100 tons per hour of dry steam which has high energy potential. It has been estimated that one ton of hot water at a temperature of 100°C can produce 4 kw/hr of energy, while one ton of dry steam can produce about 100 kw/hr, i.e., a 25-fold increase in output of electricity, as compared with hot water.

The Zhirovskiy geothermal springs have not been studied in detail and, consequently, their data are limited. It is estimated that the discharge rate is about 15 kilograms per second, and the steam mixture has a temperature of 100°C at the surface. During 1968 to 1970, some prospecting and research activities were to take place for the purpose of producing a number of wells having a total length of 19,000 meters. From these activities, it has been established that the maximum temperature ranges between 100° and 130°C, with the natural heat capacity between 25 and 60 million calories per second with a production capacity of about 36 Mw.

OTHER APPLICATIONS OF GEOTHERMAL RESOURCES

Geothermal resources are also being considered for diverse exploitation and most possible uses, whether of low- or high-temperature sources, appear to be economically viable with great advantages to domestic and industrial development.

In the USSR, hot water is found in over 20% of the nation, although the present explored potential amounts only to 1%. It has been estimated that during 1970, the total fuel cost saving arising from the use of hot geothermal waters amounted to one million dollars. By 1980, this saving is expected to be approximately ten million dollars.

Besides generating electric power, geothermal energy has broad application in several fields of the national economy. There are presently many engineering projects of varying scale and purpose in various stages of completion across the USSR, and describing all of them would be beyond the scope of this study. Therefore, only the major projects of basic interest to the respective fields will be outlined in order to illustrate the degree and magnitude of multipurpose exploitation of geothermal resources in the USSR.

Domestic and Industrial Applications

Heating: The largest user of geothermal energy for heating is the Soviet Union, where consumption of low-temperature waters from thermal springs or drill holes represents an annual savings of about 15 million tons of conventional fuel. A ten times greater saving is expected by 1980. It is estimated that 50 to 60% of the USSR territory contains thermal waters which are economically exploitable, with the total heat value being comparable with the total coal, oil and peat resources of the country.

Considering the annual heating demand, several Soviet scientists have investigated the possibility of using geothermal water for district heating in conjunction with fossil fuel boilers operating during peak heating months. They stress the importance of seasonal demands on the overall geothermal load factor and also indicate the limitations which apply to the direct use of cooler waters for heating. To overcome the low temperature disadvantage, the Soviets are stressing the great potential of the lithium-bromide absorption machine operated with geothermal energy as a heat pump for winter heating and air conditioning in the summer.

The economic use of thermal water depends on the selection of appropriate heating systems, the discharge capacities of wells, and the temperature and chemical composition of the waters.

In case the chemical compositions are beyond permissible limits, before entering the heating system the thermal water should be channeled through a degasifier and heat exchanger. In the heat exchanger, the thermal water releases its heat into low-mineralized and chemically nonaggressive water, and then is used for the heating system.

For highly mineralized thermal waters, a centrifugal separator is required to separate the steam from water before it is introduced into the heating system. In general, geothermal water should not have more than 0.1 mg/l of oxygen, 5 mg/l of sediments, and 700 meq/l of hardness (carbonate residuals).

With regard to heating systems, among the most interesting is the system used by the town of Paratunka, Kamchatka, to heat three 48-unit apartment houses. The maximum load is 0.55 gigacalorie per liter, and geothermal water at 80°C is used for heating both the tap water and the apartments. A conventional radiator system or radiant heating with pipes embedded in the floors or ceilings is used, and the heating system is designed for a temperature drop from 80° to 40°C.

Utilizing a heat pump, part of the 40°C return water can be reheated to 60°C by extracting heat from the remainder of the return water, which in turn is cooled to 10°C. The 60°C water is mixed with the 80°C geothermal water. The temperature of this mixture can be increased as desired by an electrical peak heating unit. Increased loads during cold spells can also be met by increasing the heat extraction from the geothermal water with the heat pump.

Pipelining of hot water will improve the present poor heating conditions of Petropavlovsk-Kamchatskiy, where only about 30% of its households have individual steam heating. The town is supplied with heat from the Paratunka geothermal fields, a distance of 45 to 65 kilometers, by different means.

In the town of Makhachkala and adjacent areas, utilization of geothermal waters for heating of domestic and industrial space was initiated 24 years ago following the discovery of thermal waters as a result of oil prospecting. From about 100 oil wells, having depths between 1,200 and 1,500 meters, 40 wells produce hot water at 60° to 70°C utilized for heating about 9,000 square meters of domestic and industrial space. The capacity of the wells has been estimated at 2,000 cubic meters daily, with plans being considered to increase the heating capacity to 10,000 cubic meters daily.

At the town of Groznyy, the Caucasus Special Geothermal Directorate (Ministry of Gas Industry) recently commissioned two powerful geothermal wells which will be utilized for heating of the town, replacing the present heating system operating on gas which during the winter months proved to be insufficient.

The largest single heating system in the USSR serves three districts of the town of Cherkassk with about 19,000 inhabitants.

Hot Water Supply: Hot water from geothermal springs and wells with moderate temperatures and of required chemical composition are being used extensively for domestic, industrial, and other hot-water supply. To meet required specifications, the chemical compositions of geothermal water should not exceed one gram per liter of dry residuals, 0.5 gram per liter of sulfate, 0.35 gram per liter of chloride, and 7 meq/l of average hardness. Besides the required clarity, taste, and odor, the content of heavy metal salts, radioactive elements and other contaminants should be within the established norms. In addition, the content of lead should not exceed 0.1 mg/l, while arsenic, fluorine, copper, and zinc should not exceed 0.5, 1.5, 3.0 and 5.0 mg/l, respectively. Hot water supply systems are constructed using conventional techniques and equipment and, in many cases, in conjunction with heating systems.

Permafrost—Mining and Construction: The Soviets foresee future large-scale use of geothermal waters in mine heating to facilitate the exploitation of mineral resources in permafrost, frozen soil, and regions subject to long and severe winters.

In Magadan Oblast', the main Soviet gold mining region, scientists are studying methods for year-round hydraulic placer mining with geothermal waters. Similar methods are under consideration in mining diamonds, lead and other minerals abundant in the Taymyr, Yakutia, Kolyma, and Chukotsk regions. A study conducted by the Leningrad Mining Institute indicates that for this type of mining, the thermal water should have a temperature between 20° and 30°C and a minimum discharge of 250 cubic meters per hour.

There are two ways to obtain geothermal waters for the above purposes: by direct tapping of thermal waters with a temperature above 25°C, regardless of the degree of mineralization; and by injecting cold water into deep, specially drilled wells, where it will be heated above 25°C by the deep regional geothermal (dry heat) field. It has been estimated that one such well can produce 400 to 500 cubic meters per hour of thermal water at 20° to 80°C for a duration of 10 to 30 years, i.e., the equivalent of 2 to 20 million kcal/hr capacity.

Possibilities also exist for utilizing geothermal waters for various types of construction in permafrost and frozen soil. Applied thermal water creates a hot slurry which, through gravity, penetrates into the frozen soil and facilitates drilling of holes for pillars and other foundation components.

Refrigeration and Air Conditioning: Geothermal waters are a very cheap source for refrigeration and air conditioning in domestic, public and industrial installations, with great savings of over hundreds of thousands kilowatts of electric energy possible. Geothermal waters with temperatures above 13°C provide better results for year-round air conditioning applications. Thermal waters of 70°C require special conversion equipment. To obtain temperatures above 0°C, special lithium-bromide machines are in use, and for temperatures below zero, the ammonia-water machines are appropriate.

Below is a schematic diagram (Figure 2.5) of a lithium-bromide absorption unit. This unit was designed by the Thermophysical Institute of Siberian Branch of the USSR Academy of Sciences and was tested at the Chernigov Synthetic Fiber Plant. It operates on thermal water between 85° and 120°C.

This unit has a capacity of 2.5 Gcal/hr, is manufactured in the USSR, and is in great demand for refrigeration in various chemical industries such as synthetic rubber, ammonia and at metallurgical plants. The arrangement and operation characteristics of the unit are similar to units used in other countries. Tabulated data on cooling-heating requirements are given in Table 2.2.

FIGURE 2.5: SCHEMATIC OF A LITHIUM-BROMIDE ABSORPTION UNIT

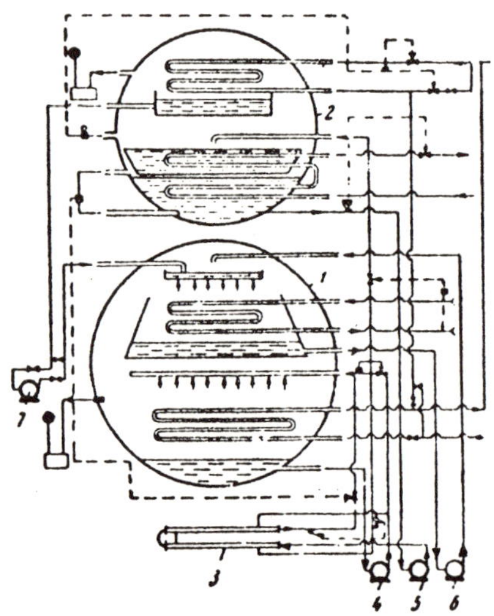

1 absorber-evaporator
2 generator-condenser
3 heat exchanger
4–7 pumps

Source: AD-754 947

Swimming Pools and Baths: Thermal waters for swimming pools and baths should have a temperature between 25° and 40°C and mineralization not exceeding 50 grams per liter. The thermal waters can be obtained chiefly from a geothermal well, geothermal electric station or geothermal heating system.

Near the town of Nal'chik, there are five geothermal wells ranging in depth between 1,600 and 2,840 meters with a free-flow discharge capacity of 9,000 cubic meters daily at temperatures between 50° and 80°C, and mineralization of 0.6 to

78 grams per liter. One well with a temperature of 80°C is supplying several swimming pools with thermal waters cooled to the required 25° to 40°C. There are many towns and resorts utilizing thermal waters for the above purposes, and only a few have been mentioned.

Agricultural Applications

Agricultural use of geothermal waters for hotbed and hothouse cultivation of vegetables has recently increased considerably. It has been estimated that one hectare of heated soil will produce year-round vegetables for a town with a population of 10,000, and for a town of 100,000 about 10 hectares will be required. The norms for above acreage are based on standards established by the Food Institute of the USSR Academy of Medical Sciences assuming an annual per capita consumption of 146 kilograms (400 grams daily) of the fresh vegetables grown in hotbeds or hothouses. It is desirable, that 25% of the above quota should be naturally grown vegetables.

In general, the water temperature varies according to the type of farming, geographic location, and season. A temperature of 60°C is required for hothousing in winter and 50°C in spring; for heated hotbeds, a temperature of 40°C is required, while for heating of soil (sheltered or unsheltered), 30°C temperatures are appropriate.

The thermal water for heating of shielded soil (by glass plates, plastics, or wooden boards) should have natural heat capacity of 2.5×10^6 cal/hr, a temperature between 35° and 40°C, and a head of at least 10 to 15 meters at the surface. Heating soil by the shielding method is said to be 4 to 5 times cheaper than by hothousing installations. This method has been used in Karnchatka, Kazakhstan and the Caucasus foreland, and has proven to be very economical.

The temperature of water used for heating soil and for thermal irrigation should be above 25°C. For thermal irrigation, the mineralization should not exceed two grams per liter.

In 1970, it was reported, that the largest hothousing complex in the world, at Sredne-Paratunka springs, was being completed and would cover an area of 60,000 square meters. Production in the initial stage is expected to be 900 to 1,000 tons of fresh vegetables annually and the final production goal is set at 2,000 tons annually.

Another hothousing complex in the planning stage is in the Pauzhetka thermal springs region, which will cover an area of 150,000 square meters and will have an estimated annual production capacity of 1,800 tons of fresh vegetables.

Near the town of Makhachkala there are several hothousing complexes with moderate production capacity. However, to increase the supply of fresh vegetables to the town of Makhachkala and adjacent areas, several hothousing projects are in the planning stages or nearing completion. It has been estimated that the production capacity of hothousing complexes in this area will amount to 3,000 tons annually.

Agricultural use of geothermal water is being considered also in Sakhalin Oblast', especially in the Kurile Islands, the Yakutsk ASSR, and Magadan Oblast', where the Soviets appear to be confronted with the immediate problem of a geothermal resources survey. In Lorino and vicinity, however, geothermal resources are already being utilized for hothousing, but on a rather moderate scale.

Medical and Health Applications

Over the years, thermal water has been used by clinics, hospitals, and health resorts for therapeutic purposes. Out of 1,500 hot water springs in the Soviet Union, over 340 are utilized exclusively by sanatoriums, hospitals, and health resorts. There are 86 mineral water bottling plants in operation.

Based on medical research and empirical experience, geothermal waters and their chemical components have a significant healing influence on various human maladies. The basic indicators for the therapeutic value of thermal waters are: mineralization, ion content, gas saturation, gaseous components, content of specific biologically active components (CO_2, H_2S, As, Fe, Br, I, H_2SiO_3), radioactivity (Rn), hydrogen ion concentration (pH), and water temperature.

To be considered for balneological use, the thermal water should meet the following norms: overall mineralization 2 g/l, content of soluble CO_2-0.5 g/l (for external application, 1.4 g/l), content of H_2S-10 mg/l, As-0.7 mg/l, Fe-20 mg/l, Br-25 mg/l, I-5 mg/l, H_2SiO_3-50 mg/l, and Rn-5mμ Curie/l. However, the above-stated content of As, Fe, Br and I pertain to therapeutic drinking water with an overall mineralization of 10 g/l.

Various therapeutic treatment and health resort facilities using low-temperature springs have been developed in the Soviet Union, especially in the Soviet Far East. The largest balneological resort center, near the town of Nal'chik, utilizes one geothermal well having a temperature of 80°C.

In the bed of dry Duzkan lake, near the town of Chardzhou (Turkmen SSR) there are many thermal water holes utilized by the local people for balneological treatment.

Considerable therapeutic value has been attributed to the Garm-Chashma thermal springs in the Anderob river valley near the town of Khorog (Tadzhik SSR). Several years ago, a balneological sanatorium was built, with an additional two presently in the planning stage.

The hot springs area in the southern part of Kamchatka Oblast' has developed considerably with modern facilities for medical treatment and recuperation. There are sanatoriums with good medical treatment facilities at Nachiki and Paratunka. Recently, several hot spring areas have been developed for medical treatment.

One such is the Magadan Oblast'. The Talaya hot springs have a well-equipped sanatorium, and the recently discovered Novoye Chaplino hot springs are in the development stage for medical treatment. Hot springs near the town of Lorino are being used for medical treatment and recuperation.

Another is the Sakhalin Oblast'. Hot springs have been discovered sporadically on the main island. For instance, at the town of Sinegorsk there is a large medical treatment facility, and the hot springs in the vicinity of Dagi have been utilized as health resorts.

However, several hot spring areas are in the planning stage for development in the near future as balneological, medical and health centers. Geothermal waters with their therapeutic potentials have been regarded as very valuable to the national economy and to the population's health.

The Hydrogeological and Hydrophysical Institute of the Kazakh Academy of Sciences recently published a comprehensive study emphasizing the curative value of geothermal and mineral waters. The study indicates that therapeutically valuable elements, such as radon, iron, hydrogen sulfide, and iodine-bromine have been found in thermal waters of southern Kazakhstan, in the Mangyshlak peninsula, and the Sary-Arki desert.

The Ministry of the Food Industry, in cooperation with the Central Scientific Research Institute for Balneology and Physiotherapy (Ministry of Health) is accelerating the production of bottled mineral waters. It has been estimated that the production capacity is over 1.5 billion bottles annually. During the current five-year plan (1971-1975) several large bottling plants are to be built and production is to be increased by about 2.4 times. One plant under consideration will have a capacity of 250 million bottles annually.

About 80% of the resources presently being utilized are located in the northern Caucasus, the Ukraine, and Transcaucasia. A considerable increase in bottling has been achieved in Soviet Central Asia, Chita and Kemerovo Oblast's, and Kamchatka peninsula. Presently, about 80 new mineral water sources are recommended for immediate exploitation.

Geothermal By-Products

Geothermal waters are enriched with about 80 different chemical elements, which can be extracted either coincident to the purification processes of thermal waters intended for geothermal electric power stations and other facilities, or specifically for the chemical industry.

The extraction of sodium and magnesium sulfates, iodine, and bromine from geothermal waters has been carried out in the Soviet Union. For example, in the Kashkadar'insk artesian basin (Uzbek SSR) for only one exploratory well, about 2,700 tons of salts (mostly potassium chloride), 9 tons of bromine, and about 100 kilograms of iodine have been extracted in one year.

Recently, the Soviets announced that techniques for the extraction of boron, alkali and alkali-earth metals, and trace elements are being developed.

Frequently, hot and superheated geothermal waters contain large amounts of salt, iodine, bromine, naphthenic acid (mixture of cycloparaffin acids), boron, strontium, lithium, fluorine, and other rare elements, which can be extracted easily for the chemical industry.

The potential benefits from extracting chemical elements from highly mineralized geothermal waters can be demonstrated with actual data as follows:

(1) Near the town of Berekey, during prospecting drilling well No. 3 produced a fountain of highly mineralized water (brine) with a discharge capacity of 70,000 cubic meters daily. Considering the discharge volume and the amount of chemical elements, in one year such a well could produce:

Elements	Tons
NaCl	1,470,000
KCl	30,000
$CaCl_2$	30,000
$BaCl_2$	14,300
$MgCl_2$	24,300
$SrCl_2$	277
Various trace elements	6,704
Total of obtained elements:	1,575,581

(2) During prospecting drilling in the oil region of Datykh, Chechen-Ingush ASSR, well number 6 produced a highly mineralized water (brine) fountain having a discharge capacity of 40,000 cubic meters daily. It has been calculated that annually, this well can produce:

Elements	Tons
NaCl	636,690
$CaCl_2$	78,930
$MgCl_2$	80,000
$MgSO_4$	4,000
Various trace elements	2,240
Total of obtained elements:	801,860

The geothermal waters of the Cheleken peninsula annually produce about 300 to 360 tons of lead, 48 to 50 tons of zinc, 24 to 35 tons of copper, and several other components valuable for chemical industry.

The western Siberian and Irkutsk basins are prospective sites for the extraction of iodine, bromine, and several other elements. In addition, the Kamchatka peninsula and the Kurile Islands are considered as potential regions for the extraction of bromine, rubidium, cesium, lithium and arsenic from available superheated water and steam.

Near the town of Zhigalovo, Irkutsk Oblast', during prospecting drilling, a well 1,125 meters deep produced highly mineralized geothermal water with brine density of 0.4 g/cm^3 and the soluble salts at 600 grams per liter. This high content of salts represents a very valuable raw material for the various chemical products for which eastern Siberia has a great need. However, the exploitation of these valuable minerals is still in the early stage, and considerable work has to be done to achieve production on an industrial scale.

EQUIPMENT AND INSTRUMENTS

The drilling equipment and instruments for the exploitation of endogenous fluids are very much the same as used in the oil industry. Experience gained by the oil industry has been applied in geothermal drilling with minor changes in the methods, machinery, and instruments.

The standard rotary rigs are used for wet geothermal drilling, and as a rule they are for medium-depth penetration, since steam producing wells are usually less deep than those producing oil. Therefore, this particular use calls for some specific adjustments and minor modifications.

During the last decade, no great changes have become apparent in the design of drilling equipment. However, the Soviets designed their own separator with rather marginal improvement over those used by other countries, and the blowout preventer (on the base of wellhead connection) has been slightly modified to withstand the higher pressure. The Soviets did not experience any blowouts during exploratory drillings, although they anticipated the possibilities in 1958 during drilling of the first rotary borehole in the Pauzhetka region.

Regarding the power plants and prime movers for drilling rigs, diesel engines and the direct-current electrical motors have been considered. The diesel engine is used primarily in the strictly exploratory phase, when electric power distribution points are at a considerable distance from drilling site. The electrical motor may prove highly economical in the subsequent stage, when a power generating plant is already in operation and may easily supply the drilling site.

To exploit deep dry geothermal resources, Soviet scientists have suggested a long-range project to drill wells to a depth of 10 kilometers and create large subterranean "boilers" which would supply a steam-hot water mixture for a geothermal electric station of more than 1 million kilowatts capacity.

The designers of the Uralmash Plant (Ural Heavy Machinery Plant), Sverdlovsk, RSFSR, have developed plans for a drilling installation called "Uralmash 15,000," which will be capable of drilling to a depth of 15,000 meters. The entire complex of the "Uralmash 15,000" consists of a power plant, an engine room, and a pumping station. This installation will have a derrick over 60 meters high, with elevators for cargo and operators. The superdeep drillholes may be bored either by turbine or rotary methods.

In geothermal investigations, the instruments for measuring heat flow, thermal conductivity variations, thermal gradients, chemical composition of water, and other elements of geothermy, are the same as those used in standard geological exploration and hydrogeological engineering.

The Soviets have developed an instrument used for evaluating such physical parameters as thermal conductivity, permeability, and heat flux in rock under dry or wet conditions. The method involves heating and pressurizing the center of blocks of different types of rock with or without addition of liquid solutions. Quite large blocks are used, up to tens of cubic meters, with internal temperatures up to 1,000°C. Temperature and pressure gradients are measured in the walls by probes. Additional technical and graphical data are lacking.

The Technical Thermophysics Institute of the Ukrainian Academy of Sciences has developed a series of instruments for measuring heat flux over a broad range; however, technical and graphical data on these instruments are presently insufficient.

An extensive list of references is given in the report. Several of the more important appear to be the following.

Kutateladze, S.S., and Rozenfel'd, L.M. The problems of geothermal energy. IN: AN SSSR. Vestnik, no. 10, 1965, 25-31.

Makarenko, F.A., and Kononov, V.I. Hydrothermal resources of the USSR. IN: AN SSSR. Geologicheskiy institut. Teplovoy rezhim nedr SSSR (Heat regime of the Earth's interior in the USSR). Moskva, Izd-vo Nauka, 1970, 199-209 (ITS: Trudy, no. 218, 1970).

Shubin, B.F. Experimental Freon geothermal power station. Elektricheskiye stantsii, no. 5, 1967, 20-22.

United Nations Symposium on the Development and Utilization of Geothermal Energy, Rapporteur's report. Pisa, 1970.

Vymorkov, B.M. First geothermal electric stations. In: AN SSSR. Vestnik, no. 10, 1965, 32-39.

Additional information on geothermal research and development in the Soviet Union is given in JPRS 59496, July 1973 entitled *Deep-Seated Heat from the Earth* which consists of excerpts from the Russian language book by I.M. Dvorov, *Glubinnoye Teplo Zemli,* 1972. The following sections are derived from JPRS 59496.

GEOTHERMICS

By geothermics (from the Greek words "geo" = Earth and "therme" = heat) is connoted a science studying the heat state of the Earth's crust and of the Earth as a whole; its dependence on geologic structure, composition of rocks, magmatic processes, and an entire range of other factors.

The criterion for the heat state of the Earth's globe is the surface temperature gradient, which permits formation of a judgement concerning the Earth's heat losses. Extrapolating the gradient to slight depths, to some extent the temperature condition in the Earth's crust can be estimated. The temperature of thermal water in the upper parts of the Earth's crust are unevenly distributed. The value corresponding to depth in meters, at which temperature rises by 1°C is said to be the geothermal gradient.

In connection with the variation in the intensity of solar radiation, the heat regime of the first 1.5-40 m of the Earth's crust is characterized by diurnal and annual fluctuations. There also occur many years' and secular fluctuations in temperature which gradually increase with depth. At any depth the temperature of rocks T_h can roughly be established on the basis of the equation $T_h = t_B + H + n/G$, where t_B = average air temperature in the given locality; H = depth for which temperature is determined; n = depth of layer of constant annual temperatures; and G = geothermal gradient. The average value for the geothermal gradient equals 33 meters and with deepening from the zone of constant temperature, the temperature rises 1°C every 33 meters. Below are listed the average geothermal gradients in certain regions of the USSR.

Region	Geothermal Gradient, m/°C
Northern Caucasus (Groznenskiy Rayon)	12.0
Dagestan	21.4
Western Caucasus foothills (Maykopskiy Rayon)	25.1
Apsheron Peninsula	27.4
Western Caucasus foothills	31.6
Donbas	32.2
Emba	33.3
Ukraine, western part	42.2
Lower Volga	49.5
Karagandinskiy Basin	62.0
Samarskaya Luka	64.3
Bashkirskaya ASSR	82.6
Belorusskaya SSR	86.5
Kamskoye Priural'ye	88.2
Baltic Shield	92.5
Krivoy Rog	112.5

The geothermal conditions are extremely diversified. This is associated with the geologic structure of any given region on Earth. The USSR has areas where an increase in temperature by 1°C occurs at a penetration of 2 to 3 meters. These anomalies usually exist in the regions of contemporary volcanism. At depth of 400 to 600 meters in certain regions, e.g., on Kamchatka, temperature reaches 150° to 200°C and more. This does not signify that at depth of 1,000 meters temperature will be 500°C. On the other hand, in the zone of permafrost, temperature is low at great depths.

At the present time data has been obtained concerning the fairly deep freezing of the upper zone of Earth's crust in the upper reaches of the Markha River (left tributary of the Vilyuy River). Geothermal observations in a deep borehole provided the possibility of establishing that the depth of frozen rocks equals 1,500 meters here. At a depth of 1,800 meters, temperature comprises a total of only 3.6°C. As is evident, the geothermal gradient will be 500 meters per degree C.

On the separate platform parts in the territory (assume there is on the Russian Platform), temperature with depth is about as follows: 500 meters, not above 20°C; 1,000 meters, 25° to 35°C; 2,000 meters, 40° to 60°C; 3,000 to 4,000 meters, up to 100°C and more.

For a calculation of the heat flow, it is necessary to know the heat conductance factor of hot rocks and the temperature gradient.

The geothermal measurements have great importance for planning the large underground structures. They are also important in the areas of permafrost where the thermal regime influences the possibility of installing water and heating lines, and also influences the stability of soils and foundations.

THERMAL WATER

Underground water with increased temperature is said to be thermal. It is found chiefly at great depths and occurs widely both in regions of recent and contemporary volcanism and also in the fault mountain and platform regions.

Temperature of underground water will fluctuate within broad limits, affecting its condition, influencing its composition and properties. Therefore, temperature as a criterion of classifying the underground water according to its qualitative indexes has found reflection in many descriptions. For the water having curative importance, temperature is determined with allowance for the heat influence on the human organism; for the potable water, it is determined with consideration of the taste qualities while for water having a heat-power utilization, temperature is determined proceeding from the value for the heat potential.

In establishing the basic temperature limit permitting differentiation of cold water from thermal water, at the comparison stage, the water temperature standard for a given locality, average annual air temperatures and maximal average air temperatures on Earth (35°C), i.e., purely geographic and climatic factors, was adopted.

A diversified approach to an evaluation of thermal water and a considerable provisionality in the selection of the temperature range led to the situation that at the present time, many classifications claim universal acceptance. The basic of these classifications has been listed in Table 2.3.

In the existing classifications, an excellent resemblance of temperature ranges is observed only in the differentiation of the cold (below 10° to 20°C) and warm (20° to 35°C) water, while in respect to the hot, very hot and thermal water, the differences are extremely great. Specifically, for the upper limit of thermal water in some instances 90°C is assumed while in other cases 100°C is assumed, while in still other cases more than 100°C is assumed.

However, in the compilation of a standardized classification for the critical temperature between the hot and very hot water, it is feasible to assume 50°C (as the transitional level for utilizing water for curative and heat-power purposes), while the upper limit of thermal water can be expanded to the critical temperature owing to the reheated water. The remaining subdivisions of underground water by temperature (cold and warm) can remain without change as having received adequate validation and recognition on the part of most authors. With allowance for the refinements introduced, the suggested classification of subterranean water will assume the following form.

	Temperature, °C
Cold:	
Supercooled	<0
Very cold	0 - 10
Cold	10 - 20
Low-thermal:	
Warm	20 - 35
Hot	35 - 50
High-thermal	
Very hot	50 - 100
Reheated	>100

With respect to the fumarolic springs and steam jets of the tectonically disturbed zones and sectors of spontaneously burning coal and shales, they can be combined under one roof under the title "Steam Hydrotherms". The supercritical water also theoretically belongs under this group.

In this manner it is necessary to distinguish the thermal water and the steam hydrotherms which differ from each other in respect to phase-aggregate stage and operating conditions.

TABLE 2.3: CLASSIFICATIONS OF THERMAL WATERS

Temperature, °C (0, 20, 40, 60, 80, 100)	Author
Cold; Subthermal; Thermal; Hypothermal	Ovchinnikov, 1947
Very cold; Cold; Warm; Hot; Very hot	Alekin, 1953
Cold; Subthermal; Thermal	Shcherbakov, 1953
Extremely cold; Very Cold; Cold; Warm; Hot; Very hot; Extremely hot	Shcherbakov, 1953
Super-cooled; Very Cold; Cold; Warm or slightly thermal; Hot or Thermal; Very hot or highly thermal; Over-heated	Makarenko, 1960
Extremely cold; Very Cold; Cold; Warm; Hot; Very hot-highly thermal; Super-heated water & steam	Group of authors (from book "USSR Health Spas") 1962
Cold; Moderate; Thermal	Lefort 1858
Cold; Invigorating; Warm; Hot; Very hot; Burning	Daffner, 1886
Cold; Warm; Hot; Very hot; Hyperthermal	Gauss, 1903
Cold; Hot	Kel'gak, 1907
Hypothermal; Thermal; Hyperthermal	Mouren, 1910
Cold; Hot; Very hot	Schneider, 1913
Cold; Hypothermal; Homothermal; Hyperthermal	Heinze, Grünhut, 1916
Cold; Moderate; Thermal	Jardin, Astruk, 1932

Source: JPRS-59496

DISTRIBUTION AND EFFICIENT USE OF THERMAL WATERS

The USSR has tremendous resources of thermal water and steam-water mixtures. Numerous deposits are known in the European part of the USSR, in Western and Eastern Siberia, in the Far East, in the Republics of Central Asia and Kazakhstan (Figure 2.6).

Prior to 1960 in the USSR, no special drilling for thermal water was conducted (with the exception of balneology). The available data were generalized, chiefly in the geologic-prospecting and exploratory operations for oil and gas. From 1956 to 1961, of the 9,280 exploratory boreholes taken into account, 3,938 (42.4%) proved to be nonproductive but some of them could be a source for obtaining thermal water.

At the present time, in many regions of the country, scientists are conducting continuous geothermal surveying, have developed efficient methods for geothermal mapping, have disclosed the basic tendencies in the regional distribution of thermal water in the entire USSR territory and have determined the thermal water deposits most promising for practical utilization. Estimation of the probable effectiveness of the natural heat of Earth will be possible just as precisely as the estimation of the reserves of oil and gas, with the sole difference that this form of heat is renewable.

The calculations of the reserves of thermal water were based on available data concerning the volumes enclosed in the beddings of gravitational water, volumes of the actual water-bearing horizons and concerning the collecting properties of the rocks forming them. In addition, there was taken into account the conditions involved in feeding the beds, composition of water, etc. All the factors were determined during the drilling of boreholes for oil, gas and thermal water. The supplies of thermal water can be subdivided into predicted and operational.

The predicted reserves of thermal water represent the total quantity of water existing in the pores and fissures of the aquiferous horizon and taken into account within the temperature limits ranging from 40° to 200°C, mineralization to 50 g/l and occurrence depth to 3,500 meters from the day surface.

The preliminary calculations conducted on the predicted reserves of thermal water permitted the estimation of their total possible output at 19,750,000 m^3 per day. However the reserves indicated are far from complete, will be made more precise through time and undoubtedly will be increased in connection with the fact that now more and more attention is being devoted by the scientific and production organizations to a study of the deep-seated heat. For example, it has been established that in 50 to 60% of the USSR territory, thermal water occurs which is suitable for certain economic purposes.

With the development of deep drilling (to 7-10-15 km from the day surface), great prospects are opened for the discovery of high-temperature heat sources. At such depths, in certain regions (excluding the volcanic ones), temperature can be 200° to 350°C. Then the predicted reserves of deep heat from Earth will increase by many times.

FIGURE 2.6: MAP SHOWING THE OUTLOOK FOR THE UTILIZATION OF THERMAL WATER IN THE USSR

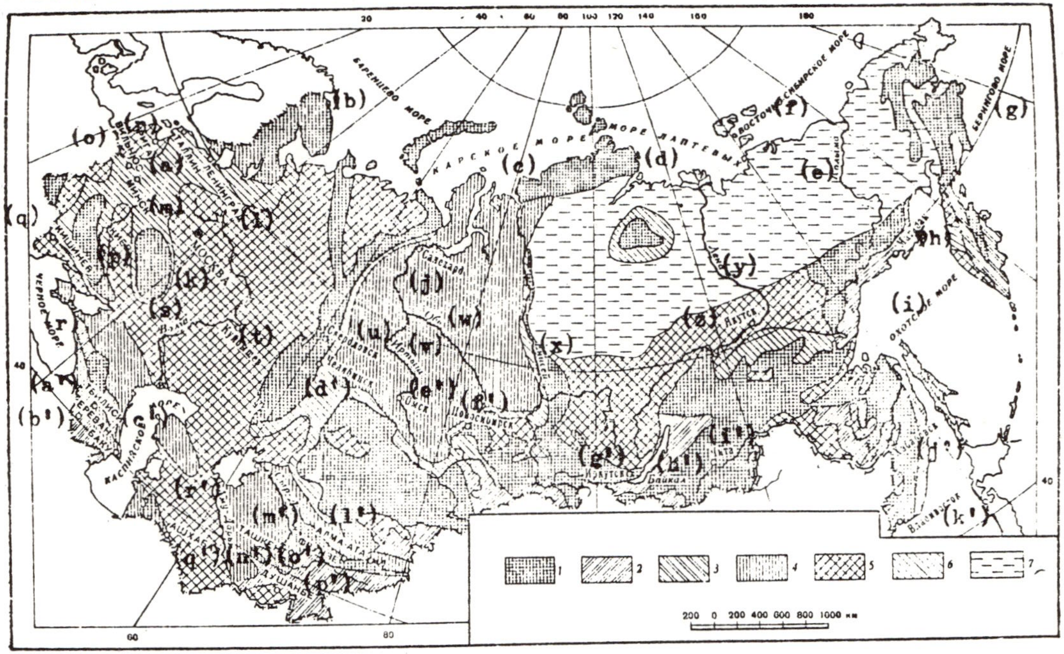

(a) Tallin
(b) Barents Sea
(c) Kara Sea
(d) Laptev Sea
(e) Kolyma River
(f) East Siberian Sea
(g) Bering Sea
(h) Magadan
(i) Sea of Okhotsk
(j) Salekhard
(k) Moscow
(l) Leningrad
(m) Minsk
(n) Riga
(o) Vilnius
(p) Kiev
(q) Kishinev
(r) Black Sea
(s) Volga River
(t) Kuybyshev
(u) Sverdlovsk
(v) Irtysh
(w) Ob'
(x) Yenisey
(y) Lena
(z) Yakutsk
(a') Tbilisi
(b') Yerevan
(c') Caspian Sea
(d') Chelyabinsk
(e') Omsk
(f') Novosibersk
(g') Irkutsk
(h') Lake Baykal
(i') Chita
(j') Khabarovsk
(k') Vladivostok
(l') Alma-Ata
(m') Syr-Dar'ya River
(n') Tashkent
(o') Frunze
(p') Dushanbe
(q') Ashkhabad
(r') Amu-Dar'ya River

1. Regions promising for the exploration and prospecting of thermal water
2. Regions with local outcroppings of thermal water suitable for utilization in heat supply and balneology (mud baths)
3. Regions of contemporary volcanism with local outcroppings of thermal water and steam hydrotherms suitable for electric and heat supply and balneological purposes
4. Regions with a real development of thermal water suitable for heat supply and balneology (in isolated sectors for extracting of chemically valuable components, i.e., iodine, bromine, boron, etc.)
5. Regions of thermal water suitable mainly for balneological purposes
6. Sectors where thermal water is lacking
7. Regions slightly studied from a hydrogeologic standpoint

Source: JPRS 59496

Among the indicated predicted reserves of thermal water, can be identified the operational, i.e., those reserves which even in the immediate future will come to be utilized for various practical purposes. The operational reserves of thermal water presented with temperatures from about 40° to 200°C, occurrence depth to 3,500 meters from the day surface, mineralization to 10 g/l, are reckoned in the level of 7,900,000 m^3/day. A considerable share of the indicated output (around 70%) can be obtained from depths up to 2,000 meters from the day surface. If these reserves are utilized in the national economy, they will be able to replace in a year tens of millions of tons of conventional fuel.

The operational reserves of thermal water in the entire USSR territory have not yet been fully estimated. It is sufficient to present only one example. From 1969 to 1970, as a result of a special prospecting of five deposits of thermal water, are obtained an output of 100,000 m^3/day with temperature ranging from 65° to 170°C. The utilization just of this heat will permit the saving of more than 200,000 tons of conventional fuel annually. At present workers have accomplished the overall regioning (division into regions) of the USSR by geologic distribution of thermal reserves, their thermal regime, chemical composition and genetic types.

At the present time in the USSR, activities are underway on utilizing the plutonic heat of Earth for obtaining electricity, heating, hot water supply and cooling supply to residential and industrial buildings, for the technological requirements of production, in agriculture for heating the structures for protected soil, livestock buildings and also for obtaining chemical materials, for balneological and other purposes.

It is necessary to point out that for obtaining electric power from Earth's natural heat, one should utilize the regions of contemporary volcanism where not far from the day surface, the steam-water mixtures outcrop with temperature up to 200°C and higher. In the USSR, such regions are Kamchatka and the Kurile Islands.

The construction of geothermal electric power plants, as has been confirmed by world experience and the first experimental-industrial Pauzhetskaya Geothermal Electric Power Plant operating in the USSR, is economically more advantageous than building electric power plants of other types.

While for obtaining geothermal electric power are utilized relatively high-temperature natural heat carriers, for heating and hot water supply to residential and industrial buildings, less heated water is required, i.e., 50° to 100°C. Thermal water with such temperatures is more widely distributed. In addition, the low-temperature underground water as a rule occurs closer to the day surface and more often is artesian. With the explored reserves of low-temperature thermal water, tens and hundreds of large cities and populated points in the country can be heated and supplied with hot water.

While thermal water has mineralization to 10 g/l and the content of harmful ingredients in it does not exceed the standard established for tap water, it can be used in hot water supply (baths, showers, etc.). If the thermal water temperature is not less than 50°C, its utilization in hot water supply systems is accomplished according to a very simple pattern (Figure 2.7).

By self-effusion, water flows from the borehole into a sectional tank (storage tank), from which it is conveyed along pipelines to users with the aid of a pump. The hot water supply installations constitute one of the most effective users of thermal water. One borehole with an output of 1,500 m^3/day at water temperature of 60° to 65°C could provide the hot water supply to a residential microregion or settlement of a municipal type with a population of around 14,000 persons.

If the thermal water's chemical composition does not meet the requirements imposed on tap water, a device is envisaged including special intermediate heat exchangers in which the thermal water heats the conventional tap water (Figure 2.8). Understandably in this case, there is no saving in tap water.

The thermal water with temperatures below 50°C can also find application in hot water supply. However, before delivery to the users, it is necessary to heat it in boilers or with the aid of heat pumping systems.

At the present time, the Soviet scientists have developed a number of basic systems for the integrated utilization of hydrotherms for various requirements (Figure 2.9). In this connection, it is necessary to keep in mind that thermal water is employed in a water supply system only once, and hence, the potential of discarded water is lost irrevocably.

Depending on temperature of the water being discarded, nature of the users of heat and on the climatic factors, one and the same geothermal borehole could provide a varying number of users and could prove equivalent to entirely varying quantities of thermal energy obtained during the year. This is reflected particularly sharply when thermal water is utilized in systems of water heating operating only part of the year and in which the dumping temperature in the very coldest period is not below 35° to 40°C. In connection with this, in developing the basic system of geothermal heat supply, it is extremely important to find the proper engineering solution.

FIGURE 2.7: DIAGRAM SHOWING DIRECT UTILIZATION OF THERMAL WATER WITH TEMPERATURE OF 50° TO 70°C

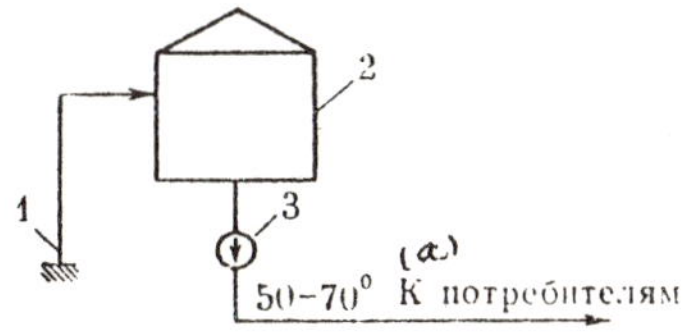

1 Borehole
2 Storage tank
3 Hot water supply pump
(a) To the users

Source: JPRS 59496

FIGURE 2.8: SYSTEM OF UTILIZING THERMAL WATER WITH TEMPERATURE FROM 20° TO 40°C FOR HOT WATER SUPPLY WITH THERMAL PUMP

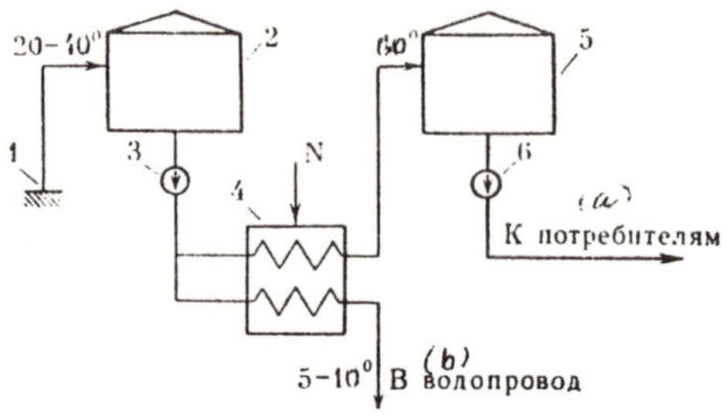

1 Borehole
2 Thermal water tank
3 Thermal water pump
4 Thermal pump
5 Hot water tank
6 Hot water pump

(a) To the users
(b) To the water line

FIGURE 2.9: DIAGRAM SHOWING UTILIZATION OF THERMAL WATER FOR HOT WATER SUPPLY WITH HEAT EXCHANGER

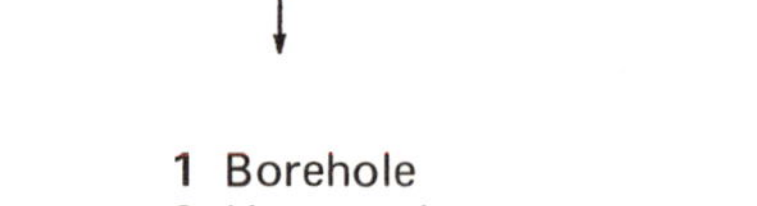

1 Borehole
2 Heat exchanger
3 Storage tank
4 Hot water supply pump

Source: JPRS 59496

Thus, based on calculations made by B.A. Lokshin, a borehole with an output of 1,500 m^3/day and water temperature at mouth of 60°C, with parallel feeding of thermal water for heating and hot water supply, permits provision of heat to seven residential 70 apartment buildings in the city of Makhachkala and saving of 1,200 tons of conventional fuel per year. Owing to the combination of peak additional heating, systems of air and water heating and open water collection, the fuel savings in this system under the utilization of the same borehole increases to 3,100 tons, while the number of 70 apartment buildings connected to the borehole increases to 16-17.

The operating principle of the system is as follows (Figure 2.10): thermal water from borehole **1**, passing the peak additional heating **2**, is fed to a water heating system **3** and to the air heaters of second heating **4** of the air heating systems. From the water heating systems, the processed water passes through the air heaters of the primary heating, **5** systems of air heating and is expelled with temperature ranging from 15° to 20°C. However, after the air heaters in the second heating stage, the thermal water is directed for the hot water supply of all buildings (equipped with systems both of water and of air heating).

The peak-time heating of thermal water can be accomplished both in conventional boiler rooms with steam or water heating boilers and also with the aid of electricity. In the latter case, the peak-time boiler room is completely automated and operates without servicing personnel. The fuel consumption is very low because the peak-time boiler room is required to operate only for a minor part of the heating season.

It should be pointed out that the traditional radiator heating systems usually do not comply with the specifics involved in utilizing the thermal water as a heat transfer agent. Most promising are the new designs, namely the panel-type (ceiling-floor) convector and air systems which permits the construction of flow-through welded systems or localization of the mineralized water at the entry to a house (air heating). These systems also greatly facilitate the installation of heating units in buildings, which at low water temperatures (50° to 60°C) encounters considerable difficulties.

FIGURE 2.10: HYDROTHERMAL SYSTEM OF HEATING SUPPLY

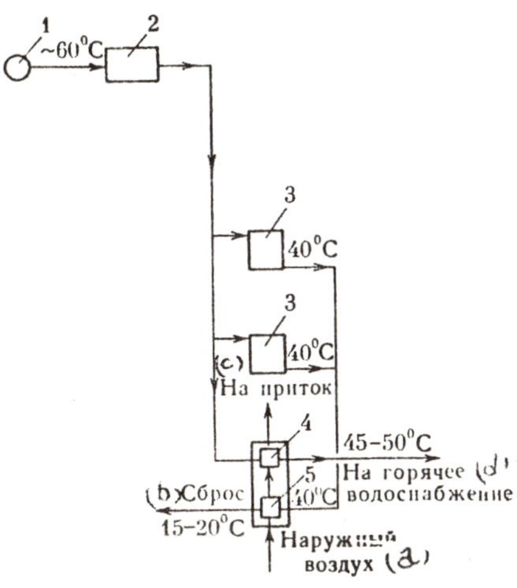

1 Borehole
2 Peak-hour boiler room
3 Water heating
4 System of air heating
5 Disposal

(a) Outside air
(b) Disposal
(c) To intake
(d) To hot water supply

Source: JPRS 59496

Thermal water can be extensively utilized not only in heat supply but also for obtaining low temperatures with the aid of applying the working substances in refrigerators.

The inventory of refrigerating equipment is increasing from year to year in the country. A number of chemical enterprises are using billions of kilocalories of cold. The tremendous amount of cooling is used for the production of synthetic rubber, ammonia, protein-vitamin preparations and so forth. The metallurgical enterprises consume a lot of coal. The great majority of them are equipped with expensive compressors, the operation of which involves tremendous expenditures of electricity.

At the present time there has been tested and put into mass production an absorption-type bromine-lithium machine with an output of 2.5 million kcal of cold per hour, which can operate not only on electricity but also on the heat from underground hot springs (Figure 2.5). This facility is remarkable in that it not only provides cold but also constitutes a transformer of heat. It can function the year round, i.e., for cooling in summer and heating in winter. Such a machine is especially valuable for cooling in southern regions of the country: Central Asia, Kazakhstan, the South of the USSR European sector, where the hydrothermal resources are fairly extensive.

Tremendous prospects of utilizing the plutonic heat from Earth are opening in agriculture, i.e., the heating of greenhouses and hothouses for growing vegetables, fruits and flowers. For this purpose, water is required with temperature ranging from 25° to 100°C. Such water is found in a considerable part of the USSR territory at slight depths.

The utilization of natural heat can become the most promising and economically advantageous aspect in vegetable raising. Protected soil occurs extensively from the subtropics to the Arctic region. By "protected soil" is understood the sectors of earth on which an artificial favorable microclimate has been created for the off-season raising of agricultural crops. Moreover, in this case the yield is many times greater than in the open ground.

Hothouses make it possible to grow in the northern regions of the country two-three crops of vegetables per year. Many of the hothouse plants are grown under artificial conditions throughout their entire life, while others are only started there (for transplanting); still others (flowers, cauliflower) finish their growth there.

Although the area of hothouses in the country is increasing with every year, the demand, especially for vegetables in winter and spring, is far ahead of the supply. In the available hothouses (with various types of heating) in winter and summer, we grow a total of only around 1 kg of vegetables per capita. In order to provide during the entire year even the minimal quantity of fresh vegetables to a city with a population of 100,000, it is necessary to have a hothouse facility with an area of 10 hectares and this will amount to 1 square meter per person. Only in this case can be solved to some degree the problem concerning the off-season supply of fresh vegetables to the population.

The prime cost of hothouse-grown vegetables is still quite high, more than 60 rubles per 100 kg. The main expenses in growing vegetables in hothouses relate to heating the facilities. They amount to at least 50 to 60% of the prime cost of production. In 1971, the total consumption of heat was expressed by more than 60 million gigacalories, which corresponds to burning 8 million tons of conventional fuel.

A hothouse combine with an area of 10 hectares in the regions of the central belt of the country requires as much heat as is used for domestic and communal requirements in a modern city with a population of 50,000 to 70,000. If a plan to increase the capability of the hothouse combines is realized millions of tons of coal or billions of cubic meters of gas would be used.

Obviously such an approach will not be economical and by the same token it will not accelerate the development of vegetable growing but will only retard it, taking into consideration the already existing crisis in the heat balance in various regions of the country and primarily in the USSR European sector. This crisis not only will not decrease but, on the other hand, in connection with the tremendous requirements for heat resources, it will grow. Therefore it is necessary to solve two problems simultaneously, i.e., providing the country with off-season vegetables, and the production of cheap hothouse output.

It is known that the cost of one gigacalorie of heat obtained from burning various types of fuel differs: in the boilers operating on gas, it is 3 to 3.5 rubles; if operating on peat fuel, 4.5 rubles; and if running on solid fuel, it is 5.6 rubles. The effective prices for heat in the form of hot water from a heat and electric power plant (HEPP) released to the hothouse combines, even with consideration of 50% discount, will fluctuate within very wide limits and often will prove quite high.

For example, the Sakhalin-Energo releases water to the hothouse combines for a price of 6 rubles 53 kopecks per gigacalorie on Kamchatka and in the Magadanskaya Oblasts, the cost per gigacalorie is still higher. It is not necessary to speak of the requirement for growing cheap vegetables in winter and spring particularly in the regions of the North and North-East of the country.

The prime cost of one gigacalorie of heat in thermal water for heating the hothouse farms is 0.5 to 1.0 rubles. However, if from the Pauzhetskaya Geothermal Station, is connected to the hothouses the hot water which is being dumped into the river, with temperature of 110°C and a discharge rate of 100 l/sec, the prime cost per gigacalorie of heat will be essentially trivial.

Thus, the deep heat from Earth can become the very cheapest form of fuel. With its proper utilization, it will be practically inexhaustible. Consider such an example: for heating one hectare of hothouses during the year, an average of 1,600 to 1,700 tons of conventional fuel is required. At the same time, a borehole with a discharge of 2,000 to 2,500 m^3/day and temperature of thermal water amounting to 65° to 70°C will permit heating around two hectares of hothouses which could grow 300 to 350 tons of fresh vegetables per year. The economic advantages of utilizing the natural heat for heating protected soil are obvious.

The unique feature of natural heat is that in most regions, it can be utilized in an integrated manner: for obtaining electricity, heating and hot water supply to residential and industrial buildings, and also for extracting chemical elements.

As is known, the hothouse enterprises heated by thermal water do not use heat uniformly, i.e., during the day they use less and on the other hand, at night they use more. In summer in the temperate and southern latitudes, the hothouses are not functioning. From the hothouse facilities, the unused heat can be directed to the refrigerating installations. Particularly promising is the concept of the combined construction of hothouse and refrigerating enterprises with a standardized power unit and common auxiliary buildings. The uniformity of utilizing the deep-seated heat completely eliminates the expensive cooling equipment and will save the national economy tremendous sums of money.

While the economic effectiveness of utilizing thermal water for heating hothouses does not cause any doubts, the heating of seedbeds and of open ground is not highly economical. This is connected with the irregularity in the functioning of cultivation-type structures during the year, since for a greater part of the time, they do not operate and from this, they have low output as compared with the hothouses. In the modern designs of large hothouses, extensive use is made of the diversified agricultural technology and automation.

Automatic devices "track" the air temperature in a hothouse, accomplish irrigation and feeding of plants, control the concentration of mineral fertilizers, etc. Here the level of mechanization and automation is higher than in the structures on open ground. This also permits harvesting greater yields here. Such in brief are the prospects for utilizing thermal water in agriculture.

With regard to the utilization of thermal water for balneological purposes, the readers already know what tremendous importance thermal water has for man's health. Almost all the thermal water having increased temperature independently of its chemical composition, conditions of formation and circulation, outcrops to the surface in the form of natural springs. They are curative just as the water from the deep artesian basins and the fissured systems.

The high curing effectiveness of thermal water and its wide application in balneology is one of the leading forms in the organization of sanitary-health resort assistance in the USSR.

DRY GEOTHERMAL RESERVOIRS

The following material is derived from AD-774 108, December 1973, *Ultra-Deep Drilling for Geothermals* by S.O. Patterson, B.E. Sabels and A. Kooharian of Tetra-Tech, Inc. This report was prepared for the Nuclear Monitoring Research Office of the Defense Advanced Research Projects Agency.

Geothermal energy offers significant promise as a source of clean and widely available power. Its general availability, however, is correlated with the three stages of geothermal energy development defined by J. Banwell and T. Meidar, in "Geothermal Energy for the Future," a paper presented at the *138th Annual Meeting of the American Association for the Advancement of Science,* Philadelphia, 1971.

The United States is currently in the first stage, using geothermal energy developed from sources exhibiting surface manifestations. The second stage will involve the development of geothermal resources where the Earth's surface is estimated to process abnormal thermal gradients. Eventually, the third stage of geothermal development will involve deeper wells, utilizing the normal thermal gradients of about 16°F/1,000 ft that occur essentially everywhere on the Earth.

The third stage of geothermal energy development offers promising potential for providing a highly dependable, independent supply of clean energy. The problems that lie ahead in developing the potential for the widespread use of geothermal energy are scientific, technological, engineering, and economic. There is generally little scientific understanding of the nature and behavior of rock at the temperatures and pressures of these greater depths. The response of materials and the basic heat transfer processes necessary for exploiting such geothermal systems must be better understood through further research and development.

Engineering developments, from drilling geothermal wells to developing underground conversion systems and circulating heated fluids back to the surface, offer problems of both an applied and theoretical nature. A critical factor in the substantial research and development program implied in the third stage is development of a practical means of drilling wells in hot, dry geological settings to the depths required for further scientific research and technological development. A prime means of drilling such holes is the adaptation of rotary drilling technology to this task. Rotary drilling systems have been developed and successfully applied to the problems of exploration and production of petroleum and gas resources. The geological factors involved in such applications are significantly different in certain critical respects.

Thus, considerable understanding and careful analysis are required to assess the potential and limitations of rotary drilling for deep geothermal system development. A substantial portion of the study undertaken by Tetra Tech, Inc. was devoted to this assessment.

The principal study conclusions for hot, dry rock with abnormally high thermal gradients are as follows.

(1) Rotary drilling technology has been proven feasible for application in developing hot, dry rock geothermal wells up to depths of 20,000 ft in terrain where the thermal gradient is several times normal, leading to reservoir temperatures of 500° to 600°F at target depth.

(2) Assuming a hydrofracturing or equivalent underground energy conversion technology, 100-Mw (or less) geothermal energy systems would definitely be more economical than nuclear power and would be competitive with oil-fired, conventional, electric-power generation at heavy fuel oil costs of $5.10/bbl delivered. This assumption drives the economic viability of hot, dry rock geothermal energy development in a threefold way. It minimizes the number of well holes required initially. It provides adaptive expansion of underground conversion. It provides long-term (20-yr) use of initial well configuration.

The principal study conclusions for hot, dry rock with normal thermal gradients (16°F/1,000 ft) are as follows.

(1) Rotary drilling technology would have to be extended somewhat beyond its proven capabilities to produce the 35,000- to 40,000-ft geothermal wells necessary to provide heat reservoirs at 600°F. There are serious risks from a material technology and operations point of view in operating drill strings, logging systems, and other subsystems which make up rotary drilling systems.

(2) Even making the same assumption of a hydrofracturing conversion capability, the extrapolated costs and risks of geothermal systems in this case (based on rotary-drilled wells to 35,000 to 40,000 ft) do not offer competitive promise at this time. As rotary drilling technology is advanced within the petroleum production industry to wells beyond 30,000 ft, this conclusion would certainly have to be reconsidered.

(3) An alternative development of geothermal energy under normal gradient conditions involves drilling more but shallower wells at a site and using a binary fluid conversion technology, e.g., five pairs of wells at 18,000 to 20,000 ft at 340°F has been proposed. Rotary drilling is feasible although the number of wells required is large and drives the initial capital costs very high. Furthermore,

the binary conversion technology does not offer the adaptive expandability of the hydrofracturing approach since the reservoir costs during the operating lifetime of a particular well configuration may be 3 to 4 yr rather than 20 yr, which would make the costs of such energy virtually noncompetitive. Indeed, based on a 20-yr life for the original reservoir, such a system would have power costs estimated at 15.5 mils/kwh, whereas a 3- to 5-yr life would drive costs to 21.5 to 28 mils/kwh.

In summary, hot, dry rock geothermal energy produced from wells in the 35,000- to 40,000-ft range offer good potential for meeting the following needs: clean energy, self-contained or independent energy source, no resupply problems nor special maintenance systems, promise economic competitiveness of up to 100 Mw, reasonable extension/improvement of rotary drilling technology, and, development and proving of underground conversion concepts.

CONVENTIONAL DRILLING

Findings

The technical problems in conventional deep drilling systems are complex. Penetration of known or unknown sequences of rock strata by a hollow drill string is an ever new race with time and obstacles, which must be kept in check by a well-planned, finely tuned, and delicately balanced drilling system that can involve a million pounds on the hook and several thousand horsepower drawworks and pumpworks.

Conventional drilling involves counteraction of pressures and transport of cuttings by the drilling fluid; transmission of power and fluid by drill strings; holding open of the well under varying pressure conditions by casing strings; and logging and completion of wells, including cementing and perforating under great varieties of pressure, temperature, and mechanical/chemical conditions.

Conventional drilling is an effort of customized mass production. Most of the 30,000 holes drilled in the United States each year have very similar characteristics in terms of diameter, depth, drilling time, and cost. Consequently, drilling hardware is virtually standard. Almost all conventional drill holes are made in sedimentary rock. However, because of varying conditions, of pressure, rock competence, porosity, etc., casing of varying strength is required. Moreover, casing strings have decreasing diameters so that they can be inserted through the preceding, emplaced casings. Thus, drill holes have a typical tapering, which can cause a serious reduction in the hole's diameter in cases of complex casing requirements. Unless the selected diameter at the head of the well is wide enough, a drill hole may be restricted in its terminal depth.

The planning and engineering of drill holes prior to spudding has, therefore, become an important part of drilling methodology. This planning is primarily based on past drilling experience. Use of computerized records and programs such as the Amoco Optimized Drilling Program allows the complete specification of drilling parameters and saves considerable time and money. During the actual drilling, calculation of hourly costs, costs for each foot drilled, and mechanical monitoring of the work plan and deviations in scheduling are measured and compared with averages of previous performance data. The computer programs and the experience on which they are based also specify and regulate the speeds of drill string movement and the pressures exerted on the casing and hole walls.

The predominant drilling method is rotary drilling, which requires the addition of a drill tool at the end of a rotating drill string. The drill bit must be changed periodically, necessitating the withdrawal and reinsertion (round-tripping) of the entire drill string. At increased depths, a large part of the rig time is spent on the withdrawl, bit change, and reinsertion of the drill string. At 10,000 ft, for example, the process requires 10 hr; at 30,000 ft, it requires 24 hr. This time is essentially unproductive.

Because of the cost (time and money) involved in this process, methods have been sought to decrease or eliminate the round-tripping necessary in changing the drill bit. Research has been conducted on materials and bit configurations as well as on methods for changing bits through the emplaced drill string. Another operational variation is to rotate the drill bit by drilling fluid pressure at the bottom of a still-standing drill string. However, all the recent modifications have had little operational significance in the United States in comparison with the proven, standard rotary drilling method using diamond or hard-metal drilling bits.

In 1971, about 159 trillion Btu's of oil and gas were produced in the world. Of this amount, 53 trillion Btu's were provided for the U.S. economy. The oil and gas produced supplied about three-quarters of the total energy demand; the balance was comprised of coal, hydroelectric and nuclear power.

A large number of oil and gas tests have been made between 10,000 and 20,000 ft; increasing numbers, however, are being made in excess of 20,000 ft; and a few have been made as deep as 30,000 ft. These tests have been achieved without any significant changes in drilling methodology. However, the deepest drillings were performed under reasonably favorable geological conditions in sedimentary formations (Amadarko Basin, Oklahoma).

From an engineering viewpoint, drilling in igneous rock is generally easier than in sedimentary rock since casing and drilling fluid requirements [to counteract formation (pore) pressures] are less rigorous, although drilling rates are generally slower. However, oil and gas are seldom found in igneous rock and drilling in these formations is usually done for mineral deposits and construction.

To date, geothermal wells have not been drilled beyond 10,000 ft. Generally, under conditions of air drilling, heavy wear on drilling equipment occurs, and temperatures in excess of 600°F present critical difficulties in logging, cementing, and completion methods. Such high temperatures have been encountered only rarely in sedimentary drilling, whereas they are encountered quite regularly in geothermal well drilling. Thus, the geothermal gradient in the terrains that are being drilled appears to be quite variable. So far, geothermal drilling has been restricted to areas of high to extreme geothermal gradients, as is reflected in the comparison of about 3,000 geothermal wells drilled to date in the world with several million for oil and gas (100,000/yr for about 20 yr).

Areas for Further Research

Conventional drilling has been specifically adapted for the needs and requirements of sedimentary rock penetration in search of oil and gas. Recently, the method is being stressed in progressing to even greater depths, without significant system changes. However, rotary drilling may have an optimum range and an optimum sector of application, 20,000 to 25,000 ft in sedimentary rock. Because of exponentially increasing drilling costs with depth, deeper geothermal systems may require other means of penetration than rotary drilling to remain economic. Thus, further analysis must be done of the complex man-machine-lithosphere system. Also, further research must be conducted on new techniques and mechanical operations.

In geothermal drilling, the cheapest means of developing geothermal energy is by emplacement of shallow drill holes in high-temperature gradient areas, which are relatively scarce. Geothermal development requires the construction of wider diameter conduits than do oil and gas. Ideally, a minimum 12- to 24- inch i.d. hole should be available to support a steam turbine. Also, emplacement of heat exchangers at depth would be possible in wide-diameter structures.

DEEP ROTARY DRILLING OPERATIONS

Findings

Only a few wells have been drilled in excess of 25,000 ft. Because of technological and geological difficulties, costs increase exponentially with depth. Thus, the suitability of rotary drilling operations for deep geothermal exploitation is questionable. The problems and limitations of deep drilling are specified in the section on Limitations of Technology. Here the general state of the art is described and the general technical problems and limitations inherent in the current engineering, testing, completion, and logging practices, techniques, and equipment used in deep drilling operations in the oil and gas explorations are outlined. To provide this overview, a variety of wells were examined in various geological locations: California, Oklahoma, southern Louisiana, southern Texas, West Virginia, and Wyoming.

For deep drilling, the knowledge of the target at depth is often limited. Seismic records of strata below 25,000 ft are weaker and less definitive than those at shallower depths. Ultra-deep drilling to date is generally carried out in sedimentary basins where extensive structural features can be predicted and defined.

In drilling deep wells for oil or gas, a sufficiently large hole at total depth is required to permit the setting of a production casing string. Such wells usually require several immediate casing strings, which in some cases reduce the hole size such that the overall objective cannot be met. These and other limitations and difficulties can be seen in the following case descriptions.

California, Great Basins Petroleum, Tenneco 31-10X Well, Total Depth 21,640 ft: The Great Basins Petroleum Company, an operator for a group of companies, has drilled the deepest well in California to date although two other wells, currently being drilled, are expected to exceed 22,000 ft in depth. All of these wells are in the Bakersfield area (southern San Joaquin Valley). In sedimentary basins such as this, formation pressures and incompetent strata require protective casing strings. These sedimentary rock formations usually consist of layers of sandstone, shale, limestone, and combinations thereof. Pores created within these materials may comprise 10 to 30 percent of the bulk volume. The fluid in these pores is under pressure, and since most of the pore spaces are connected, fluid can move, under pressure, from one point to another. Under normal conditions, the fluid in the pore spaces is under a pressure equal to the freshwater or saltwater gradient for that depth.

For example, the normal pressure of a formation at 10,000 ft is 4,333 psi or 4,650 psi for freshwater or saltwater gradients, respectively. Some formations have been known to have abnormal pressure gradients of up to 1.2 psi/ft of depth, or 24,000 psi at 20,000 ft. In oil fields, such pressures are usually expressed in equivalent heads of drilling mud in pounds per gallon (ppg). If at 30,000 ft a fluid in a pore is under pressure of 25,000 psi, a 30,000 ft drilling mud column with a density of 16.03 ppg would be required to balance this formation pore pressure. Thus, high-pressure formations must be cased and cemented off to permit further drilling.

Another factor that necessitates protective casing is induced formation fracturing. When drilling fluid densities become too high, certain formations with low-formation pore pressure tend to fracture on each side of the well bore. When this occurs, drilling fluid will enter the formation (lost circulation). The fracture gradient of certain formations is so low that the formation will not support the head of drilling fluid above it. A casing string must be set at such zones of lost circulation and cemented.

These drilling problems requiring the setting of casing strings are well illustrated in the deepest well drilled to date in California. Figure 3.1 shows the conditions encountered while the well was being drilled. The curve to the left shows the magnitudes (in equivalent mud density heads) of the formation pressures encountered. The actual drilling fluid density used is shown by the middle curve. The mud weight curve is always at a higher value than the formation pressure at the same depth; if it is not, the well could blow out of control. Formation fracture gradients are also shown. If the mud weight should become greater than the fracture gradient curve at some depth, fracturing and lost circulation would occur.

Casing strings were set and cemented in this well. The strings are less than ideal, but this well was originally planned to have a total depth of 12,000 to 14,000 ft. At 12,000 ft, 9⅝ inch casing was set because hole conditions were unstable and the drill pipe was almost struck. The decision was made to set the 9⅝ inch casing and drill deeper. At 17,000 ft, the required mud weight to prevent a blowout became so high that lost circulation and sticking drill pipe were experienced. These problems were solved by setting the 7 inch drilling liner, and drilling continued. Below the 7 inch liner, the hole was drilled to 21,640 ft without setting additional casing. The 4½ inch casing was set as a gas production casing string. Eventually the casing restricted the size of the drill hole to such a degree that greater depths could not be reached.

FIGURE 3.1: PRESSURE CONDITIONS IN DEEPEST CALIFORNIA WELL

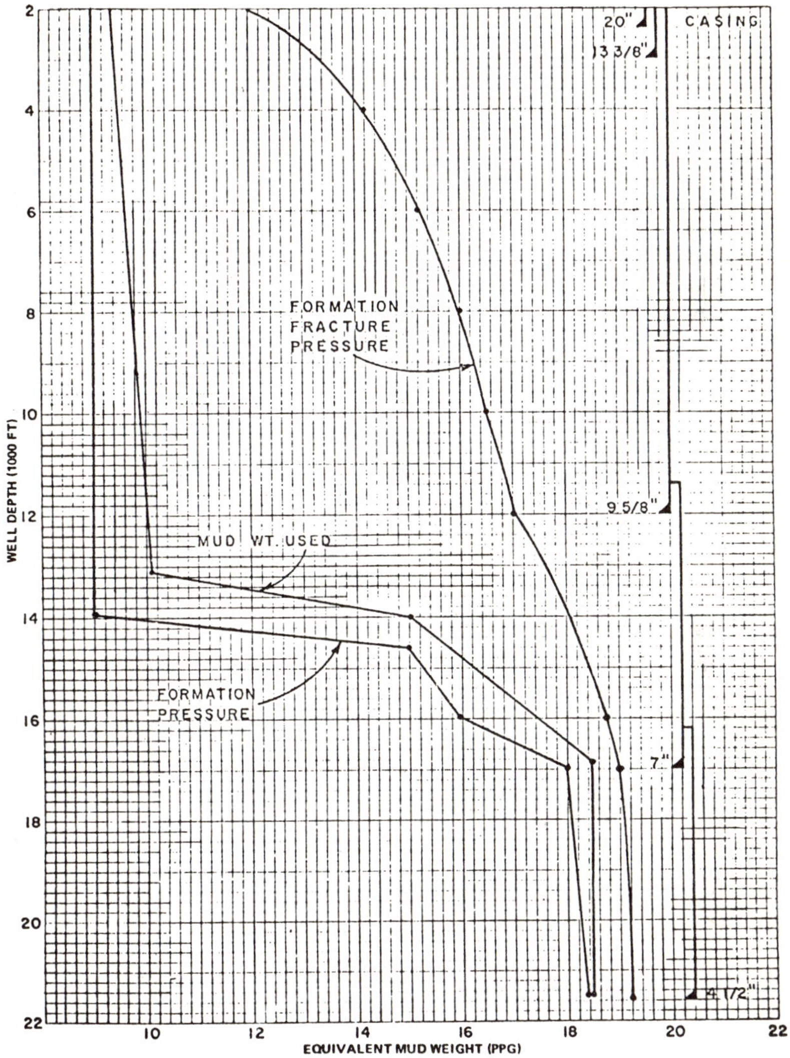

Source: AD-774 108

Another problem in these high-pressure situations is the collapse of casing when emptied of drilling mud. Figure 3.1 shows that the pressure at total depth is about the equivalent of a 21,640-ft head of 18.4 ppg drilling fluid, equal to about 20,705 psi. If the 4½ inch casing were completely emptied, approximately 20,000 psi of pressure would be applied to the outside of the casing, sufficient pressure to collapse all commercially available casing except small diameter casing such as a 4½ inch outer diameter (o.d.) thick-walled pipe.

Temperatures encountered at great depth can create severe problems with drilling fluids, cements, and logging tools. Figure 3.2 shows the temperature versus depth for the California well. Formation temperatures are normally expressed as temperature gradients in degrees Farenheit per 100 ft of depth.

FIGURE 3.2: CALIFORNIA TEMPERATURE VERSUS DEPTH (BAKERSFIELD AREA)

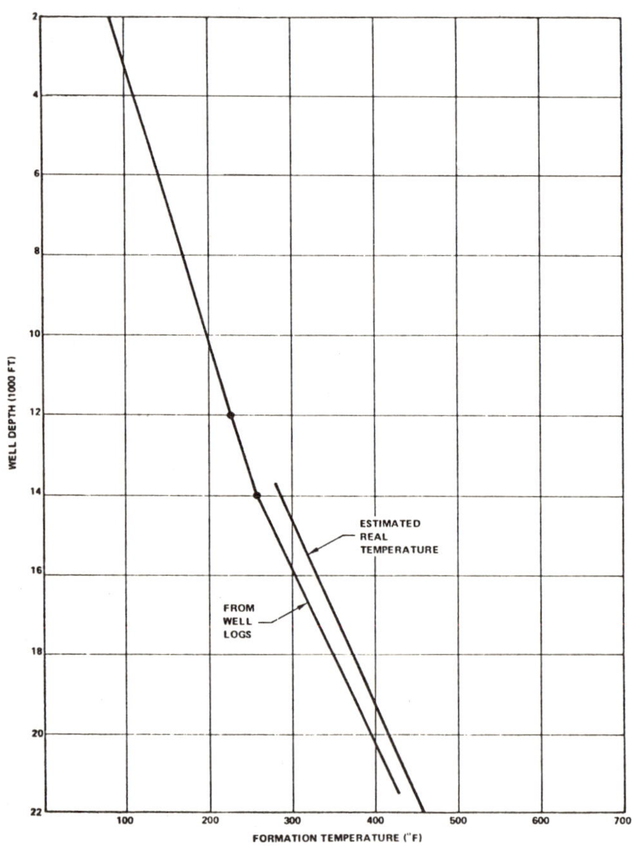

Source: AD-774 108

In the figure, the temperature gradient from 0 to 14,000 ft is 1.3°F/100 ft, while at the 14,000 to 22,000 ft level it is increased to about 2.1°F/100 ft. Extension of such plots to depths of more than 50,000 ft reveals that temperatures in excess of 1,000°F could be encountered. Use of present-day muds, cements, and logging tools would be impossible in such hot environments.

Oklahoma; El Paso Natural Gas Company; Easley No. 1; Total Depth 27,500 ft: This particular well is within a few miles of the world's deepest well. The geology, pressures, and temperatures for both wells are much the same and are shown in Figure 3.3.

FIGURE 3.3: PRESSURE CONDITIONS IN EASLEY NO. 1, WASHITA COUNTY, OKLAHOMA

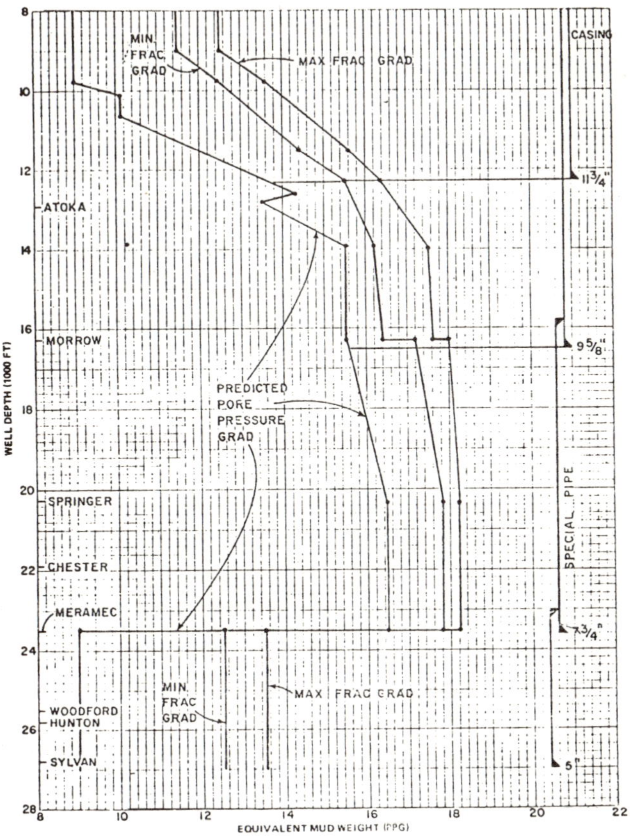

Source: AD-774 108

The important formation tops are shown on the left side; pore pressures are shown by the left-hand curves; and a range of fracture gradients are shown by the next two curves. Again, the mud density must always be just slightly higher than the pore pressures, but less than the fracture gradients. Therefore, all of the casing strings shown were required. A 16-inch string of surface casing was also set at 4,000 ft.

Nearby, the Glover-Hefner-Kennedy Green 1-1 was drilled earlier. Figure 3.4 shows the mud weights used in this well. The pore pressure magnitudes were verified by the well kicking and trying to blow out at just below 21,000 ft.

FIGURE 3.4: MUD WEIGHTS USED IN GREEN 1-1, BECKHAM COUNTY, OKLAHOMA

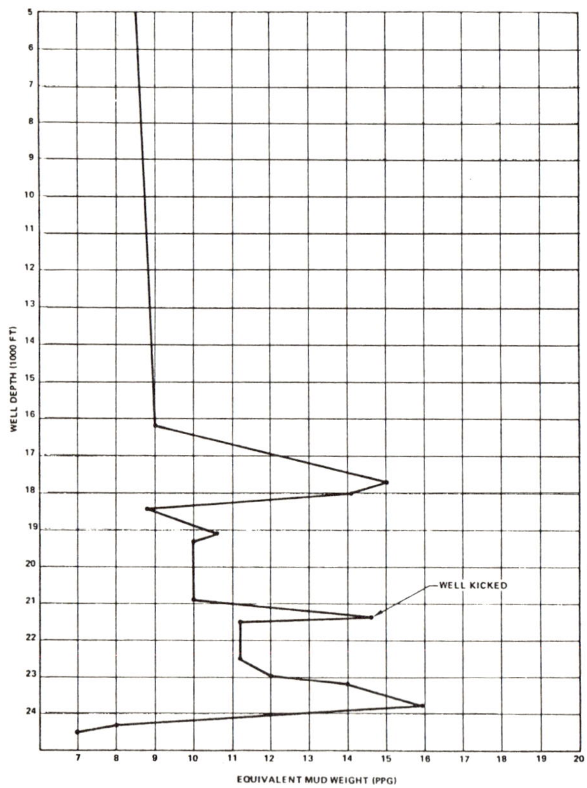

Source: AD-774 108

Temperature gradients in this area are not extremely high. However, a temperature of about 500°F was encountered in the Lone Star Gas Producing Company's Baden No. 1 well at 30,000 ft. The company was unable to obtain logs at the total depth of 30,050 ft because of the high temperature.

Louisiana (Southern); Dow Chemical Company; Felix Bacque No. 1; Lafayette Parish, Louisiana; Total Depth 17,500 ft: Drilling deep wells in this area presents problems somewhat unique and more troublesome than those of other areas. These problems result from relatively young sediments, which become very plastic with depth. Rock failure is of a plastic or ductile nature rather than brittle, as in hard-rock country.

Formation pore pressures sometimes become as high as the overburden or lithostatic pressure, which is created by the weight of all overlying sediments and the weight of the fluids contained in those sediments. For example, in this area the bulk density of sediments is about 2.0 g/cc near the surface and increases to about 2.60 g/cc at about 20,000 ft. Under these conditions, the lithostatic pressure equals about 17,300 psi at 17,500 ft. The pore pressure equals that of a head of 18.5 ppg drilling fluid, as shown in Figure 3.5. Thus, the pore pressure is about 16,840 psi at 17,500 ft, only 460 psi less than the lithostatic pressure at the same depth. Further, the fracture pressure equals about 17,300 psi at the same depth, about the same as the lithostatic pressure.

Again, the main problem is that in such pressure situations, the weight of the drilling fluid must be slightly greater than the pore pressure gradient (18.5 ppg) or a blowout will occur. However, under these conditions, when the drilling fluid pumps are turned on while drilling, the equivalent circulating density becomes greater than the fracture gradient at the bottom of the hole because of the friction as the drilling fluid flows up the annulus. This situation may cause fracturing, lost circulation, and eventually a blowout.

To prevent this situation from occurring, numerous casing strings are required, as shown in Figure 3.5. For example, the mud weight for a static mud column at 16,600 ft is 18.3 ppg. The fracture gradient at the 9⅝ inch casing seat is about 18.8 ppg. Therefore, as the 18.5 ppg mud is circulated up past the 9⅝" casing seat, the equivalent circulating density at that depth is very close to the fracture gradient. If the 9⅝ inch casing had not been set and cemented, fracturing and lost circulation would have occurred somewhere above the depth of the 9⅝ inch casing seat. The lower casing strings were required for the same reason.

In this area, it becomes impossible to drill by conventional methods below about 18,000 to 20,000 ft. The pore pressure gradient, fracture pressure gradient, and lithostatic gradient all become equal. Extreme care must be exercised with the mud density. In many cases, the mud density is so critical that the well attempts to blow with the pumps off and lost circulation is then experienced when the pumps are turned on using the same mud weight.

Temperatures in this area are known to be high, although since no well has approached 30,000 ft in depth, ultra-deep temperatures are still undetermined. However, temperatures and pressures are sufficiently high to cause the formations to be very plastic. In this situation, if a well-bore or casing is emptied of drilling fluid, the formations tend to flow into the open well-bore or collapse the casing, where the hole has been cased.

FIGURE 3.5: PRESSURE CONDITIONS, SOUTHERN LOUISIANA

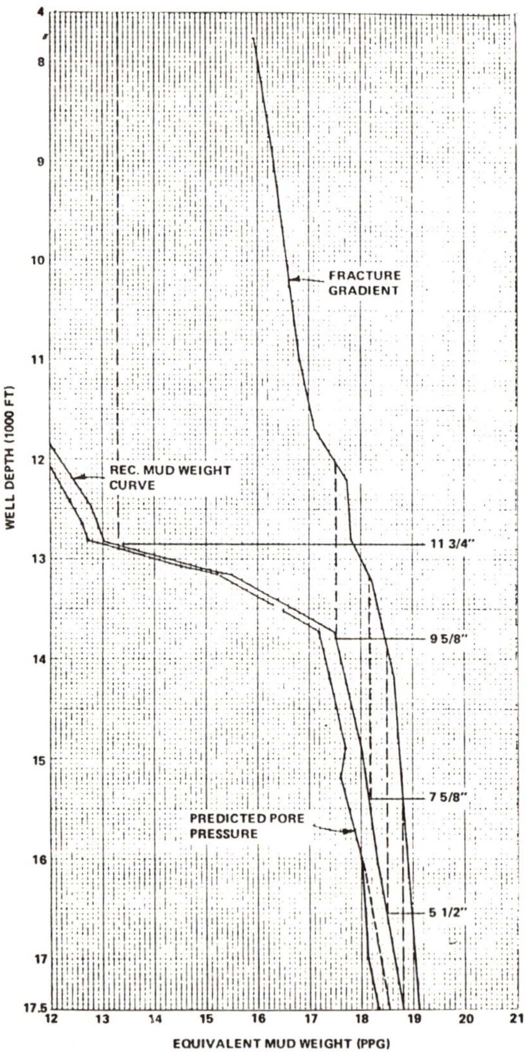

Source: AD-774 108

Texas (Southern); Shell Oil Company; Webb County, Texas (tight hole); Total Depth 22,000 ft: Because this well is a tight hole and no information is released, no actual data are available. However, it is known that this is an area where high-pressure problems are severe and are similar in nature to those in southern Louisiana. Abnormally high formation pressures begin at relatively shallow depths and increase to greater than the equivalent of an 18.0 ppg head of fluid. The pressure profile, mud weight curve, fracture gradient curve, and casing program are all thought to be very similar to the southern Louisiana well.

However, one additional pressure problem occurs here because of the vast difference in pressures in various sandstone formations. For example, suppose that a sand exists at 12,000 ft and has a pore pressure equal to a 12.0 ppg head of fluid (7,488 psi). As this sand was drilled, it would require 12.3 to 12.5 ppg fluid to control it while making trips for new bits. As drilling progressed to greater depths, sands and shales penetrated would require 18.0 ppg fluid to control the pressure.

Given this situation, a new problem exists. The sand at 12,000 ft is exposed to a head of 18.0 ppg fluid. The pressure of this static head is 11,232 psi, while the pore pressure of the sand at 12,000 ft is 7,488 psi. A differential pressure of 3,744 psi is directed from the well-bore into the sand, causing mud filtrate loss into the sand formation. When the drill pipe is stationary (as it is while making a connection for another joint of drill pipe), this differential pressure can cause the drill pipe to stick against the hole wall, referred to as differential sticking of the pipe. This is a severe problem and has caused many holes to be abandoned.

High temperatures are also severe in southern Texas. Temperatures exceeding 500°F have been encountered at the 18,000- to 20,000-ft level, with an average temperature gradient of about 2.5°F/100 ft. If this gradient is extended with depth, the temperature should be about 1000°F at 40,000 ft. These high pressure and temperature conditions are such that plastic deformation or flow of the rocks is likely to occur, possibly causing the drill pipe to become stuck.

West Virginia; Columbia Gas Transmission Corporation; Well 9674T, No. 1; Mingo County, West Virginia; Total Depth 19,591 ft (20,000 ft test): This Columbia Gas Transmission Corporation well is the deepest well in the eastern part of the United States. During the drilling of this well, there were four major problems: crooked hole, high pore pressures, very hard rocks and slow penetration rates, and near blowout. This well was drilled using the following casing sizes and depths:

Conductor casing: 26 inches at 54 ft,
Surface casing: 20 inches at 1,641 ft,
First intermediate casing: 13⅜ inches at 6,020 ft,
Second intermediate casing: 10¾ inches at 10,910 ft,
Third intermediate casing: 7 inch liner at 16,338 ft.

Air and/or natural gas were used as the drilling fluid in the well down to about 13,000 ft. However, if there is much water influx into the well bore, air is not desirable for drilling. Therefore, the shallow water-bearing strata had to be cased off. Using air or natural gas allows a bit to penetrate much faster than it will if water or drilling mud is in the hole. However, air or gas does not exert much static pressure on the drilled formations. For this reason, this particular well attempted to blow during drilling at 13,000 ft.

Fortunately, the 10¾ inch casing was set deep enough to provide the fracture gradient required to control the well. Also, the 10¾ inch casing had sufficient burst resistance at the top to prevent an accident. A drilling mud with a density of 18.2 ppg was required to control the gas flow.

In this area, the formation pore pressures are normal down to the 13,000 ft level, increase rapidly to the equivalent of an 18.0 ppg head, and remain equal to a 16.0 to 18.0 head to depths of 20,000 ft or more.

Because of the high mud weight required and the very hard rocks, penetration rates are very low. The following data (Table 3.1), a breakdown of some of the latter bit runs, indicate that penetration rates are very low and diamond bits drill faster than tricone insert bits.

TABLE 3.1: DRILLING DATA

Bit	Size (in.)	Type	Depth Interval	Time (hr)	Footage (ft)	Rate (ft/hr)
88	6½	Diamond	18972-19265	149	293	1.96
89	6½	Insert	19265-19372	79	107	1.35
90	6½	Diamond	19372-19554	92	182	1.98
91	6½	Insert	19554-19582	21¼	28	1.32

Source: AD-774 108

Figure 3.6 shows the formation encountered in drilling this well. Depths of each age are also shown. Because of the hardness of the rock, considerable bit weight is required to drill, leading to crooked hole problems. This particular well was so crooked that it was plugged with cement and redrilled very slowly and vertically. (Ultra-deep wells must be vertical or nearly so to prevent severe pipe wear.)

FIGURE 3.6: FORMATIONS ENCOUNTERED IN THE COLUMBIA 9674T NO. 1 WELL, WEST VIRGINIA

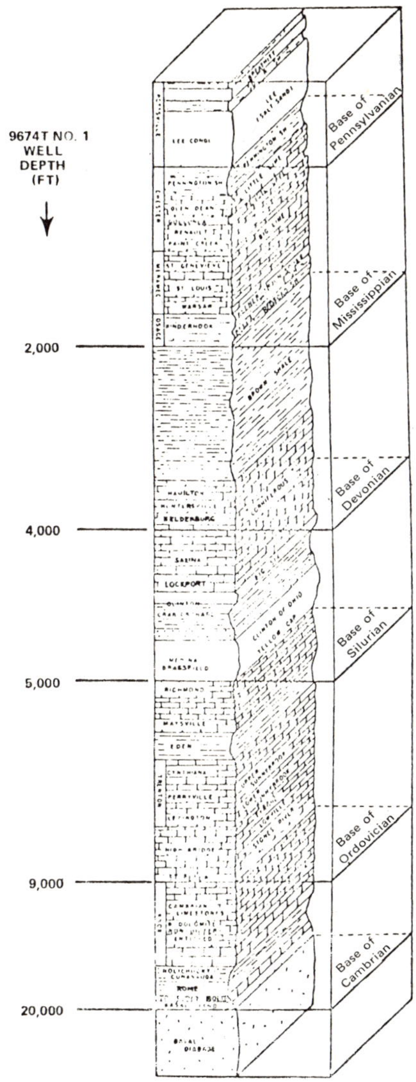

Source: AD-774 108

Temperatures of formations in this area exhibit a fairly low gradient. Figure 3.7 shows temperatures obtained while logging in West Virginia. The recorded temperatures fall on a straight line with depth. Actual formation temperatures are usually slightly higher than those recorded while logging because of mud circulation.

FIGURE 3.7: TEMPERATURE PROFILE, COLUMBIA WELL 9674T NO. 1, WEST VIRGINIA

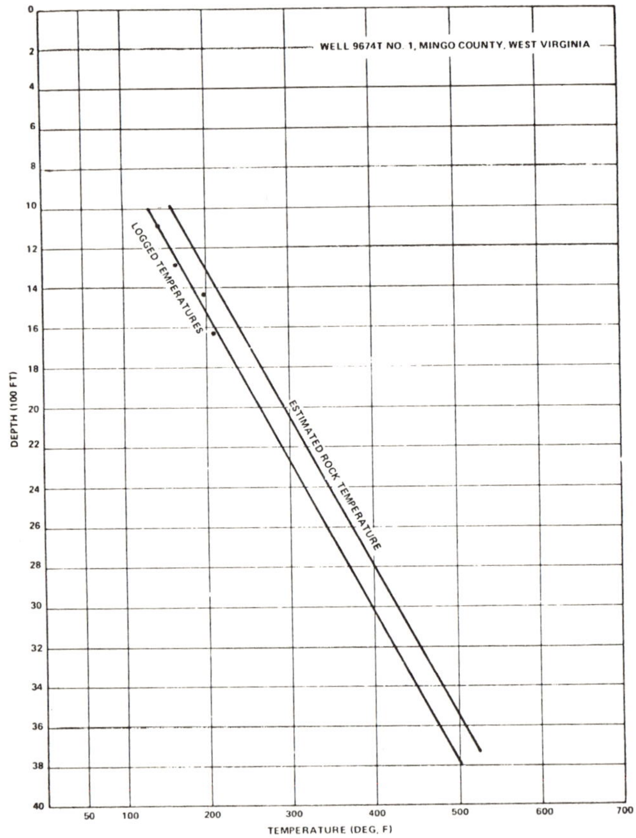

Source: AD-774 108

Wyoming; El Paso Natural Gas Company; Wagon Wheel No. 1; Sublette County, Wyoming; Total Depth 23,000 ft: Relatively few drilling problems occurred in drilling this well. Lost returns did occur but were caused by going in the hole too fast with a core barrel. Later, stuck pipe occurred with the collars at about 6,400 ft while attempting to regain circulation. The major loss zone in this well was between 6,000 and 6,600 ft. The hole was drilled extremely straight by running light bit weights. However, this resulted in low penetration rates and required many additional drilling days. The hole was drilled, logged, and then reamed out to 14¾ inch prior to running the intermediate pipe. This also required considerable additional drilling time. The only problem that occurred in drilling the intermediate hole was fluctuating mud weight. Sonic logs from this well were interpreted and plotted, as shown in Figure 3.8. Pore pressure gradients were determined from these plots and then plotted, as shown in Figure 3.9 which also shows a plot of actual mud weights used in drilling the pilot hole.

FIGURE 3.8: SONIC LOG PLOT, WAGON WHEEL NO. 1, EL PASO NATURAL GAS COMPANY

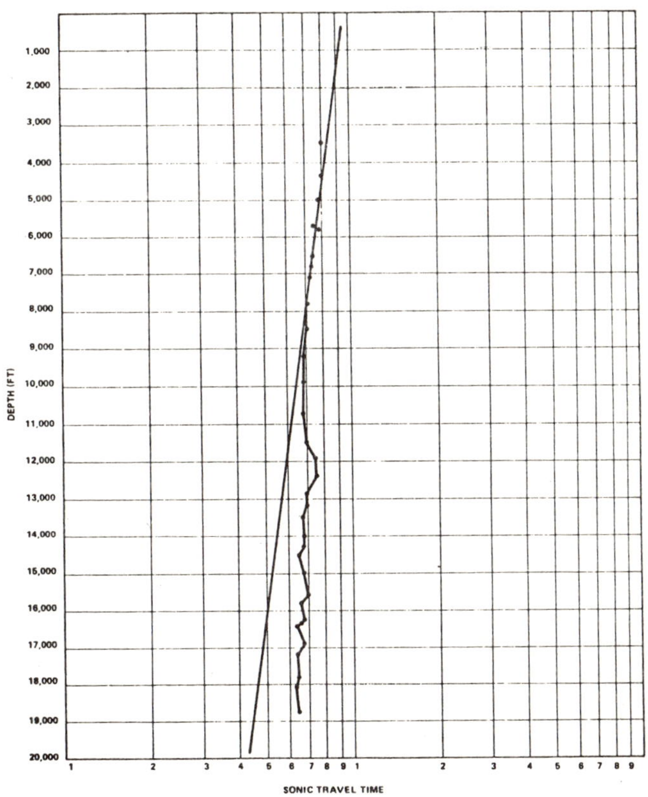

Source: AD-774 108

FIGURE 3.9: PORE PRESSURE GRADIENTS FROM SONIC LOGS AND ACTUAL MUD WEIGHT USED

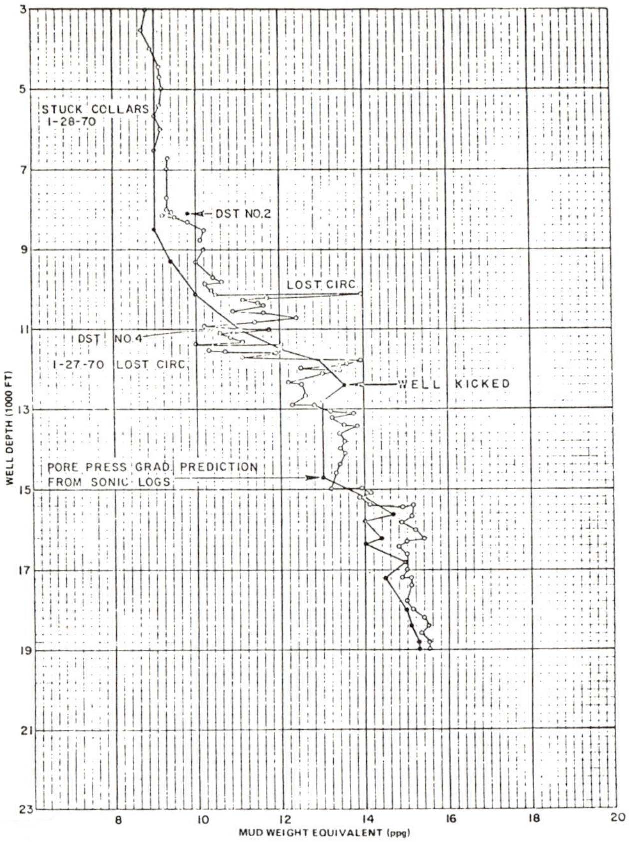

Source: AD-774 108

Note that the well was drilled under "under-balanced" or "near-balanced" conditions at all times, and that the drill stem test pressures agree with the sonic log pressure prediction. The mud weight fluctuated as much as 3.0 ppg or more at a specific depth. For example, at 10,200 ft, the mud weight changed from 10.5 ppg to 13.9 ppg over a very short interval and returns were lost.

These wide mud weight changes were attributed to (a) water entering the well bore and (b) water entering the mud system through the cooling and lubrication system on the mud pumps. However, since the mud head over-balanced the formation pressures, it seems unlikely that water could enter the well bore above 10,800 ft, the area in which the weight variations occurred. Thus, the dilution probably occurred through the pumps. Regardless, the shales were considerably washed out and additional cement was required. However, these washouts were not the primary contributing factor to the cementing failure on the 10¾ inch intermediate casing, the first major problem encountered in drilling. The job was to run in three stages since lost circulation had occurred previously in the 6,000- to 7,000- ft depth interval and while placing the casing in the hole when the casing shoe was between 6,800 and 8,300 ft. The casing was equipped with such auxilliary equipment as guide shoe, float collar, stage tools, centralizers, and cement baskets.

On April 29, 1970, with the casing in place, the hole was circulated with 12.8 ppg mud that was cut with gas to 11.8 ppg in the flow line. The sonic log shows pore pressures equal to 13.2 ppg at 12,100 ft. Mud weights were still being cut from 12.6 to 11.6 ppg with gas just prior to the first-stage cementing operation. The first stage went as planned and the 392 bbl of slurry filled the calipered annulus to about 9,150 ft, although according to calculations, the cement should have filled to only about 9,400 ft. However, when hole washouts exist, all drilling mud is never displaced, and fill-up is higher than expected. Also, some contaminated lead slurry might have gotten above the first-stage tool at 8,996 ft.

The plug was bumped on the first stage at 0145 hr on April 30, 1970 with the proper displacement. At 0300 hr on the same day, circulation was established through the stage tool at 8,966 ft. The returning mud was reported to be uncontaminated by cement. However, the returning mud viscosity was 105 sec compared to 76 to 80 sec prior to the first-stage job. Since circulation was ceased only for short times, cement seemed to be the cause of the increased viscosity. The opening plug had dropped readily in the mud (inside the casing) to the lower stage tool.

At 0755 hr on the same date, the second-stage mixing began. The cementing volume was 2,012 ft^3, or 360 bbl. The operation was normal except that when displacing, the closing plug never bumped and the second stage was over-displaced by about 86 bbl. Full returns were observed throughout the job and the displacement. At 1030 hr, the mud in the pits had gelled to some extent. There was enough cement and displacement to circulate cement from the second-stage collar (at 5,996 ft) to the surface, if this collar had been open. Had this been the case, cement returns would have been observed.

As a result of this cementing failure, the casing became buckled and holes were worn in it by later drilling. Later, when cementing the 7⅝ inch liner, the cement flash set in the liner, causing another cementing failure.

Figures 3.10 and 3.11 show all the data used in planning such a well, for a total depth of 23,000 ft. Note that the objective is to have a 6½ inch hole at total depth and be able to set a 5 inch casing liner. Temperatures at 20,000 to 23,000 ft are in the 425° to 450°F range, which is not excessively high. As previously stated, the only problems encountered in drilling this well were cementing, casing wear, and abnormal pressure.

As indicated by the previous examples, the general methodology of deep drilling in sedimentary rock is to simply extend standard drilling procedures; however, the setting of casing strings, excessive pressures and temperatures, and resulting effects on the system components become more severe with increasing depth. Yet, there are other methods of achieving deep, wide-diameter, untapered, underground construction. These methods include wide-diameter drilling as developed in volcanic rock at the Atomic Energy Commission's Nevada Test Site and ordinary shafts sunk by manned, semiautomatic or fully automatic systems. These latter methods were also investigated and found to offer some promise for deep-hole environment. However, these latter methods lead by necessity to large-scale endeavors which require multiple uses to justify the increased expenditures.

FIGURE 3.10: WELL PLANNING GRAPH, WAGON WHEEL NO. 1

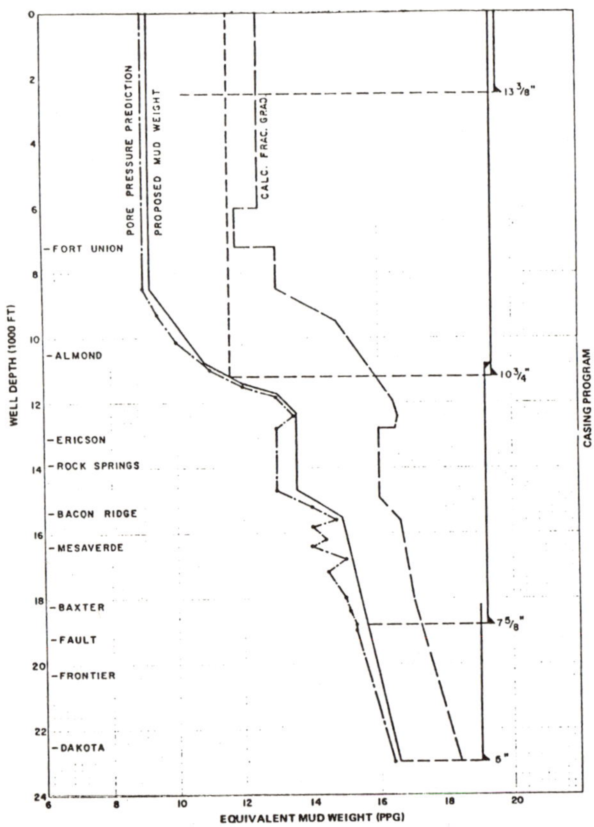

Source: AD-774 108

FIGURE 3.11: PROPOSED WELL PLAN OUTLINE

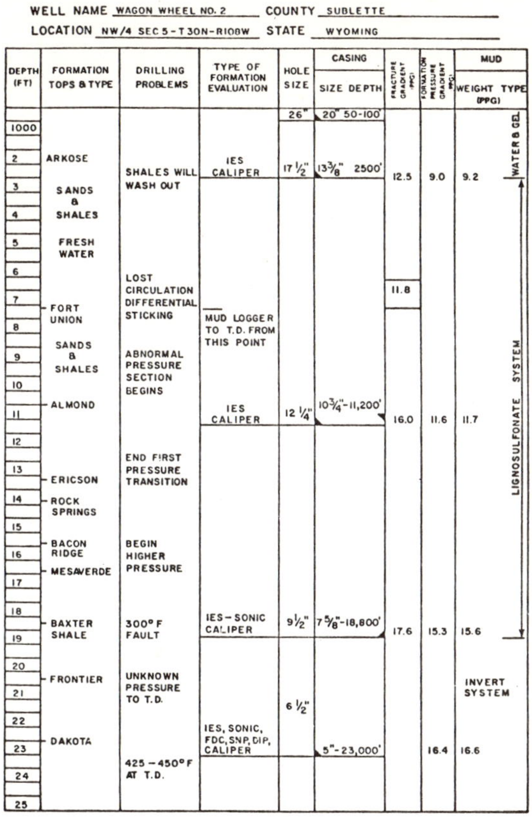

Source: AD-774 108

Areas for Further Research

In terms of geothermal exploitation, the question is whether to construct many shallow wells or a few deep wells or a few wide-diameter, deep wells in igneous rock. Essentially the answer is one of economics.

Since the deep drilling has been based primarily on oil and gas explorations, one facet of deep drilling not reflected in the records is drilling in igneous rock. A study, *Deep Hole Drilling Feasibility Study,* May 1969, sponsored by the Atomic Energy Commission and conducted by Fenix & Scisson, Inc., Tulsa, Oklahoma, was reviewed within this program. Based on it and other considerations, it is felt that deep drilling for geothermals should be concentrated in igneous terrain. This opinion is shared by Los Alamos Scientific Laboratory investigators, whose geothermal explorations center on granitic rocks and whose target depths are approximately 15,000 ft.

LIMITATIONS OF TECHNOLOGY

Findings

Depth: One depth constraint in drilling a well by conventional methods is the length of the drill string which transmits the drilling energy and emplaces/retrieves the rock-breaking tool. This length has been proven to exceed 30,000 ft and may be as long as 50,000 ft under favorable geological conditions. However, the efficiency of drill string manipulation decreases rapidly with extended depth, causing exponential escalation of drilling costs with increased depths.

Temperature: The temperature limitation of all drilling operations to date has been approximately 600°F, with the exception of a few geothermal drillings which have exceeded 700°F. Deep wells that depend on downhole measurements for monitoring and control are especially vulnerable to high temperatures. Furthermore, drilling fluids, cements, and sensors become unstable at temperatures about 600°F.

Well Diameter: The drill hole diameter is generally small but should be large enough to allow the setting of casing strings required to withstand zones of high pore and formation pressure. The deeper the well, the more varied the lithologic column, and the greater the multiplicity of pressure zones, the larger the well diameter should be when spudding.

However, the ability of a hole wall to withstand collapse under static pressure is a direct function of the casing diameter/thickness ratio along with the grade of steel used in the casing (Figure 3.5). To avoid a collapse in high-pressure zones in various rock types, several thick-walled casing strings may be required, which can reduce the well bottom diameter to zero.

As long as a well is filled with appropriately heavy drilling fluid, the pore and formation pressures can be counteracted. With the replacement of weighted drilling fluid by water or gases, however, the pressure can become prohibitive. For example, a pore pressure of 20,000 psi cannot be withstood by any commonly available, empty well casing except the highest grade steel, V-150, at a diameter/thickness ratio of 13. The more common P-110 steel needs a diameter/thickness ratio of about 9 to do the job (Figure 3.5), and such configuration is not generally available (U.S. Steel, for example, markets P-110 steel with a diameter/thickness ratio of 20:13 in outside diameters of 4½ to 13⅜ inches).

Since pressures up to 50,000 psi must be expected in 50,000 ft deep wells in sedimentary rocks, the conditions of wide hole diameter, versatility of usage (removal of drilling fluid), and resistance to collapse are conflicting, creating a major problem in preserving adequate well diameters.

Pressure: Just as the temperature level and gradient limit the depth of a well, the pressure affects its diameter. High pressure can be handled using the technology currently available, but only in rather narrow (4½ inch) cased wells, for example, the well plan shown in Figure 3.6. From 10,000 to 16,000 ft the pore pressure and formation pressure increase with increasing depth, requiring the setting of several interlocking casing strings. From 16,000 to 24,000 ft, the high pressures hold steady. At this depth, the well must be run on heavy, 18 ppg drilling fluid to counteract the pressure. Below 24,000 ft, the pore and formation pressures decrease to 13 ppg equivalent mud weight, and drilling can continue on a normal basis. Such abnormal pressures are considered normal in almost all sedimentary drilling.

These pressures have limited rotary well diameters and forced precautions and constraints. Pressure containment technology in large wells and excavations is only beginning. Presently, bulk construction plastics of exceptional mechanical properties are being investigated. Also, since high pore and formation pressures are believed to exist only in sedimentary rocks, drilling in igneous rocks is being further investigated.

Rock Form: High pore pressure in sedimentary formations represents a technological boundary to date. Until methods of pressure containment in gas-filled wells have been devised, deep drilling or excavation will be limited to igneous formations.

Structural Barriers: In igneous rock the primary impediment at any depth is jointing and fracturing of the rock, and it is highly unlikely that 50,000 ft granitic monoliths exist. However, convective water in fracture systems would be favorable for geothermal utilization of a deep well, even though its existence might endanger the drilling performance.

Economic Barriers: In any excavation process, the intended use of the structure will dictate its allowable cost. For example, a geothermal well should not cost so much that it makes the resulting electrical energy system prohibitively expensive. However, a structure intended for multiple purposes, such as drilling research and development, hardened communications systems, and the special needs of national defense, would allow the application of larger economical scales.

Technological Barriers: Drill steel strength is established both by the requirements of the drilling industry and the limitations of the current state of technology. Although the present drill steel length of 30,000 ft appears to be marginal and capable of breaking under its own weight, some drillers believe that 50,000 ft could be reached under favorable geological conditions. However, although depth does not seem to be a limitation at this time, there is no indication of plans to extend the design limit beyond 50,000 ft. The U.S.S.R. and some European countries are attempting to develop new alloys, but so far these alloys have not exceeded the capabilities of drill steels.

Available casing strengths are limited to the configuration of the most common casings required by the oil and gas industries. The ratio of outside diameter and wall thickness determines the collapse resistance, or strength, of a casing. This ratio is normally limited to the range of 40 to 10, with the most common ratios being about 20. However, to keep open wells under conditions of extreme formation pressure, as in ultra-deep drilling, casing with a diameter/thickness ratio below 10 would be required. Such gun-barrel equipment is technologically feasible but its custom manufacture makes it time-consuming and expensive to obtain. Currently, there is no effort to change this situation.

The hook load of current drill rigs limits the maximum casing weights to approximately 1 million pounds. This limit may be raised gradually as designs for new and specialized rigs are improved.

The drilling fluid temperature barrier remains unbroken and discussions with leading research and development laboratories in Texas and Oklahoma indicate that no imminent breakthrough is anticipated. To date, all water-based or oil-based drilling fluids become useless at temperatures over 600°F.

As with drilling fluid, cement also becomes useless at temperatures over 600°F, and a breakthrough is not anticipated to date. Currently, at temperatures over 600°F, the water in the cement boils and evaporates before setting can occur; the cement powder is then blown out the steam wells. In California, some wells have been successfully cemented at temperatures above 700°F by saturating the cement with modifying chemicals. However, an application at extended depths, with the inevitable delay between preparation and emplacement, is highly questionable.

The logging tool industry has been trying to extend the reliability of logging tools in temperatures over 500° to 600°F. Currently, it is being claimed that reliable performance in temperatures up to 1200°F, at least within several hours of tool insertion, is within reach.

Other barriers dictated by the equipment, such as energy deliverable at bottom and drilling cost, are also areas for further research and development activities. To date, energy deliverable at bottom decreases with depth, while drilling cost increases exponentially.

Expansion of technological limitations cannot be expected to occur rapidly or without a demand to fill a need in the drilling industry. In terms of geothermal development, the best approach seems to be circumvention, using current technology and carefully selecting locations, depths, and rock formations.

Areas for Further Research

In addition to the extensions of the current technology that have been mentioned in this discussion, new techniques must be investigated. Some advances can be achieved within the current technology, but 50,000 ft would remain a maximum depth. The ultimate research effort, then, must be in drilling beyond that depth. In addition, geothermal development remains even more limited than oil or gas drilling. These research areas for geothermal development have been indicated in previous sections.

EXTENSIONS OF CURRENT TECHNOLOGY

Findings

Several methods of deep hole making are reviewed and introduced in the following paragraphs. These methods are shown in Figure 3.12.

FIGURE 3.12: CONCEPTS IN DEEP HOLE MAKING

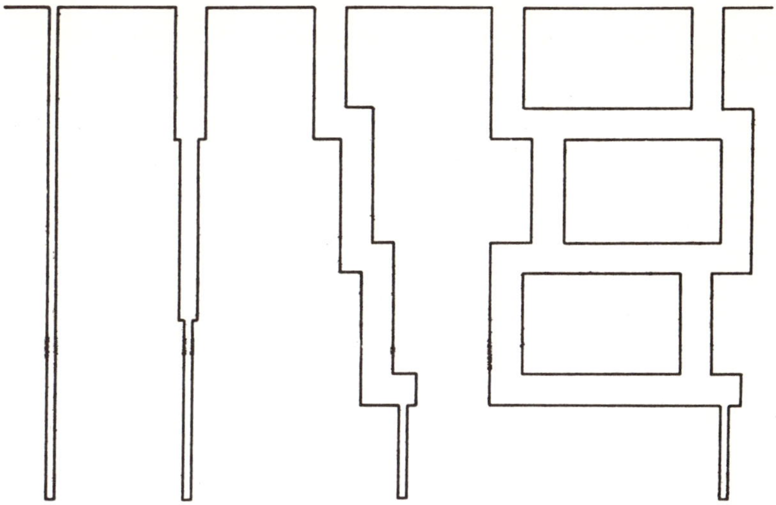

Source: AD-774 108

Single Drill Hole: The typical deep excavation structure is a single drill hole, generally rotary drilled in sedimentary rock (Figure 3.12A). Although the single drill hole using the rotary mode is the most economical means of deep penetration, it suffers from a number of drawbacks. In sedimentary formations, the diameter of the drill hole is generally not constant, but decreases with depth because of the casing required. The planned terminal depth usually cannot be exceeded by a change in drill plans because the diameter becomes too narrow. Furthermore, the hole must remain filled with drilling fluid at all times to counteract the pressure on the well bore, thus restricting the utilization of the well. Also, any control over the well and the penetrated formations is limited by the confinement of access through the casing string and its attached safety and operational mechanisms.

However, there are several ways to increase the drilling efficiency in igneous rock. The rotary method of rock disintegration has been modified by hydraulic water jets, hydraulic and turbine bottom hole motors, and wide-diameter rotary drills. These modifications result in increased rates of hole making, but without changing the basic drilling system, which is still at 30,000 ft and ultimately limited to 50,000 ft. Also, some recent development efforts have focused on improved, novel drill bits, which are expected to extend the bit life and reduce round-trip requirements for the drill string. However, none of these improvements will extend the depth of a single well beyond 50,000 ft within the foreseeable future.

Staged, Single Drill Hole: Staged, single drill holes (Figure 3.12B) allow the use of different technologies in completing the hole, including allowing for time lapses between stages. This approach is being pursued in the U.S.S.R. First, conventional rotary methods are used, followed by automated rotary drilling to 30,000 ft, which entails a more rapid handling of the drill string above the ground. The well may then be completed and reentered when more advanced turbine drills are available to bring it to target depth.

Using wide-diameter drilling or shaft sinking during the first stage of such a drilling operation would allow a reduction of drill string round-trip time and, through increased efficiency, of drilling cost. The costs of a wide-diameter first stage are no higher than ordinary rotary costs, and the staging might allow the cost increments to be kept more linear than exponential. Other advantages of the wide-diameter hole are discussed below.

Staged, Wide-Diameter Hole/Shaft: A staged, wide-diameter hole/shaft (Figure 3.12C) is one step beyond the staged hole with a wide-diameter first phase that was just described. In this case, the wide-diameter hole continued throughout the shaft and stages.

A wide-diameter hole can be produced by rotary drilling, semiautomated or automated drilling, or blasting and mucking techniques. The advantages of the wide-diameter hole are: manned access to the well bottom, the possibility of horizontal tunnels, staging/stacking of several shafts on top of one another, extended depth range in add-ons, and a constant, large diameter.

Staged, Wide-Diameter, Tandem Holes/Shafts: A staged, wide-diameter, tandem system (Figure 3.12D) offers the greatest possibility of utilization and development. About 50 percent of the broken rock can be hoisted mechanically, after collection downhole, rather than by flotation in drilling fluid. The system allows for dual access to most points and ventilation and refrigeration.

Most important, every level can be established as a new working level, thereby reducing hole-making costs as compared with one-hole deep drilling. Since construction methods can be employed and most equipment in the shaft can be man-operated, this approach offers a high probability of completion in accordance with plans.

Ultra-deep drilling by rotary methods becomes prohibitively expensive. For example, under the right geological conditions, a well could be rotary drilled to 50,000 ft, and would cost about $20 to 30 million. Of this cost, the following percentages would be spent on the respective depth increments:

Depth Interval	Cost %
0 - 10,000	3
10 - 20,000	6
20 - 30,000	13
30 - 40,000	26
40 - 50,000	52

However, if a dual-shaft system with staging and wide diameters were used, 10,000 ft wells could possibly be stacked below 50,000 ft while keeping drilling costs low. Although this approach could conceivably decrease the drilling costs of ultra-deep drilling, life-support systems would also be necessary at the various staging levels, which could increase the cost of the total system. Thus, detailed cost effectiveness analyses of such a system should be conducted as part of any further research efforts. The implications and conclusions derived from an examination of these drilling methods are summarized as follows:

(1) No one practical and operational hole-making techniques should be ruled out as a potential possibility for ultra-deep penetration. Techniques include rotary drilling and its modifications—turbine, wide-diameter, and high-pressure jet and rock-melting drilling; automated, semiautomated, and manual techniques; and manned stations, combinations of drilling and blasting, and hoisting of rock.

(2) No individual hole-making technique known is believed to be capable of economical, ultra-deep penetration; combinations of techniques are believed necessary to provide the performance and economic levels required in a competitive market.

(3) The total hole-making process must be planned and defined prior to beginning the operation. If staging is to be used, each stage must be integrated with every other section.

(4) Initially the well should be as large as possible in diameter and the width should be decreased as little as possible so that the hole is not restricted at great depths.

(5) Tandem wells should be advanced simultaneously and within a short distance of each other so that they can be linked at any level. Under certain conditions, the advancement of two communicating wells to extreme depths can cut costs, allow for special techniques useful in geothermal development, and increase the utllization of the well.

Current Operational Drilling Rigs

There are, at the present time, rotary rigs, capable of drilling to 40,000 ft. One such rotary rig has been used to drill a well to 30,050 ft and is currently drilling a second ultra-deep well (Rig C in Table 3.2). Table 3.2 shows the rig specifications for three of the world's most powerful rotary drilling rigs (Loffland Brothers Drilling Company). Undoubtedly, these specifications must and will be increased for ultra-deep drilling in the future.

TABLE 3.2: TYPICAL BIG RIG HARDWARE

Item	Rating		
	Rig A	Rig B	Rig C
Mast	147 ft, w/1.3 million-lb GNC	147 ft, w/1.3 million-lb GNC	142 ft, w/2 million lb GNC
Draw works	3000 hp	4000 hp	3000 hp

(continued)

TABLE 3.2: (continued)

Item	Rating		
	Rig A	Rig B	Rig C
Substructure	32 ft high, w/750,000-lb setback capacity	32 ft high, w/750,000-lb setback capacity	28 ft high, w/1.1 million-lb setback capacity
Rotary	37½ in., 500 ton	37½ in., 500 ton	37½ in.
Primary mud pumps	2 at 1400 hp each	1 at 1750 hp 1 at 1000 hp	2 at 1650 hp each
Traveling block, hook	525 ton	525 ton	750 ton
Swivel	400 ton	400 ton	650 ton
Wire line	1½ in.	1½ in.	1-5/8 in.

Source: AD-774 108

Areas for Further Research

As stated previously within this section, the tandem shaft system with staging and wide-diameter holes should be investigated further, with emphasis on a cost-effectiveness analysis. Drilling techniques and advancements and life-support requirements and systems should also be investigated, including the effects of various geological conditions on these.

NOVEL TECHNIQUES

Findings

The most significant problems in ultra-deep drilling are high temperature and pressure. These problems, the state of the art, and new developments (potential and currently under investigation) have been discussed in detail. A synopsis of the novel techniques being developed in rotary drilling, particularly in reaching greater depths, follows.

Tubular Goods: Oilfield tubulars that are currently available could be used to case a well to a depth of 40,000 ft or deeper. Special techniques, such as running liners to bottom on drill pipe, cementing it in place, and then tying the liner back to the surface, will be required. This is due to the high tensile loads imposed upon the upper portions of full, long casing strings while they are being run. Special sizes of casings have been developed to accommodate more standard bit sizes in deeper drilling. Undoubtedly, additional casing sizes and high-strength grades of steel will be developed in the future.

Drill Pipe: Drill pipe for conventional drilling has been developed to the point that a tapered string would drill to more than 40,000 ft. However, if a bit should become stuck at great depth, little additional strength would remain in the drill pipe to pull the bit loose. Again, higher strength steel has been developed for drill pipe, and additional grades and sizes will be developed as the need arises. Table 3.3 shows the maximum length of common drill pipe that can be suspended before deforming under its own weight; Table 3.4 shows how different grade and size drill steels must be combined to make it possible to attain depths in excess of 30,000 ft, and in addition to deliver 100,000 pounds of pull at the drill bit.

TABLE 3.3: MAXIMUM LENGTH OF COMMON GRADE E DRILL PIPE SUSPENDED IN AIR BEFORE PERMANENT DEFORMATION OCCURS

Pipe Size (in.)	Grade	Weight (lb/ft)	Tensile Strength (1000s lb)	Maximum Length Before Deformation (ft)
2-7/8	E	10.40	214	20,577
3½	E	15.50	323	20,838
4½	E	20.00	412	20,600
5½	E	24.70	497	20,121

TABLE 3.4: SINGLE-SIZE DRILL STRINGS CAPABLE OF DRILLING AT 30,000 FT (100,000 lb of Pull Available at Bit)

Pipe Size (in.)	Weight (lb/ft)	Grade	Length (ft)	Attainable Depth (ft)
4¾ x 2 Drill Collars			600	
3½	9.5	E	9,520	
	9.5	G	9,550	
	15.5	E	3,550	
	15.5	G	3,260	
	15.5	S	8,340	34,820
7½ x 2½ Drill Collars			600	
4½	16.6	E	10,500	
	16.6	G	3,270	
	16.6	S	7,970	
	18.1	S	3,550	
	20.0	S	4,960	30,850
8½ x 3 Drill Collars			800	
5	16.25	E	5,300	
	16.25	G	9,430	
	16.25	S	8,080	
	19.50	S	7,450	31,060

Source: AD-774 108

In ultra-deep drilling, because of the instability of the rock, it may become difficult, if not impossible, to remove the drill pipe for bit changes, logging, or casing. One of the novel technology developments foreseen is one-way drill pipe, which is advanced to the greatest attainable depth and is then cemented into place, after removal of the drill bit from the inside of the pipe. The drill pipe then takes on the role of the casing string for the protection of the next length of drill string, which eventually becomes a casing string. Such technology, if developed and made operational, may permit drilling even in plastic, incompetent formation, and the advancement of a telescoping hole to the required ultra depth. However, at present this is merely a suggestion, even though some producing wells have operated using a drill pipe that was modified for the purpose.

Drill Bits: Many novel drill methods are being developed, tested, and refined. Among these methods are: erosion (spark), explosives, forced flame, jet piercing, electric disintegrating, pellet, turbine (1 stage), plasma, electric arc, high-frequency electric, fusion, lasers, electron beams, and ultrasonics. Various combinations of these drilling methods are also under investigation. Figure 3.13 shows the rates at which wells can be drilled in medium-strength rock using these various methods and tools.

FIGURE 3.13: ESTIMATED NOVEL DRILLING RATES IN MEDIUM-STRENGTH ROCK

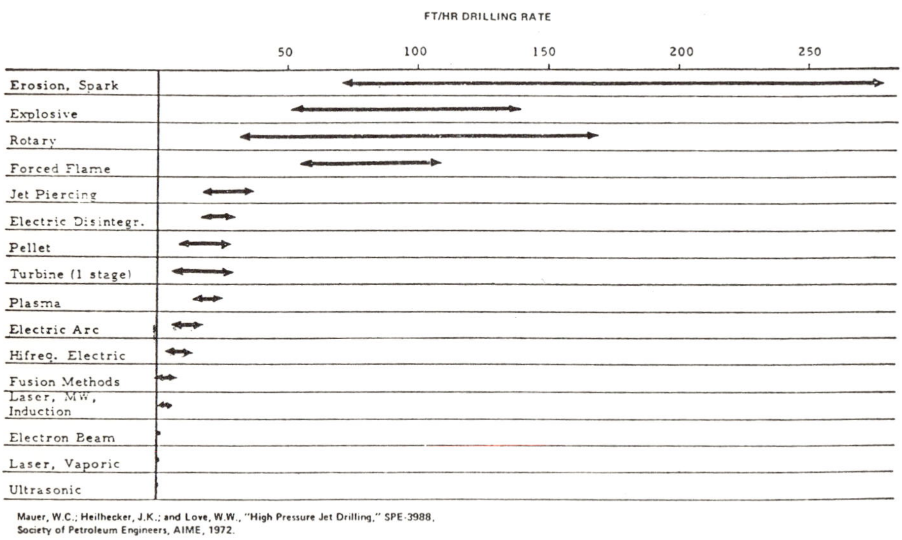

Source: AD-774 108

Today's drilling bits are much improved over those of only a few years ago. The new journal bearing bits with tungsten carbide inserts for cutting have provided reasonable penetration rates at the 25,000- to 30,000-ft level. The bearings allow for 50 to 60 hr of on-bottom drilling per bit, even in such hostile environments. It is believed that currently available bits could drill to 50,000 ft. Average penetration rates, footages, and other data taken from the 520°F Falcon-Seaboard "Tally" well are shown in Table 3.5.

TABLE 3.5: BITS USED AND HOW THEY PERFORMED

Interval (ft)	Number of Bits	Size (in.)	Average Penetration Rate (ft/hr)	Average Footage (ft)	Weight on Bit (lb)	rpm
0- 119	1 fishtail	27				
119- 3,675*	2 rock	15	70.5	1,778	12,000-15,000	225
3,675-12,520	10 rock	14¾	24.3	884.5	10,000-50,000	100-175
12,500-19,750	31 rock	9½	9.32	197.5	15,000-40,000	80-125
	2 diamond	9½	2.56	547	35,000	120-140
	1 diamond core	7-13/16				
	1 rock	9½		Used for drilling at top of line		
19,750-23,884	1 rock	6½	1.92	25	25,000	90
	1 diamond core	6-7/16	2.42	175	10,000-12,000	100
	5 diamond	6½	3.92	774	12,000-15,000	150
	1 junk mill	6-3/16		19		

*15-in. hole opened to 22 in. with one pilot bit.

Source: AD-774 108

Bit Trip Times: Time spent in pulling a worn bit out of a well, replacing it, and running the new bit back in the hole, is wasted as far as actual drilling is concerned. Such round trips represent a large percentage of the overall time spent in drilling any well, as shown in the data in Table 3.6 taken from Falcon-Seaboard's Tally well.

TABLE 3.6: OPERATION ALLOCATION PERCENTAGES OF THE 228½ DAYS ON THE TALLY WELL

Operation	Percentage of Total Time
Drilling and reaming	50
Trips	20
Maintenance and revision of equipment	8
Logging, coring, etc.	8
Circulating, conditioning hole and mud	5
Running, casing and cementing	8
Fishing	1

Source: AD-774 108

Further, a great deal of downhole problems are caused, or at least occur, while round tripping. Therefore, the ultimate goal is to eliminate round trips. To date, this has not been possible. However, there are many ideas as to how trips could be decreased both in number and time. Among these are: (1) better bits that drill faster and longer, (2) bits that could be replaced by wire line through the drill pipe, and (3) a means of pulling longer stands of drill pipe. For example, trips could be faster if the drill pipe could be rolled up on a spool, or pulled in 1,000 ft segments rather than the conventional 93 ft stands. The actual trip times versus depth for the world's deepest well are as follows:

Depth, ft	Trip Time, hr
10,000	8
15,000	13
20,000	18
25,000	20.5
30,000	22

These round trips were performed with the same rig and crew. Therefore, they should be indicative of the length of round-trip time at 40,000 ft. Extrapolating the data, it takes 26 to 28 hr for 40,000 ft, assuming that drill bit life and other parameters can also be extended linearly (which, according to the literature, is not certain). Typically, the drill pipe is pulled and run in triple lengths of about 90 ft total, so that eleven stops and accelerations are required for every 1,000 ft moved. Since the drill collars act on the hole walls like a piston in a cylinder, every dynamic effect on the drill pipe ultimately results in a force on the total exposed well.

It has been pointed out that the wells drilled in the shortest time not only cost the least but are usually the least troublesome. Extending the drilling time often compounds the severity of other problems, further increasing well costs. Therefore, one of the foremost goals in deep well drilling must be to shorten drilling time.

Drilling Rate versus Overall Drill Time: There are several ways in which the rate of progress can be improved. This becomes clear when one realizes that operational drill speeds in penetration tests, using rotary and other methods, are about 30 to 200 ft/hr and more in medium strength rock. Speeds of 10 to 30 ft/hr are attainable even by spalling and other techniques (Figure 3.13). Water jets have drilled in controlled field tests with rates of 50 to 280 ft/hr (Figure 3.14). Effective drill speeds of API classified drill bits range from 135 to 700 ft/hr. Yet, despite such high rates of drilling penetration, actual effective hole completion rates are much slower.

The Lone Star well in Oklahoma took 540 days to complete, with an average drilling rate of 2.2 ft/hr. Some estimates are that a 40,000- to 50,000-ft well would take 5 to 7 yr to complete, with an average drilling rate of 0.8 ft/hr. Even if such an estimate should prove wrong, drilling would probably take at least half that time, depending on the location. Obviously, in ventures of such large drill time and distance, even minor improvements of the effective drilling rate can lead to enormous savings.

FIGURE 3.14: WATER JET DRILLING

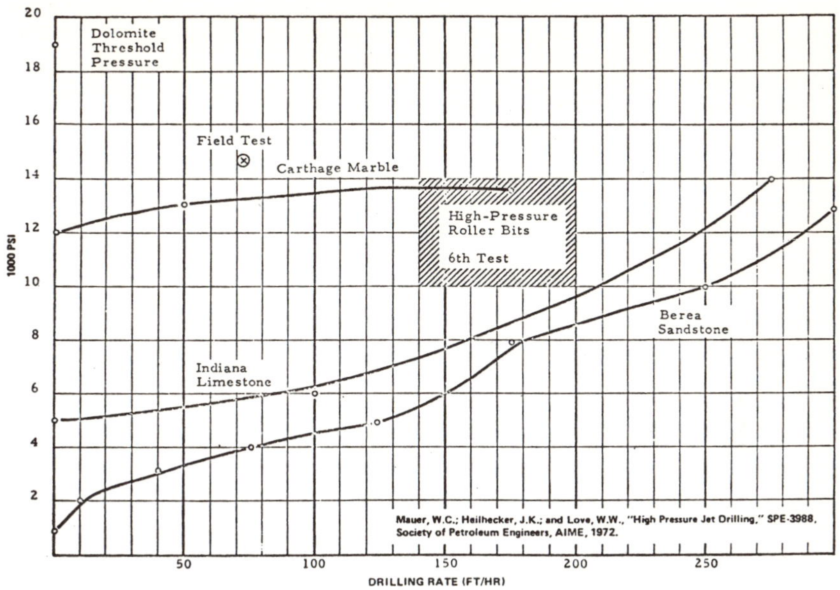

Source: AD-774 108

The N x 1,000 ft Drill Pipe Segment: In deep ocean drilling, it was suggested that long drill strings could be hung in the moor pool of the ship, thus minimizing trip time. In land drilling, large segments of drill pipe can be pulled intact, provided the derrick is tall enough; the pipe can be flexed or rolled; or the derrick can be extended at depth by a shaft that is accessible at bottom to man or equipment. To date these have not been done. However, mud head problems could probably be overcome by mud density manipulations or by mechanical means.

However, increased trip speed is dependent on pore pressure, viscosity of the drilling fluid, suitability of drill collars, and related parameters which must be coordinated to avoid dynamic disturbances affecting the stability of the hole.

A time line of the effect of shortened trip time in a drilling operation will be given later. At 30,000 ft, one round trip constitutes a time demand of about one day (22 hr). At greater depth, the time is likely to increase to 1 to 2 days. Since bit life decreases with increased temperature and pressure, causing an even greater need for increased tripping, it is conceivable that about 50 percent of the total rig time at depth near 40,000 to 50,000 ft will be spent in conventional tripping. Therefore, a redesign of the drill pipe handling may well be the most decisive factor in affecting total drill time and cost. A suggested solution is offered in Figure 3.15.

FIGURE 3.15: DRILL STRING TRIP TIME AT A DEPTH OF 45,000 FT

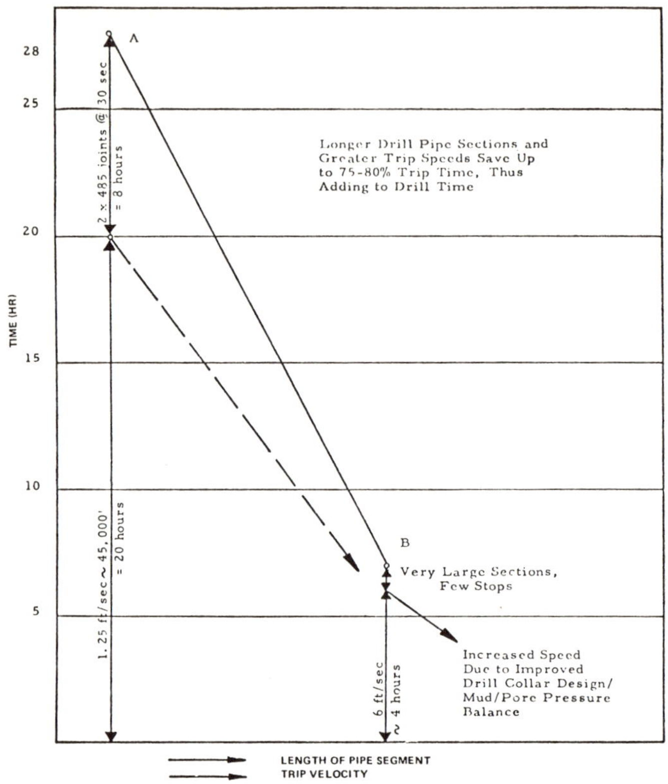

Source: AD-774 108

At well depths of 45,000 ft, about 28 hr would be spent per trip, assuming average pipe motion. Handling long, multiple lengths of pipe can decrease the trip time. Very large sections of about 1,000 ft (since they could be moved in deep water, assuming automatic coupling to be available, or by using a shaft) may decrease the trip time to 8 to 16 hr by saving coupling/decoupling time and allowing constant speeds for intervals of thousands of feet (assuming mud problems can be solved). (This constant driving speed may also be beneficial for the entire well.) Further savings in trip time can only be realized by increased trip speed. Trip surge calculations for one well are in the range of 1 to 6 ft/sec with resulting, acceptable pressures in equivalent mud weight.

For example, to move 45,000 ft of string, with a few neglibible stops, in 6 hr would take a constant velocity of 2.1 ft/sec. A 5 hr trip time appears attainable, assuming improved drill collar designs and matching mud/pore pressure balances. At that time, 80 percent of the total trip time, or 40 percent of the rig time, potentially can be saved. If only a fraction of this savings could be converted into active drill time, rotary total drill yields could be increased tenfold, from 2 to 20 percent of total rig time, or from 2 to 20 ft/hr overall effective drilling. Such increases appear entirely feasible without any major breakthrough. The potential cost savings per ultra-deep well are in the millions to ten millions of dollars.

A well-organized 228-day drilling operation, drilled by Falcon-Seaboard Drilling Company for Amerada-Hess in Victoria County, Texas, reached 23,884 ft in soft to medium hard, hot rock and passed into 520°F temperatures. The average drill speed was 4.3 ft/hr, and the time table (Table 3.6) shows that only 20 percent of the rig time was spent on trips; 50 percent drilling and reaming; 8 percent each on maintenance and revision of equipment and on logging and coring; and another 8 percent on casing and cementing. The high-temperature drilling operation utilized refrigerated mud. A combination of a high-capacity derrick and a wide-diameter well or shaft could be utilized to increase the effective drill time of a modern drill rig about tenfold, thereby reducing costs and technical drilling problems.

Drilling Muds: Currently, there are two basic types of drilling muds. The most common is the water-base fluid. However, in many deep, hot wells, an oil-base fluid is used because of its stability at higher temperatures. Research is currently being performed at the three major mud company research centers in Houston, Texas, as well as in major oil company research centers. The objective is to develop drilling fluids to be used in deep wells with bottom hole temperatures in excess of 700°F.

The main problem with drilling fluids at high temperatures is that they tend to thicken or solidify at the bottom of a deep well during a round trip. Some of the various possible solutions to this problem are: new highly stable polymer water-base muds, new oil-base muds, surface refrigeration of the drilling fluids, and faster trip times.

Any new drilling fluid developed for drilling to 40,000- to 50,000- ft must have the qualities that enable it to stabilize the well bore, remove the cuttings from the well, retain rheological properties for optimum hydraulics, and withstand contaminants.

The role of drilling muds and drilling fluids in deep wells has been discussed by G.R. Gray and F.S. Young of Baroid/National Lead Company in "Drilling Fluids of Deep Wells in the United States," a paper presented at *Proceedings of the Eighth World Petroleum Congress,* 1971. They are advocating ways to maximize penetration rate and minimize nondrilling time. Drilling fluids of 2.34 g/cm^3 specific gravity, stable at temperatures of 750°F and higher, are proposed using suspensions of heavy minerals in temperature-stable oily liquids. Oil mud will also cut down corrosion of drill pipe, and reduce stress corrosion failure.

Pore pressure to be found at maximum depth may range from 22,000 to 50,000 psi. Temperatures may range from 500° to 1000°F. Rocks, under these conditions, may have begun metamorphism and recrystallization. Tests on various sedimentary rocks indicate that a reduction of effective stress (formation pressure approaching overburden pressure) and an increase in temperature reduced the ultimate rock strength [J. Handin, et al, *American Association of Petroleum Geologists Bulletin* 47(5):717-55, (1963)]. (Furthermore, tests on marble at higher temperature showed a pronounced decrease in ultimate strength between 570° and 930°F.) At greater depths, borehole stability will become an increasingly more difficult problem. Rocks become weaker, but also more ductile and difficult to drill.

The long intervals of open hole made necessary by deep drilling place special emphasis on the problems of high torque and stuck pipe. Both problems are related to mud properties, although no specific measurements are made routinely to define the optimum characteristics. Torque reduction in water muds has been brought about by the addition of extreme-pressure lubricants.

Because of the severity of differential sticking, which depends on the magnitude of the difference between the mud pressure and the pore pressure and on the area of contact and the friction between the pipe and the mud cake, efforts have been directed toward preventive measures. The area of contact between the pipe and the mud cake has been reduced through the use of special drill collars and stabilizers. Several additives have been proposed to reduce the friction between the pipe and the filter cake of water muds. The use of oil mud virtually eliminates the problem of stuck pipe because of the thin filter cake and the low coefficient of friction between the cake and the preferentially oil-wet steel.

Corrosion of drill pipe deserves special attention in deep holes because of the difficulties of fishing operations. Corrosive attack is accelerated at high temperatures. Not only do corrosive substances enter the mud from the formations drilled but also thermal degradation of common constituents of water muds may create a corrosive environment. A consistent program on monitoring and counteracting corrosive attack of drill pipe is necessary.

The temperature requirement eliminates consideration of water-base muds for ultra-deep drilling. The critical temperature of water is 705°F. The reactions of the normal components of water mud with one another and with the formations drilled definitely limit the usefulness of water mud at even lower temperatures.

The forced selection of a nonaqueous continuous phase for the mud narrows the search for a stable liquid for high-temperature lubricants, hydraulic fluids, and heat-transfer liquids. Such liquids fall into the general classes of aromatic of polyphenyl ethers, aromatic or phosphate esters, and aromatic silicates, silanes and silicones, and possibly, certain naphthenic and paraffinic oil and halogenated hydrocarbon esters. The required suspending and sealing qualities may be supplied by carbon black and gellants for high-temperature greases. Current facilities for testing such compositions for use as drilling fluids are inadequate and these speculations are unsupported by any experimental performance tests.

Cements and Cementing: The equipment is available at present to cement any casing string in an ultra-deep well. However, the cementing materials will have to be improved. High temperatures cause two very serious problems with current cements. The first problem is that of flash setting or hardening of the cement before it is totally in place. A second problem is also caused by high temperatures. The high-temperature environment causes the cement to lose most of its compressive strength, referred to as strength retrogression.

Logging Tool Capability: The Falcon-Seaboard well provided interesting insights into the limitations of logging tools. Below 21,000 ft all logging tools were inoperable because of the high temperature, although high-temperature difficulties actually started at 15,000 ft. All-Teflon insulated cables were used below 12,000 ft to avoid trouble. Below 19,000 ft, where temperatures could not be controlled, even with the mud cooling system, logging was done by insulating the tool completely.

The inverted oil emulsion mud used at this depth restricted the tool choice to gamma ray-neutron logging. But formation logging was not affected alone, even the deviation survey tools had to be altered to withstand high temperatures. It is believed that the highest temperature reached in drilling the Falcon-Seaboard well was 520°F and that logging tool difficulties became serious at temperatures agove 400°F. Thus, considerable progress will be required in logging technology to make proper monitoring of ultra-deep wells possible, particularly since logging time availability will become shorter and shorter, in the interest of maintaining an open well.

Table 3.7 shows a typical recommended optimum drilling bit program. The bit series conforms to the new API classifications. Amoco's *Op Drilling* program, developed through nearly a decade of research and field application, can minimize the capital risk in drilling a well. Field results indicate that the actual homemaking costs were reduced 15 to 20 percent when optimization programs were properly applied.

TABLE 3.7: OPTIMUM RECOMMENDED BIT PROGRAM

Bit No.	Size	Bit Series	Depth Out	Footage	Hr	Ft/hr	Weight	rpm
2	9.875	1-1	6,200	2200	16	135	40	180
3		1-1	7,800	1600	15	110	40	150
4		1-2	8,300	500	10	50	35	150
5		1-1	8,900	600	13	45	40	140
6		1-1	9,350	450	11	40	30	140
7		1-1	9,700	350	11	32	30	130
8		1-1	10,100	400	10	40	40	100
9		5-2	11,100	1000	60	16	35	50/55
10		5-2	12,100	1000	60	16	35	50/55
11		5-2J	12,900	800	65	12	35	45/50
12		5-2J	13,500	600	65	9	35	45/50
13		5-2J	13,950	450	65	7	35	45/50
14		5-2J	14,300	450	65	7	35	45/50

Source: AD-774 108

When a well is to be optimized, the *Op Drilling* program draws on an extensive data bank of drilling and geological information from many of the nation's oil productive areas.

This extensive and expanding data bank is the key; it is the experience base on which optimized drilling is built. The data bank allows computer programs to calculate and optimize the maximum number of variables in drilling a given well.

From this data bank, Amoco's program selects wells and conditions as close as possible to the proposed well. The selected data are then computerized to prepare a predrilling plan outlining the most economical and efficient methods to be used for drilling a given well.

The program recommends a combination of drilling variables such as weight on the bit, rotary speed, mud characteristics, and bit selection. The driller, by following the prescribed plan, can cut costs and time and substantially increase drilling efficiency.

Areas for Further Research

It is recommended that research on alloys, bits, temperature and pressure effects, lasers, high-pressure jet drilling, and dual shafts be continued and that controlled testing be conducted in hard rock.

A STATUS REVIEW

In *Geothermal Well Technology and Potential Applications of Subterrrene Devices, A Status Review* by J.H. Altseimer, 34 p, LA-5689-MS, August, 1974, the past, present, and some future aspects of the geothermal energy (GTE) industry are reviewed.

Special attention is given to geothermal well-drilling problems. Geothermal wells can be produced with present equipment and methods, mostly derived from the oil and gas industry, but costs are relatively high.

Short-term improvements are needed in drilling rigs and auxiliary surface equipment, drill bits, bit-bearing lubrication systems, tubular goods, high-temperature muds and cements, logging and downhole sampling equipment, directional control equipment applicable to geothermal conditions, and in the use of a data bank for GTE wells to help optimize drilling programs.

A PROPOSED COMMERCIAL SYSTEM

Using available drilling techniques a method for economically recovering the thermal energy contained in the much more numerous reservoirs of hot rock that are nearly impermeable to circulating ground water is described by D.W. Brown, M.C. Smith and R.M. Potter in LADC-72-1157, September, 1972, *A New*

Method for Extracting Energy from "Dry" Geothermal Reservoirs from which the following material was derived. It is postulated that a commercial energy source, based on the Los Alamos dry geothermal energy concept illustrated in Figure 3.16, would be developed in the following sequence of operations.

FIGURE 3.16: PROPOSED SYSTEM FOR DEVELOPING A SHALLOW, DRY, GEOTHERMAL RESERVOIR AS A COMMERCIAL ENERGY SOURCE

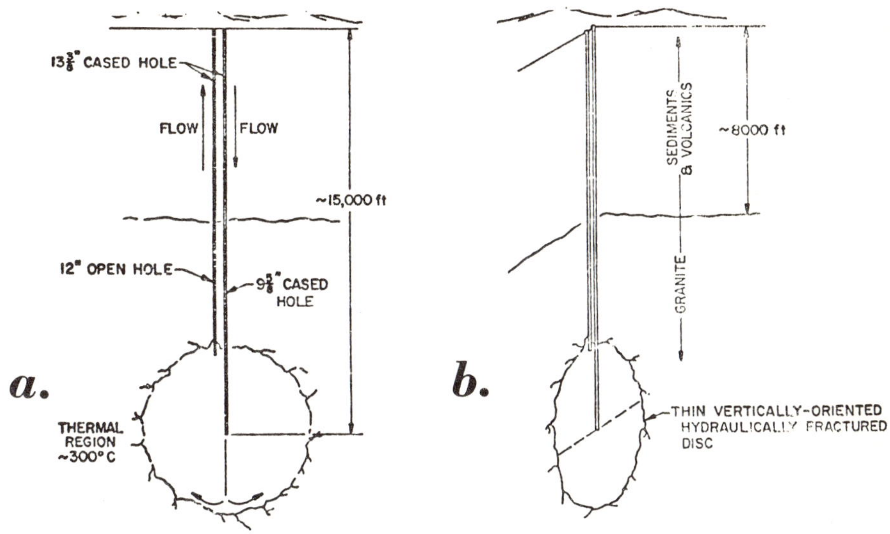

Source: LADC-72-1157

Drilling, Hydraulic Fracturing and Water Circulation

Site Selection: An accessible site (or sites) would be selected from within a region where the geology is favorable (free of major faults or other unknown structural complexities, and with reasonably competent rock at the projected reservoir depth), and where the heat flow has been determined to be adequate for the planned application. Since measured heat flow values are reasonably constant for surface displacements of many hundreds of feet (most definitely not the case when exploring for petroleum or geothermal steam), the plant location would be determined more by topography and ease of accessibility than by slight variations in heat flow.

This measured heat flow might range from an acceptable value of 1.5 HFU (1 HFU = 1 μ cal/cm^2-sec) for the eastern United States to from 3 to 6 HFU in the western United States—west of the eastern slope of the Rockies.

Drilling the Deeper Hole: The first (deeper) hole would be drilled to the projected depth, and then logged to obtain geologic, stratigraphic and reservoir rock physical property and diagnostic data—which would of course include a measurement of the relatively undisturbed bottom-hole (reservoir) rock temperature. The hole would then be cased with steel pipe to within the planned upper limit of the reservoir, and the casing cemented in place.

Initial Hydraulic Fracturing: Following post-cementing hole cleanout operations, the uncased portion of the hole below the casing would be hydraulically fractured using methods that are common in the oil well service industry. This would be done, using a high-pressure pump at the surface, by pumping water down a high-pressure line extending through a seal (packer) inserted in the annulus between this line and the bottom of the casing string. Hydraulic pressure (of the order of 7,000 psi above the hydrostatic head) would be exerted against the rock adjacent to the well bore, developing tensile stresses in this rock sufficient to cause fracturing.

The stress required to extend a crack is in general much less than that needed to form it. Therefore, once breakdown (cracking) had occurred at the well bore, pumping would continue at a reduced pressure until the principal cracks had been extended to the desired radius. Mathematical analysis agrees with field experience in indicating that the resulting crack would be in the form of a thin, vertically-oriented disc of elliptical cross section, as suggested by the sketch of Figure 3.16.

Following the initial fracturing operations, the crack system would be allowed to collapse by slowly reducing the system pressure by back-flowing through a pressure control valve. This procedure is necessary to preclude the dangers inherent in intersecting a large volume of superheated water during the air drilling of the second (shallower) hole.

From diagnostic measurements obtained during the hydraulic fracturing operations, the orientation (azimuth) of the fracture system would be determined. Typical of the diagnostic measurements that might be used for this purpose would be impression packers, down-hole pressure transducers and geophones, and surface arrays of seismometers and/or tiltmeters.

Drilling the Second Hole: The second hole would be located several hundred feet away from the first (deeper) hole, on a line normal to the orientation of the fracture system formed at the bottom of the first hole. This hole would be essentially the same size as the first hole, but several thousand feet shallower. It would be drilled parallel to the first hole until adjacent to the upper portion of the fracture system, and then slanted (whipstocked) to intersect this fracture system using directional drilling techniques. During the directional drilling operations, the deeper hole would be periodically pressurized to test for communication between the two holes.

Final Hydraulic Fracturing: After sufficient communication has been established through the fracture system connecting the two holes (which may require additional fracturing from the bottom of the second hole), the fracture system would be enlarged to its final dimensions by additional pumping down the deeper hole.

Pressurized Water Circulation: Following the final fracturing operations, the previously filled and pressurized surface system; heat-exchangers, water-treatment system, valving and surface piping; would be pumped into the flow loop, and circulation initiated by an auxiliary pump. Circulation would be down the deeper hole, through the fracture system, up the shallower hole, through the heat-exchangers and water-treatment system, through the flow-control valve, and finally back down the deeper hole.

Once a moderate temperature difference had been established between the ascending and decending legs, the auxiliary pump would be bypassed and then valved off. From this point on, the earth-loop flow would be maintained by natural convection, no circulating pump being required. However, for some eastern U.S. systems working on a smaller ΔT; say an earth outlet temperature of 150°C (302°F) and an earth inlet temperature of 40°C (104°F); some auxiliary pumping might be required to obtain adequate water flow rates.

If water entered the underground system at 65°C and left it at 280°C, the difference in water density in the two legs would produce a pressure difference sufficient to maintain natural convection circulation at the rate of about 610 lb/sec (4,500 gpm) for the hole sizes shown in Figure 3.16. If permeability of the hot rock for this system was low enough to permit it, a pressure of approximately 2,000 psia would be maintained in the heat-exchangers, which is safely above the vapor pressure of water at 280°C; 930 psia.

This overpressure would keep the flow system liquid throughout, maintaining a negative temperature coefficient of viscosity in the heat-transfer medium, which relative to steam, offers a very large heat-transfer advantage in extracting thermal energy from thin cracks in the hot rock; and greatly increasing the rate at which energy can be transported through a pipe of a given diameter. It would also permit the operation of the system at higher water temperatures if spontaneous extension of the cracks occurred and hotter rocks were encountered.

Commercial Power Plant

It was realized many years ago that naturally-occurring hot geothermal waters were very unique energy sources; so unique, in fact, that no attempts were made to utilize these low-enthalpy waters for power generation. More recently, however, with a growing interest here and abroad in binary cycle power plants and the apparent availability of more suitable down-hole submersible pumps, the situation appears to be changing. However, common to all proposed (or operating) geothermal energy systems based on circulating pressurized water as the heat source, and using heat exchange to a separate vapor-cycle working fluid, is the following unique problem: A large temperature drop in the circulating geothermal water is required to vaporize a significant amount of the power cycle working fluid.

The advantages of a low-boiling-point working fluid over water are obvious for such a geothermal water heat source, particularly if the earth outlet temperature level is below about 200°C (392°F). This would be the general case for dry geothermal energy applications in the eastern U.S., if electric power geothermal power generation was the objective.

For a typical commercial man-made geothermal energy source for the western U.S. as shown in Figure 3.16, and with an earth outlet temperature level of 280°C as discussed above, a dual-cycle power generation system would be appropriate. Such a proposed system, using a conventional steam cycle for the higher source temperatures, and an isobutane vapor cycle for the lower source temperatures, is shown in Figure 3.17.

FIGURE 3.17: POSSIBLE DUAL-CYCLE POWER GENERATION SYSTEM

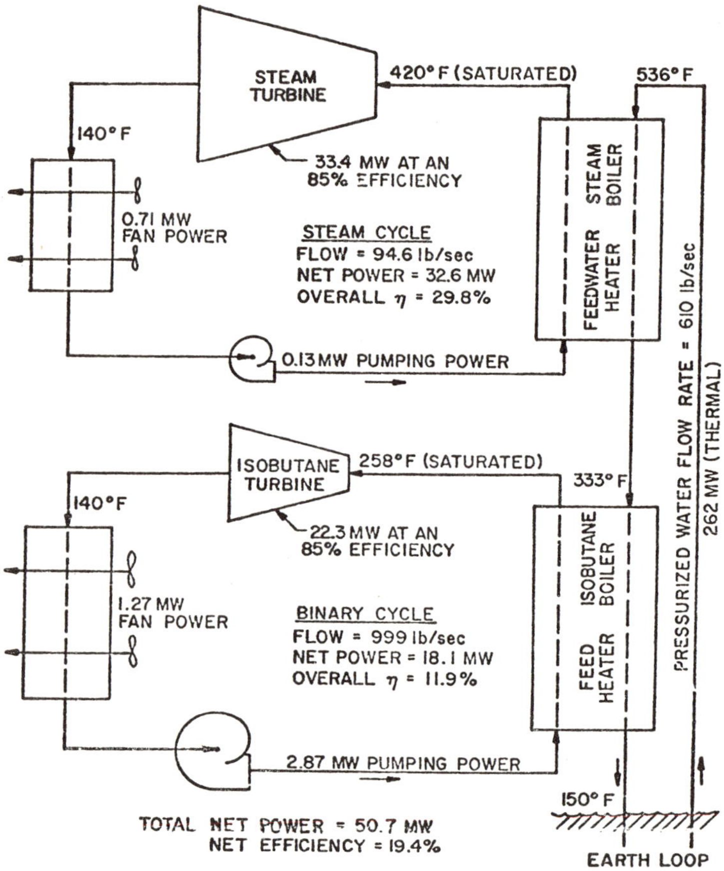

Source: LADC-72 1157

For applications in the western U.S., where water availability is already becoming critical, forced-draft air-cooled condensers were specified. Although adding 10 to 20 dollars per kilowatt to the capital costs of such a plant, the resulting freedom in site selection and the absence of thermal pollution problems (at least as now generally applied only to the heating of bodies of water or rivers) appears to justify this choice. Further, the economics for this type of geothermal power plant are so favorable, that the additional 0.2 to 0.4 mills per kilowatt hours in generating costs associated with the air-cooling requirement can easily be absorbed.

Feasibility Experiment

This experiment would be performed in a region referred to as the Jemez Plateau, in northcentral New Mexico. This area, on the western side of the Jemez Mountains and about 20 miles from Los Alamos, is within the Santa Fe National Forest and therefore under federal control. Detailed geological and geophysical investigations of this area over the past year, and the drilling of an exploratory hole into the basement crystalline rock during May and June of this year, confirm the suitability of this site for an initial experiment.

The planned experimental configuration and operating conditions are shown in Figure 3.18. For this experiment, the crack size (a 1,500 ft radius) and the pressurized water flow rate (315 lb/sec, or about 2,300 gpm) have been so selected that the artificial geothermal reservoir should be depleated in less than a year, if no reservoir enhancement due to thermal stress cracking were to occur.

FIGURE 3.18: EXPERIMENTAL CONFIGURATION AND OPERATING CONDITIONS

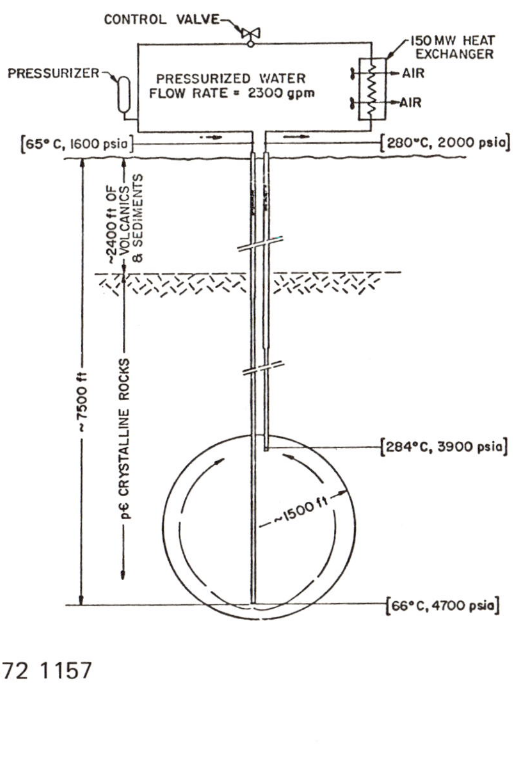

Source: LADC-72 1157

Hydraulic Fracturing and Thermal Stress Cracking

Hydraulic fracturing is a very common stimulation technique used in oil and gas fields to improve reservoir flow characteristics, by creating a set of cracks in the producing formation(s) adjacent to the well bore. This is normally done by inserting temporary seals above and below the zone to be fractured, and then using a high-pressure pump at the surface and a high-pressure line extending through the upper seal, to produce a hydraulic pressure (normally with water) in the isolated zone of the order of a few hundreds to a few thousands of psi above the existing hydrostatic pressure. A crack system is created which may extend for many feet from the well bore, the resulting increase in volume being accommodated locally by natural porosity and by elastic compression of the uncracked rock.

Although there does not appear to be any depth limitations, hydraulic fracturing has normally been done only in sedimentary rocks. However, the strengths and elastic properties of some of the sedimentary formations that have been successfully fractured closely approach those of igneous rocks.

For example, Halliburton Services, *The Frac Book, Design/Data Manual for Hydraulic Fracturing,* Duncan, Oklahoma, 1971, cites hydraulic fracturing at depths of 12,000 to 15,000 feet in the Ellenburger formation of West Texas, a strong massive limestone having properties very similar to those of a granite. The theory and practice of hydraulic fracturing is well developed, and representatives of two commercial service companies specializing in fracturing have predicted that the required fracture system for this application can be produced without difficulty, using standard techniques. Hydraulic fracturing evidently offers a method of producing a large surface area for heat transfer even in hard, competent rock, and is proven and relatively inexpensive.

Extraction of heat from rock penetrated by the hydraulic fracture will create thermal-contraction stresses that should eventually be sufficient to extend the initial crack system in three-dimensions. Initially, the cooler rock next to the fracture surfaces will be restrained from contracting by the adjacent hot rock. These restraints will cause tensile stresses to be developed which, after sufficient cooling has occurred, will exceed the tensile strength of the rock. New cracks will then form along the surfaces of the initial hydraulic fracture system that will undoubtedly propagate far beyond the cooled region. Both the amount of heat-transfer surface and the total amount of heat available to a fluid circulating through the crack system should therefore increase continuously as energy is withdrawn from the geothermal reservoir.

This very important potential of the proposed development method has recently been verified by computer modeling of the reservoir behavior (F.H. Harlow and W.E. Pracht, "A Theoretical Study of Geothermal Energy Extraction," *J. Geophys. Res.,* December 1972). These calculations indicate that the new crack volume and heat-transfer surface opened by thermal-stress cracking will, following an initial reservoir thermal drawdown period, make additional heat available to the circulating water more rapidly than it is removed by the cooling that creates the thermal stresses. Figures 3.19 and 3.20 graphically show the results of one specific reservoir calculation, where the thermal-stress crack spacing was about two inches. Figure 3.19 shows the geothermal power variation with time, indicating the power recovery phenomenon discussed above. Figure 3.20 shows reservoir contours of equal porosity for several selected times which, as expected, show the reservoir growing preferentially downwards and sideways with increasing time.

FIGURE 3.19: POWER vs TIME AT CONSTANT CIRCULATION RATE

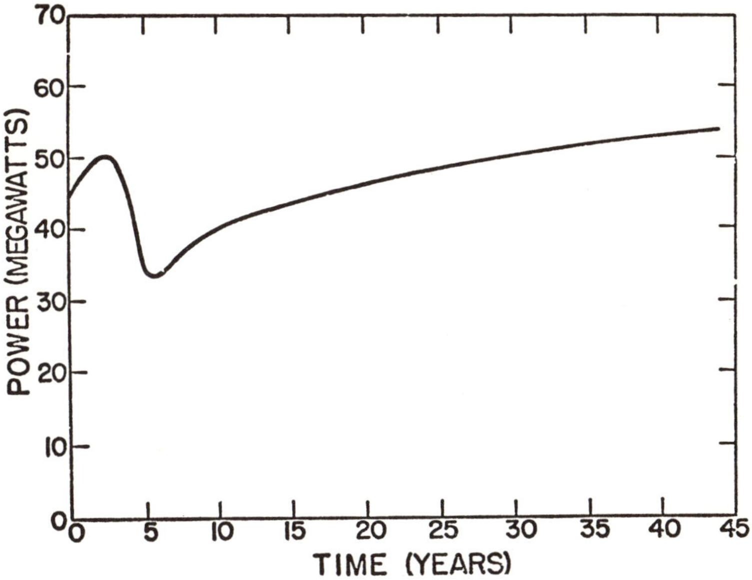

FIGURE 3.20: CONTOURS OF EQUAL POROSITY FROM COMPUTER CALCULATION

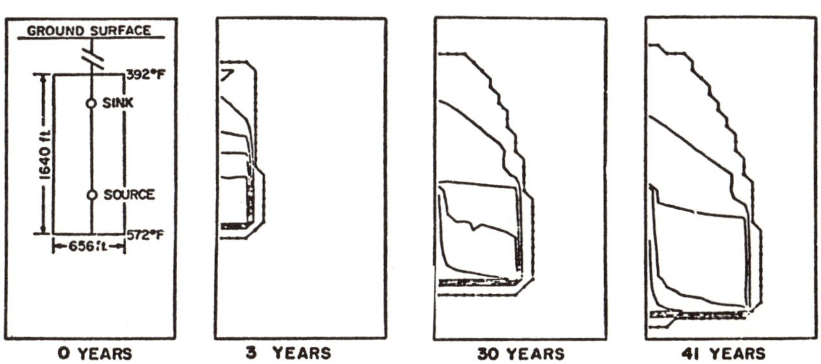

Source: LADC-72 1157

Economics

The economics of a power generating system based on the Los Alamos dry geothermal energy concept appear to be favorable now, and may possibly improve in the years to come. These improving economics may result since the only fuel costs for such a man-made geothermal power system would be in the form of maintenance charges, while conventional power plants (both nuclear and fossil fueled) are expected to experience continually increasing fuel costs, especially for oil- and gas-fired systems.

50 to 60% of the capital charges for a power plant based on the Los Alamos dry geothermal energy concept would be attributable to the drilling, casing and hydraulic-fracturing operations associated with the formation of a sufficiently large thermal reservoir. Since the costs associated with the drilling and casing of conventional-sized holes are very well-known, the importance of the more speculative estimates required, relative to the total capital charges for such a power plant, are correspondingly reduced.

Table 3.8 compares the cost for power plants based on the dry geothermal energy concept, to conventional nuclear- and coal-fired power plants to be installed near the New York load centers during the mid 1970's.

TABLE 3.8: COMPARISONS OF PLANT AND GENERATING COSTS

Plant Type	Plant Size MW (e)	Plant Cost dollars/kW	Generating Cost mills/kWH
"Dry" Geothermal Energy Source:			
300°C Rock, dual-cycle, 4 holes	100	186	4.7
175°C Rock, isobutane, 10 holes	100	316	8.0
New York Area for the Mid 1970's:			
Nuclear (PWR)	950	350	11.8
Coal-Fired Steam	950	250	13.3

Basis: 0.17 Annual capital charge rate
Use Factor = 80% of rated capacity

Source: LADC-72 1157

FEASIBILITY EXPERIMENT SITE DATA

The following information in regard to the feasibility experiment discussed above is given in LADC-72 669, *The Induction and Growth of Fractures in Hot Rock: Artificial Geothermal Reservoirs* by R.L. Aamodt and M.C. Smith of the University of California's Los Alamos Scientific Laboratory.

On the basis of geological information so far collected and heat-flow measurements made in a series of holes drilled for that purpose, it appears that an area in the Jemez Mountains; about 2 to 5 miles west of the Valles Caldera, and about 30 miles west of Los Alamos Scientific Laboratory; is particularly suitable for a field experiment. Here the geology is relatively uncomplicated, the depth to the basement granite is only about 2,500 ft, and the geothermal gradient is about 6 to 10 times normal. An exploratory hole in this area has recently been drilled and cased to 2,430 ft depth, to examine heat-flow lithology, hydrology, and drilling problems.

It is intended to extend this hole a few hundred feet into the granite, with continuous coring, to investigate the character and probably competence of the granite, to measure temperatures and heat-flow in it, and to permit preliminary hydraulic-fracturing and pressurization experiments. If the basement rock is found to be capable of containing a pressurized system, then a specific site will be selected for development of an experimental underground circulation system.

Preliminary heat-flow measurements indicate that in this area a rock temperature of about 500° to 600°F can be expected at a depth of only 7,500 ft; which is not unusual in regions of geologically recent vulcanism. It is proposed to create, at approximately this depth, a circulation loop of the general type sketched in Figure 3.16b but with a relatively small hydraulic fracture; perhaps 1,500 ft in radius; and with the extracted heat wasted to the atmosphere by an air-cooled heat-exchanger.

Simply creating such a system and establishing convection circulation in it will represent an important feasibility demonstration. Operating the loop for approximately 10 months should; with relatively small initial heat-transfer surface; indicate whether the system simply decays, or whether instead it perpetuates itself by thermal-stress cracking. Continued circulation for another 2 or 3 years will then permit detailed studies of the thermal, chemical, and mechanical behavior of the system, and of its economics. Problems are, of course, expected to develop in such areas as fluid loss through the natural void system in the rock, and plugging of circulation channels by mineral deposition. However, it is believed that there are straightforward engineering solutions to such problems.

USE OF NONAQUEOUS FLUID IN RESERVOIR

P.L. Paull and P.F. Kerr; U.S. Patent 3,827,243; August 6, 1974; assigned to Texaco Development Corporation describe a method of heating a nonaqueous fluid in a dry, geothermal reservoir formation. This process relates to a method of heating a nonaqueous fluid in a dry, geothermal reservoir formation penetrated by an injection well and a production well which comprises introducing a nonaqueous fluid into the formation through the injection well; forcing the nonaqueous fluid through the formation thereby heating the fluid; recovering the heated nonaqueous fluid through the production well and wherein the temperature of the geothermal reservoir formation is substantially above the temperature of the fluid introduced into the formation.

The heated nonaqueous fluid recovered from the production well can be passed through a boiler or heat exchanger where it is employed to heat and/or vaporize a second fluid or working fluid which can be, for example, water, steam, isobutane, pentane, isopentane, ethylene glycol, kerosene, crude oil, or, alternatively, the heated fluid recovered from the production well may be utilized directly for heating purposes or as a reactor input stream, and where the recovered heated fluid is a vapor it can be employed directly to drive a turbine, etc.

The working fluid leaving the heat exchanger which can be a vapor, a liquid or a mixture thereof, may be employed in a variety of ways as a heating medium for such process requirements as steam raising, evaporation, reactor heating or preheating of reactor input streams. The heated working fluid may also be employed in agricultural production to heat fields or greenhouses, thus permitting a number of crops to be grown in areas with short growing seasons by preventing frost damage and freezing of the ground.

After the heated nonaqueous fluid withdrawn from the production well has been employed directly, for example, in driving a turbine or has been passed through a suitable heat exchanging device, it is, preferably, recirculated through the system by returning the fluid to the dry geothermal formation via the injection well for reheating or, if desired, it may be used in other ways for processing purposes, and fresh nonaqueous fluid feedstock forced into the formation for heating.

The process offers a number of distinct advantages over processes previously proposed for utilizing geothermal energy. Since the heated nonaqueous fluid withdrawn from the production well contains an extremely low dissolved salts content there is no brine disposal problem. Likewise, noxious gases associated with steam produced from geothermal formations in other processes are not present in any appreciable quantity.

THE PLOWSHARE CONCEPT: USE OF NUCLEAR EXPLOSIVES

PNE-1550, *A Feasibility Study of a Plowshare Geothermal Power Plant,* April 1971 explores the factors involved in extracting heat directly from a heat source composed of hot rock using nuclear explosives to fracture the rock. The following information is from this study which was a joint effort of American Oil Shale Corporation and the Atomic Energy Commission. Battelle-Northwest and Westinghouse Electric Corporation aided American Oil Shale in its portion of the work while the AEC assigned its part to the Lawrence Radiation Laboratory and the Nevada Operations Office.

THE PLOWSHARE CONCEPT OF GEOTHERMAL HEAT EXTRACTION

In a natural steam field, the steam is obtained by drilling into an underground reservoir of steam or pressurized hot water. The hot water or steam is contained by an overlying, relatively impervious cap rock, and it gets its heat by circulation from a heat source deeper in the earth. This heat source may be a relatively shallow mass of intruded rock, a recently extinct volcano, or a living volcano. When a hole is drilled into such a reservoir, steam may be produced at the surface. If the steam is available in sufficient quantity, and if it is of acceptable quality, it can be used to produce electric power.

On the other hand, even though an adequate heat source is available, the circulation system and the geologic reservoir may not be good enough to supply the necessary amount of steam at the surface. The Plowshare concept of geothermal heat extraction is directed toward extracting heat from the source itself and does not rely on the reservoir or the circulation system. The basis of the concept is the use of a nuclear explosive to fracture large amounts of the hot source rock. The heat is extracted from the fractured region by piping water to it and creating steam that is then piped to a turbine-generator (see Figure 4.1). The system is closed by piping the condensed steam from the power plant back to the fractured region.

FIGURE 4.1: THE PLOWSHARE CONCEPT OF GEOTHERMAL HEAT EXTRACTION

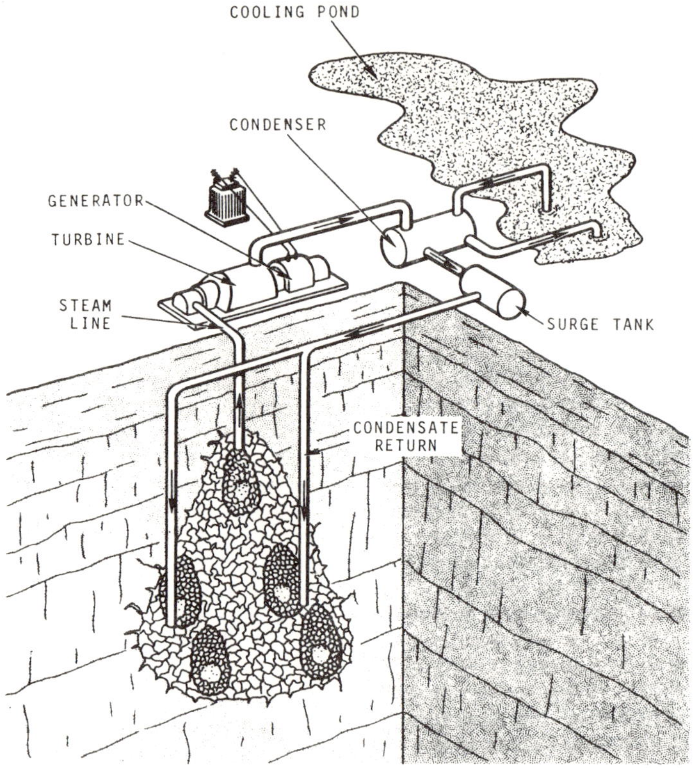

Source: PNE-1550

The heat source is expected to be deep enough so that even very large explosions could be completely contained. Much of the radioactivity created by the nuclear explosion will be trapped in the immediate vicinity of the explosion. Also, the closed design of the system will prevent any radioactivity in the steam from being released to the environment.

The heat energy contained in one cubic mile of rock is impressively large. If the rock had an initial temperature of 350°C and was cooled to 177°C, as is contemplated for the power plant described in this study, the heat energy extracted would be equivalent to that available from 300 million barrels of oil. Thus, if the energy could be efficiently recovered, it might support many tens of thousands of megawatts of electric power. The realization of such an energy source will depend on the cost of fracturing the rock, the efficiency with which the heat can be extracted, and the quality of the steam that can be produced.

Two geologically different geothermal situations are amenable to heat extraction via nuclear explosives. One is a steam field that does not have enough permeability to produce the steam volume necessary for conventional steam wells. The large well bore that could be created by nuclear explosives might increase the steam production enough to permit the economic development of marginal areas in existing steam fields as well as areas whose development by conventional methods is not now considered feasible. This concept is known as Plowshare geothermal stimulation.

The second geothermal situation consists of hot (about 350°C), relatively dry rock at a moderate depth (about 3 kilometers). The fluid nature of basaltic magmas causes them to form relatively thin intrusive bodies or to be extruded as lava flows, but intermediate and granitic magmas are very viscous and hence form plug-like plutons. A solid but still hot intrusion of granitic rock close to the surface is a likely candidate for a dry geothermal resource.

In the western United States, the volume of granitic rock that intruded during the middle Cretaceous period (110 to 800 million years ago) is several hundred times the volume that has intruded since then. However, many important intrusions are found, and they form a broad band that passes from north to south through western Utah and eastern Nevada (see Figure 4.2; age assignments in the figure are based in part on all of the radiometric ages published up to July 1962).

Gilluly has indicated a continuous history of intrusions, with the only variation being the volume of material intruded (J. Gilluly, "The Tectonic Evolution of the Western United States," *Geol. Soc. London Quart. Jour.*, volume 119, pages 133 to 174, 1963). Thus, it is reasonable to expect that numerous granitic intrusions have occurred west of the Continental Divide during the past 2 or 3 million years.

The purpose of this study is to determine the technical, economic, safety, environmental, and administrative considerations that are involved in using Plowshare geothermal power plants to produce electricity. An additional purpose is to develop data and recommendations for developing the concept further.

The basic assumptions adopted are:

Plant size	200 megawatts electric
Linear thermal gradients	125° and 100°C per kilometer (the thermal gradient of the Cerro Prieto geothermal field was substituted for the latter)
Geology	Competent granite overlain by 1,000 feet of alluvium, lava, and/or sedimentary rock
Rock temperature	350°C
Explosive sizes	200, 500, and 1,000 kilotons
Steam system	Boiling and/or pressurized water

In addition, the geothermal field is assumed to be both impermeable and dry. An increase in the permeability or the addition of water will enhance the recovery of the heat.

FIGURE 4.2: GRANITIC ROCKS OF MESOZOIC AND CENOZOIC AGE IN THE WESTERN UNITED STATES

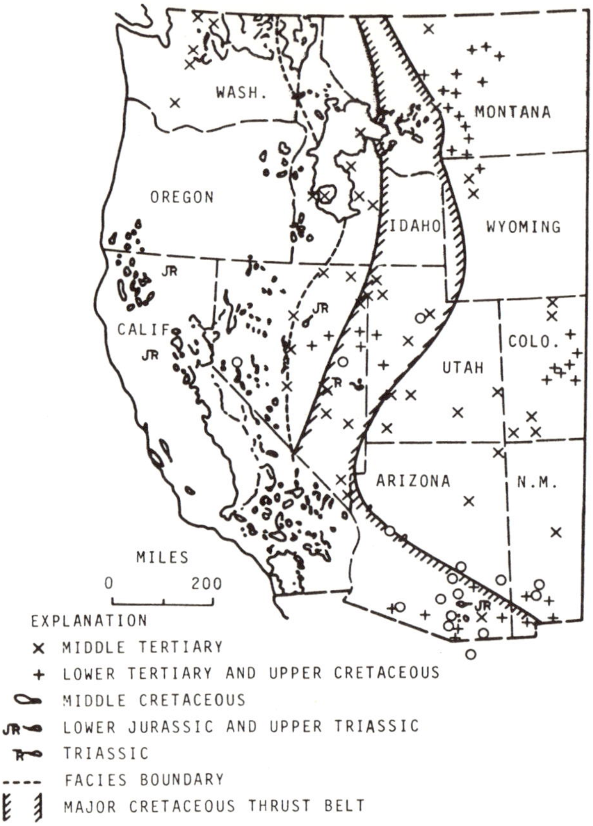

Source: PNE-1550

Three different methods of developing geothermal power plants are considered in this study. They differ according to the way in which the plant is protected against the seismic motions created by fracturing explosions. The reference method assumes a 200-megawatt plant operating on a boiling-water cycle, with the geothermal field being completely developed before the plant is constructed. This means that the plant and its equipment do not need to be built to withstand seismic shocks from subsequent explosions.

The alternate hardened-plant method does not require the geothermal field to be completely developed before the plant is built, which could be a distinct economic advantage. The disadvantages of this method are the additional investment necessary for hardening the plant and the increased losses associated with transporting steam over comparatively long distances. For example, a field providing a 10-year heat supply would be fractured and a hardened plant constructed close by. After 10 years, a second field would be fractured far enough away to minimize damage to the plant. A long, hardened pipe would then be laid on the surface between the plant and the new field. After another 10 years, a third field adjacent to the second one would be fractured and a short, unhardened pipe laid between it and the hardened pipe.

The third method consists of moving the plant every time a field is fractured. After each explosion series, power production would be continued by moving the plant back and reconnecting the steam and electrical-transmission lines. This method sounds simple enough, but in practice it would require the movement of massive pieces of equipment. There are a number of techniques for accomplishing this task–railroad cars, skids, helicopters, or trucks, for example. In order for these techniques to be used, the plant would have to be constructed of transportable modules. To circumvent this requirement, the waterway-and-barge technique might be useful. The plant would be completely built on a barge in a canal leading from the field being fractured to a point far enough away to prevent seismic damage. The plant would then be moved away from the area while the nuclear explosions were conducted. If the terrain of an actual site proves suitable, this method would provide the most attractive power costs.

SITE GUIDELINES AND GEOLOGY

The ideal site for a geothermal power plant is a sparsely populated area, for the economic benefit from this concept increases with the yield of the explosives used. In turn, the size of the nuclear explosives that can be used is strongly dependent on the number and kind of structures in the immediate area. In addition, the site must be seismically stable, and it should have soft, reasonably level ground on which a cooling pond or tower (and possibly a canal) can be built.

An adequate water supply, (more than 125 cubic feet a minute) must be available, either on the surface or in the water table. Also, a power-transmission line should be located nearby in order to minimize transmission costs. Finally, the geothermal heat source should lie at a depth of 5,000 to 10,000 feet.

A likely candidate for such a heat source is a solid, but still hot, intrusion of granite rock close to the surface. Using reasonable approximations (see Table 4.1) and neglecting the effects of convection or escaping gases, Jaeger has shown that a granitic intrusion takes 0.016 T^2 years to solidify where T is the thickness of the intrusion in meters (J.C. Jaeger, "The Temperature in the Neighborhood of a Cooling Intrusive Sheet," *Amer. Jour. Sci.*, volume 255, pages 306 to 318, 1957).

For 1-, 2-, and 3-kilometer thicknesses, the values are 0.2, 0.6, and 1.4 million years, respectively. At that time, the core is just congealing (500°C) and temperatures of more than 350°C extend upward from the center of the intrusion for a distance that is about equal to the thickness of the intrusion. As the center of the intrusion continues to cool, the temperature of the igneous pluton and the surrounding rock remains above 350°C for several times as long as the initial solidification took.

TABLE 4.1: REASONABLE APPROXIMATIONS OF THERMAL DATA FOR GRANITIC INTRUSIONS

Intrusion temperature	700°C
Final solidification temperature	500°C
Latent heat of fusion	80 calories per gram
Thermal conductivity (liquid and solid)	0.005 calorie per centimeter per second per degree centigrade
Specific heat (liquid and solid)	0.28 calorie per gram per degree centigrade
Density (liquid and solid)	2.7 grams per cubic centimeter

Source: PNE-1550

It is reasonable to expect numerous granitic intrusions to have been emplaced west of the Continental Divide during the last 2 or 3 million years. A well studied example of such intrusions is the Cornelia quartz-monzonite pluton near Ajo, Arizona. This pluton, which is of Tertiary age, outcrops over 9 square miles in the Little Ajo Mountains.

Gilluly estimates that its temperature of crystallization was 800°C and that its depth of burial is about 2 kilometers (J. Gilluly, "The Ajo Mining District, Arizona", *U.S. Geol. Survey Prof. Paper 209*, page 112, 1946). This temperature and depth also fit the conditions necessary for the hydrothermal alteration that took place as the magma cooled. Most other Tertiary plutons, particularly those with porphyritic textures, probably occur at depths of 2 to 4 kilometers. The problem is to locate still hot magmatic bodies at these depths.

A method of exploring for hidden targets has been developed in the search for phorphyry-copper deposits. After a region has been selected on geological grounds, aeromagnetic and regional gravity measurements are used to select sites for more detailed ground-magnetic and local-gravity studies. Both gravity and magnetic anomalies are clearly related to the heat sources in known steam fields. Microseismic measurements may also prove effective in locating geothermal anomalies.

Abnormal heat flows and high geothermal gradients at the ground surface can be used as a guide to buried dry heat sources, but they are not necessary. Evidence of recent surface displacements and local geologic variations may be a better guide to recent intrusions.

Although geothermal heat sources may not contain significant amounts of minerals, they give off trace amounts of many elements as they cool. Thus, both chemical and physical expressions of buried plutons can be detected. Dispersion halos of mercury may move the greatest distance and hence be detectable before abnormal amounts of the more common elements are found.

HEAT SOURCE DEVELOPMENT

Geothermal rock fractured by a nuclear explosive represents more than just a source of energy. It is, by analogy to a nuclear reactor, the core, the fuel, the containment system, and the pumps. To a large extent, the technical and economic feasibility of the concept hinges on the efficiency with which steam can be generated in the fractured region.

Nuclear Explosion Effects

A nuclear explosion in rock proceeds in the following manner: the energy released by the nuclear reactions creates a plasma having a temperature of millions of degrees and a pressure of many megabars. The heat produces a shock wave that successively vaporizes, melts, and fractures the surrounding rock before it decays to a seismic wave at relatively large distances from the explosion. The sphere of vaporized rock created by the shock wave expands until its pressure is balanced by the weight of the rock above it and by the strength of the surrounding rock. The result is a large, spherical cavity surrounded by melted and fractured rock.

The fractured rock above the cavity normally collapses in from seconds to hours after the explosion (except that for certain special materials, such as salt, the cavity remains standing). The collapse eliminates the cavity, whose volume is then distributed in the fractured region above the explosion point. For a crushable medium like alluvium, the porosity introduced by collapse is small, and collapse can continue until the surface is reached and a subsidence crater forms. In hard rock, however, the induced porosity is significant, and collapse continues only to a height at which the cavity volume is used up or a stable arch forms. The complete structure is called a chimney.

Estimates of the effects of nuclear explosions under conditions that are of interest to geothermal development are given in the following sections.

Cavity Size: The cavity radius (r_c) is given by:

$$(1) \qquad r_c = C \frac{W^{1/3}}{P_f^{\alpha}}$$

where W is the yield of the explosive, P_f is the final pressure in the cavity at the end of cavity growth, and C and α are constants. This pressure is related to the overburden pressure by:

$$(2) \qquad P_f = KP_0$$

where K is a constant and P_0 is the initial cavity pressure. Constants C and α are determined by theoretical calculations for adiabatic expansions of the silica-water system, and K is determined from nuclear test data. For r_c in meters, W in kilotons, and P_0 in bars, these constants are approximately C = 46.6, α = 0.277, and K = 2.0, but they depend somewhat on variations in P_0.

In actual experiments, there is some error involved in measuring r_c and W. Table 4.2 shows the results from experiments on granitic rock and compares them with results calculated with Equation 1. The 0.5 meter uncertainty in the cavity radii for the Piledriver and Hardhat Events is realistic, but the same uncertainty for the Shoal Event is probably larger, although it is not known how much larger.

TABLE 4.2: RESULTS OF EXPERIMENTS ON GRANITIC ROCK

Event	Measurements: Density (grams per cubic centimeter)	Depth of explosion (meters)	Device yield (kilotons)	Cavity radius (meters)	$\frac{r_c}{W^{1/3}}\left(\frac{\text{meters}}{\text{kilotons}^{1/3}}\right)$ Measured	Calculated
Piledriver	2.66	462.8	61 ± 10	40.1 ± 0.5	$10.2^{+0.5}_{-0.4}$	10.23
Hardhat	2.66	286.2	4.9 ± 1.5	19.2 ± 0.5	$11.3^{+0.9}_{-0.7}$	11.69
Shoal	[2.66]	367.3	13.1 ± 1.2	[25.6 ± 0.5]	$10.8^{+0.2}_{-0.1}$	10.38

Source: PNE-1550

To this data should be added the results of 13 French underground experiments at the Hoggar test site in the Sahara Desert. These tests were conducted in a uniform granitic rock and is apparently stronger, dryer, and less naturally fractured than the American medium. Their measured cavity radii are considerably smaller than those in American tests. The ratio is:

$$\frac{r_c \text{ (American)}}{r_c \text{ (French)}} = 1.73.$$

Thus, the cavity volumes in the American tests are five times larger than those in the French tests. This difference is confirmed by computer calculations in which cavity radii for a 5-kiloton yield (Hardhat) were computed for granites having different water contents and natural fracturing. The results are:

Medium	Cavity Radius
Dry, virgin granite	12.7 meters
Dry, fractured granite	15.0 meters
Saturated, fractured granite	20.4 meters

Estimates of cavity radii and volumes at two depths of geothermal interest and for several yields are given in Tables 4.3 and 4.4. The estimates are based on the adiabatic-expansion calculation mentioned above, and they are representative of experience with granitic rocks having some water and natural fracturing. Dryer and less fractured rock could result in smaller cavities, as explained above.

TABLE 4.3: CAVITY RADII

Depth (kilometers)	Cavity radius (in meters) for a yield of:		
	200 kilotons	500 kilotons	1000 kilotons
2.64 (P_0 = 688 bars)	36.96	50.16	63.20
3.3 (P_0 = 860 bars)	34.68	47.07	59.30

TABLE 4.4: CAVITY VOLUMES

Depth (kilometers)	Cavity volume (in 10^6 cubic meters) for a yield of:		
	200 kilotons	500 kilotons	1000 kilotons
2.64	0.2115	0.5286	1.0569
3.3	0.1748	0.4369	0.8734

Source: PNE-1550

Extent of Fracturing: The number of fractures created in the rock surrounding the explosive diminishes with radial distance from the explosion point. Rock-mechanics calculations with the stress-wave-propagation code SOC give the following fracture distances for a 5-kiloton explosion in granitic rock at a depth of 286 meters (Hardhat).

Material	Fracture Distance (meters)	$\frac{d_f}{r_c}$
Dry, virgin granite	40 to 50	3.14 to 3.98
Dry, fractured granite	55 to 65	3.67 to 4.33
Saturated, fractured granite	100	4.9

In these calculations, the fracture distance (d_f) is defined as the distance from the explosion point to the zone where fracturing due to compressional failure of the rock ceases. The third column above gives the fracture distance as a function of cavity radius.

An examination of core samples from drill holes in the vicinity of the explosion point yields an indication of the extent of the fracturing. With this method, the French experiments have a fracture distance of 3.7 r_c, whereas core samples from Hardhat give a fracture distance of about 2 r_c. However, the latter value is derived from small samples of the larger core samples, so large-scale effects may not have shown up. Even in the chimney, where the large-scale permeability is known to be infinite, measurements on chimney rubble samples show a low permeability. The permeability determined from laboratory tests of Hardhat core samples is plotted in Figure 4.3.

FIGURE 4.3: HARDHAT PERMEABILITIES AS DETERMINED FROM CORE SAMPLES

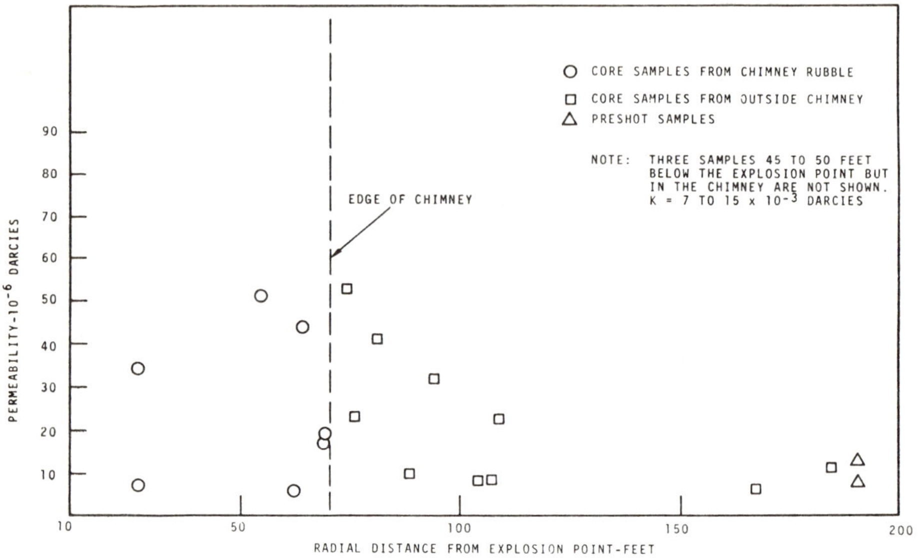

Source: PNE-1550

The in situ permeability in Hardhat was also measured by pressurizing a series of holes drilled into the walls of a drift that was mined through the Hardhat chimney. The permeability determined with this method (see Figure 4.4) is much greater than that determined from the core samples. The measurements show a misleading increase in the permeability beyond 3 r_c. These values are probably artificially high because the drift was mined to a radial distance of 3 r_c before the shot. Consequently, a free surface existed beyond this point, and the permeability along this drift could have been enhanced by spalling of the drift walls. However, the measurements do show a region of greatly increased permeability out to at least 3 r_c from the shot point.

Information on fracturing has also been obtained from cliper measurements. A cliper consists of a transducer cable rigidly grouted in a vertical hole that has been drilled parallel to the emplacement hole. The deformation produced by the explosion reaches the bottom of the cable first and causes sections of the cable to successively break off. The length of the cable is continuously measured via microwave reflectometry. The length of cable remaining after the explosion is an indication of the extent of the fracturing, and the distance between cable breaks is a measure of the density of the fractures. The results of such measurements in a number of tests show that the fracture pattern is dependent on the layering and planes of weakness in the rock.

FIGURE 4.4: HARDHAT PERMEABILITIES AS DETERMINED FROM PRESSURIZED DRILL HOLES

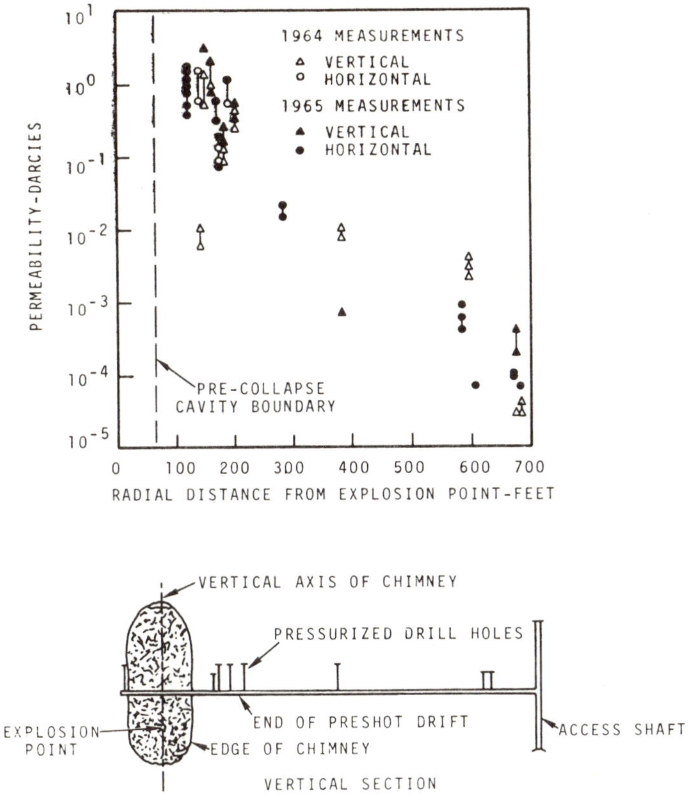

Source: PNE-1550

The zone of increased permeability in the gas-bearing shale in which the Gasbuggy Event was detonated has been estimated from production tests of the Gasbuggy chimney (see Figure 4.5). The extent of increased permeability is about 3.7 r_c.

In summary, fracturing depends on the water content, the pre-existing fracture pattern, and the planes of weakness in the rock surrounding the explosion point. For a medium like that in Hardhat or Piledriver, the permeability is assumed to be increased in a sphere surrounding the explosion point and to extend to a maximum of 4 r_c where the permeability reaches its original value. The permeability near the edge of the chimney is on the order of 1 darcy. The extent of the fractured region is reduced for dryer granite having fewer pre-existing fractures. Also, calculations indicate that the extent of the fractured region may be reduced for deeply buried explosives such as those contemplated here. This combination of factors might result in a fractured region as small as 2.5 r_c.

FIGURE 4.5: ESTIMATED PERMEABILITY INCREASES IN THE GASBUGGY EVENT

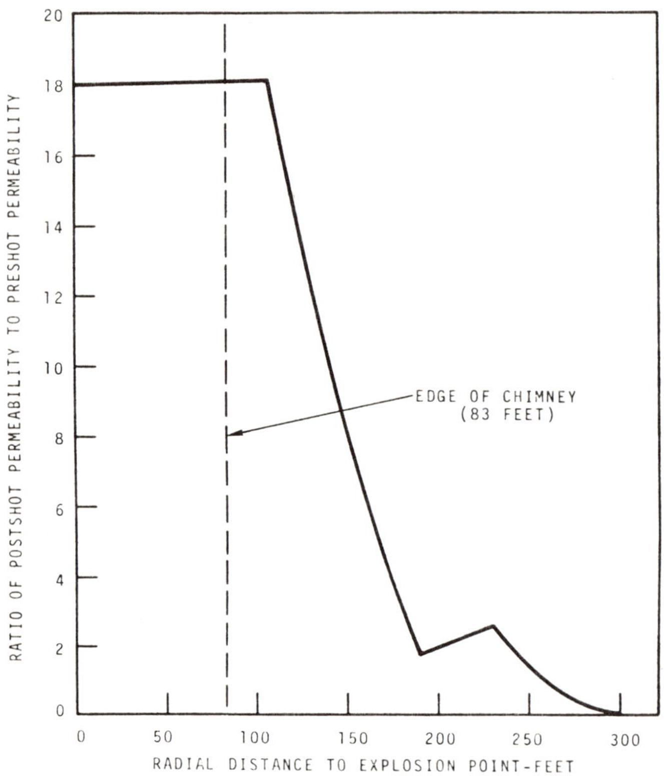

Source: PNE-1550

Chimney Characteristics: A chimney is formed by the collapse of a cavity and the subsequent introduction of the cavity's volume into the surrounding rock. Chimneys are of interest because of their large permeabilities, relatively small blocks of rubble and consequent ease of heat extraction.

Collapse takes place because of instabilities in the cavity, and it can occur at various times after the explosion. For example, Hardhat collapsed 11 hours after the explosion and Piledriver collapsed almost immediately, although they were shot within one quarter mile of each other in the same kind of rock. All tests in granitic rock (including the French tests) have collapsed and formed chimneys.

Chimney collapse need not be symmetric. The final shape of the cavity depends largely on the medium's original fracture pattern, its layering, and its zones of weakness.

The volume of a chimney depends on its geometry. Chimneys are often idealized as hemisphere-cylinders or as hemisphere-cones, and the volumes of such models can vary by as much as a factor of 2. The minimum volume for a given cavity radius and chimney height is that in a hemisphere-capped cylinder. This is the model that is assumed in the following calculations. It can be shown that any heat available in the chimney (see "Available Energy" below) is relatively little. Thus, any error made by assuming the above model has only a small effect on the final results.

The average chimney porosity, which is defined as the ratio of the cavity volume to the chimney volume, has been calculated on the basis of measurements from the Piledriver, Hardhat, and Shoal Events. The values range from 18 to 29%. In addition to these three tests, there have been several tests on Pahute Mesa at the Nevada Test Site. This data shows a tendency toward larger chimney volumes at large yields, but the trend is not conclusive because of the complicated layering of the tuffs in Pahute Mesa and because of the lack of knowledge about the tuffs' relative strengths.

In summary, there is considerable uncertainty in chimney volumes due to lack of knowledge about chimney geometries and about the extent of chimneys formed from large cavities (high yields) in hard rock. A rough estimate is provided by a hemisphere-capped cylinder whose ratio of height to cavity radius is about 4.5.

Distribution of the Explosive's Heat: A large amount of the energy from an explosive is trapped as heat in the vicinity of the explosion point. Nuclear explosions in salt (the Gnome and Salmon Events) exemplify one extreme of the way in which the explosive's heat is distributed. Except for a small partial collapse, such cavities remain standing and give the melted salt enough time to collect in a thick puddle at the bottom of the cavity. The explosive's energy is contained in a thin region of increased temperature in the cavity's walls and in the puddle of melted salt. Removing heat from the puddle or the walls would be a slow process indeed since the regions of increased temperature are impermeable and heat transfer can occur only by conduction.

A more typical distribution of the heat from an explosive is indicated by evidence from drifts that were mined near the explosion point in the Ranier Event. When the cavity in this test collapsed, blocks of chimney rubble and glassy melted rock were mixed together. The water in the melted rock caused it to froth during the collapse.

Such water is available, as steam, to percolate up the chimney and transfer the heat in the hot rock. The temperature log that was taken in the Shoal exploratory hole 2.5 months after the explosion is shown in Figure 4.6. Almost all of the chimney has an increased temperature. There is a sharp rise to a maximum of 600°C just below the explosion point and extending through the lower cavity hemisphere. It is reasonable to assume that in a nuclear explosion, 1,000 metric tons of rock per kiloton of yield is raised to the melting point of the rock or higher. This means that for the 13.1-kiloton yield of Shoal, the melted rock should occupy a volume of about 5,000 cubic meters. If this melt were distributed in a spherical segment at the bottom of the lower cavity hemisphere, it would attain a height of 8.76 meters. However, the zone of greatly increased temperature is actually about 21 meters thick. The difference, about 12 meters, is far too thick for heat transfer to have occurred by conduction alone.

Also, the temperature of the material in this zone is well below the melting point of the rock. Consequently, the melted rock must be distributed in a larger volume and mixed with rubble. Furthermore, the region must be permeable enough to have allowed the steam in the rock to smoothly distribute the heat in the cavity.

FIGURE 4.6: TEMPERATURE PROFILE FROM THE SHOAL EVENT

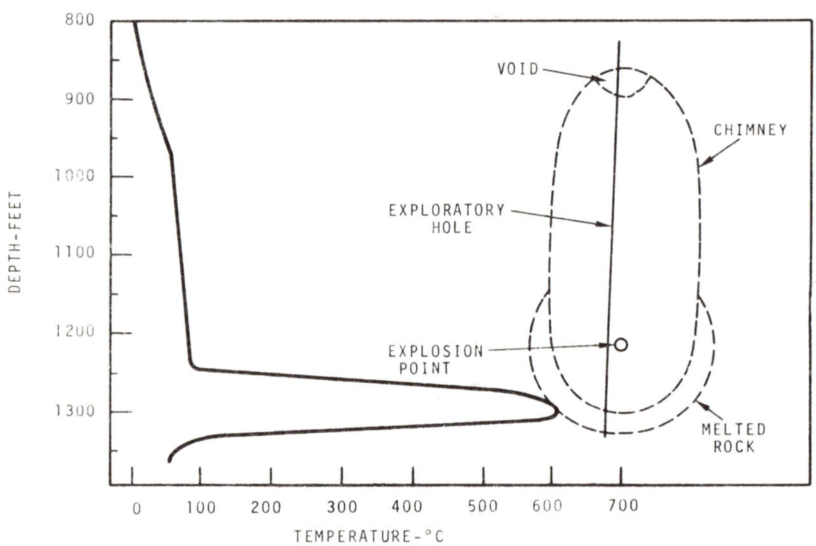

Source: PNE-1550

Core samples from the Hardhat chimney show that most of the melted rock is concentrated in the bottom of the lower cavity hemisphere in a layer about 0.25 r_c thick (about 70% of the core samples are melted rock). The one drill hole that entered the Piledriver chimney shows melted material where it penetrated high in the cavity region at about 0.23 r_c below the explosion point.

Extensive core samples from the French tests show that melted material is present throughout the collapsed cavity, with the melted material comprising 15 to 50% of the core samples. The more concentrated melt distribution observed in Hardhat may be connected with the long time that elapsed before the cavity collapsed (11 hours for Hardhat and a few seconds for Piledriver).

The permeability of the mixed melt and rubble is difficult to assess. The Shoal temperature profile shows that heat convection occurred from the lower hemisphere, presumably by means of water in the rock turning to steam and permeating through the region. A similar temperature log for Hardhat that was taken 11 months after the explosion (see Figure 4.7) shows a temperature increase throughout the chimney and a maximum temperature below the 70% melt region, where the melt was presumably injected into cracks. The highest temperature recorded is about 85°C, indicating considerable heat convection.

FIGURE 4.7: TEMPERATURE PROFILE FROM THE HARDHAT EVENT

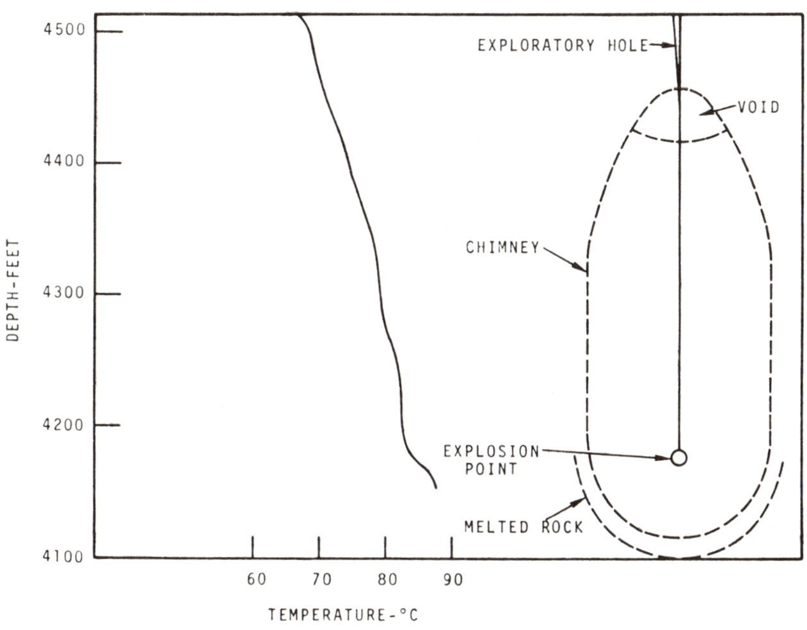

Source: PNE-1550

Thus, it appears that the melt is distributed throughout the lower cavity hemisphere, that it is mixed with rubble, and that region is permeable (although less permeable than the chimney). About 90% of the explosive's energy is trapped in this region in the form of increased temperatures. This energy may be available because of the distribution of the hot material and the permeability of the region.

Available Energy: Table 4.5 gives the relevant fractured volumes that can be produced by various sizes of nuclear explosives at a depth of about 3 kilometers, subject to the qualifications stated in the preceding sections. Figure 4.8 shows the estimated electric energy that is available from these volumes. An initial rock temperature of 350°C, a final rock temperature of 150°C, a conversion efficiency of 21% and an extractable-energy constant of 45 kilowatt hours per cubic meter are assumed. This figure clearly shows that the amount of power available from the chimney alone is too little to be of interest for a 200-megawatt power plant. Thus, the process is attractive only if heat can be removed from the fractured volume surrounding the chimney (excluding the unlikely possibility of increasing the yield, which is limited to about 1,000 kilotons for seismic reasons).

TABLE 4.5: FRACTURED VOLUMES PRODUCED BY VARIOUS SIZES OF EXPLOSIVES

	Fractured volume (in 10^6 cubic meters) produced by an explosive having a yield of:		
Region	200 kilotons	500 kilotons	1000 kilotons
Chimney	0.72	1.8	3.6
Volume fractured between r_c and $2r_c$	1.25	3.13	6.25
Total volume fractured to $2r_c$	1.97	4.93	9.85
Volume fractured between $2r_c$ and $3r_c$	3.65	9.12	18.25
Total volume fractured to $3r_c$	5.62	14.05	28.10
Volume fractured between $3r_c$ and $4r_c$	7.18	17.95	35.9
Total volume fractured to 4r	12.80	32.00	64.00

Source: PNE-1550

There appears to be no reasonable way to extract heat from the fractured region produced by a single explosive. However, if two or more explosives are set off so that their fracture zones overlap, then water or steam can be flowed through one fractured region into the other in a systematic manner. In this way, the heat in the fractured rock may become available.

The firing of an array of explosives is therefore essential to extracting enough heat for geothermal power generation. This scheme has the further advantage of lowering the per-explosive cost of the large exhaust pipe that is required to efficiently deliver the steam to the surface. The explosives could be emplaced in smaller, less costly holes, with perhaps more than one explosive in each hole, and only one return pipe would be needed.

FIGURE 4.8: ESTIMATED ELECTRIC ENERGY AVAILABLE FROM THE FRACTURED VOLUMES PRODUCED BY SINGLE EXPLOSIVES

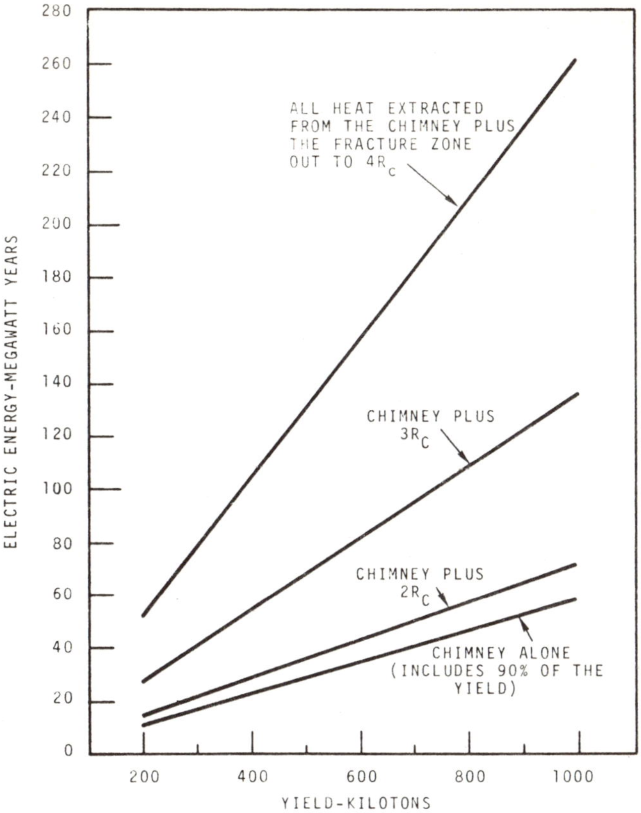

Source: PNE-1550

Multiple Arrays: As pointed out above, the fractured region produced by a single explosive has unfavorable economic and heat-transfer characteristics. Multiple arrays, on the other hand, have the advantage of spreading the piping costs over a larger energy base. In addition, the volume through which steam or water can be flowed is larger for a multiple array than for a single explosive. Lastly, the total fractured volume in a multiple array may be greater than that in a number of isolated explosions.

The enhancement of the amount of fractured rock produced by multiple charges of high explosives is well known. The enhancement factor is defined as the fractured volume created by a multiple HE array divided by the fractured volume that would be created if each HE charge were fired separately.

For bench blasting the enhancement factor varies from 1.25 to 4, for cratering it ranges from 1.1 to 3, and for tunnel digging it goes up to 4. The HE situation however, is different from the underground nuclear explosions being considered in this study. In HE operations, several charges are detonated simultaneously, they are usually line sources, and they are placed close to a free surface. In contrast, nuclear explosives having large yields are fired sequentially, they are point sources, and they are placed close to a "region of different properties" that is produced by the previous explosion and that may or may not be like a free surface. Consequently, the applicability of HE experience to multiple nuclear explosives is difficult to determine.

Holzer has calculated that the fractured volume resulting from two 50-kiloton nuclear explosives placed two failure radii apart is 25% greater than that of a single 100-kiloton explosive (F. Holzer to G.C. Werth, letter, "Arguments for Using Two or More Explosives per Emplacement Hole for WASP," *UOPK* 69-25, September 1969). He suggests that a spacing of more than two failure radii may be optimal. Selby has calculated that with a tetrahedral array, in which the explosives are placed in such a way that the critical shear stress of the medium is just equaled, an enhancement factor of 2.9 can be realized (*BNW*, "Engineering Economics Analysis of U.S. Patent 3,465,818 as It Applies to a Geothermal Plant," March 1970).

For Plowshare geothermal applications it is necessary to produce a large amount of broken rock. If a single hole can be used to sequentially fire two explosives at their maximum yields, the cost of one emplacement hole can be saved. It also appears that there is an increase in rock fracturing if two explosives are fired simultaneously. That is, the fracturing produced by two 500-kiloton explosives is expected to exceed that of a single 1-megaton explosive. However, it is not known if the increase in the fractured rock justifies the cost of an extra explosive. (The cost of an explosive depends only slightly on its yield, so the cost of two explosives is roughly double the cost of a single explosive.)

Drilling Considerations

There are numerous uncertainties and possible problems that should be taken into account in designing and constructing deep holes in a hard-rock thermal area. Some of these problems affect the costs, and some are technical considerations that come to light during the study. They are:

(1) For the feasibility study, the casing thickness for the 12- to 36-inch diameter holes was arbitrarily established at ¾ inch. This is likely to be adequate when the casing is cemented into the rock. Where cementing is not possible and for the above-ground portions, this thickness is probably not adequate for the full range of steam pressures that may be used. The final wall thickness will depend on the pipe size and the steam pressure. Thicker casings will increase the cost of cased holes.

(2) Cementing techniques at 660°F and depths to 10,000 feet should be investigated. The cement may set quickly or the mixing water may flash to steam at such temperatures, unless suppressed. Removal of the heat of hydration of cement is said to be a problem in cementing casings in normal-temperature holes.

In a high-temperature hole, it may not be significant but should be investigated.

(3) Cementing rarely provides for complete filling of the voids behind a casing. The effect of voids behind the casing should be considered where the casing is expected to resist high internal or external pressures.

(4) Jacks for lowering the casing are required in all cases for ¾-inch-thick casing, 12- to 36-inch diameter, 8,600 to 10,800-foot depths. Special lift rigs are required to prevent collapse of the pipe during the lowering operation. Casing weights for those diameters range from about 1,000,000 pounds to above 3,000,000 pounds, which is beyond the capacity of drill-rig hoists. The 1.5-inch-thick casing for the 54-inch hole 10,800 feet deep would weigh more than 9,000,000 pounds. These weights will require the use of high-strength steel for the tension loads encountered during the casing-lowering operation.

(5) The 12-, 18-, 24-, 30-, and 36-inch inside diameter used in the hypothetical cases are not the exact inside diameters of the pipe. For pressure-drop calculations, etc., actual commercially available pipe sizes should be used.

(6) Accurate estimates of hole-drilling costs will require identification of the actual site and geological information from exploratory holes. This point was stressed by the hole-drilling companies.

(7) Drilling may use air or mud for cuttings removal. If the hole is dry, air will be best. If the hole is wet and it is too costly to case it to below the aquifers, mud will have to be used. One company that has drilled several holes in granite said they hit water in nearly every hard-rock hole. They have not drilled in thermal areas, however.

(8) Temperatures to 660°F may affect the drilling equipment and might require special thread lubrication and special muds, if mud is used. Drilling has been done up to 550°F, and invert muds are said to perform at this temperature.

(9) For the 12- to 36-inch holes, it would be more economical to drill the emplacement hole the full size required for the steam line than it would be to drill it out to a larger diameter after the shot. A reaming operation costs about as much as drilling the full-size hole.

Environmental and Safety Considerations

It is essential to anticipate the potential hazards of Plowshare geothermal heat extraction, evaluate the likelihood of such hazards, and then take steps, through a safety program, to control or eliminate them. The safety program also entails making measurements for implementation of safety plans and documenting the fact that the program has indeed been carried out safely.

Radiological Considerations: At the depth contemplated for the location of the hot rock and with proper cementing and stemming, no release of radioactivity to either the atmosphere or the ground water would be expected. In the operation of the plant, radioisotopes that may be brought to the surface in the steam will be recirculated to the region fractured by the explosion, making a self-contained system.

Probability of Radioactivity Release to the Atmosphere: Results from more than 300 underground nuclear explosions show that with proper and careful deep emplacement and stemming, it is unlikely that any release of radioactivity resulting from the detonation will occur. Some causes (and cures) of unexpected releases experienced in the past are discussed below.

(1) Faulty and/or insufficient cementing and stemming. This can be controlled to prevent such an accident.

(2) Insufficient depth of burial. Here it should be noted that this study is concerned only with burial depths that are far in excess of those found sufficient for complete containment (roughly 400 feet multiplied by the cube root of the explosive's yield in kilotons).

(3) Seepage through fissures or faults in the area surrounding the emplacement hole. Thorough geologic mapping and avoidance of suspect areas will minimize this possibility.

In those past cases when small leaks have been noted, further investigations shows that only a minute fraction of the radioactivity diffused through the complicated path to the surface.

In the unlikely event that release of radioactive effluent occurred, weather data would be required to enable tracking of the effluent and implementation of appropriate safety plans. Meteorological predictions could be made on the basis of available data from established weather-bureau stations plus data collected at the site.

In summary, the probability of escape of radioactivity into the biosphere can be made vanishingly small, and the project can be conducted with safety and confidence.

Groundwater Contamination: A thorough preshot evaluation of the hydrologic regime would be made to assure that mobile-groundwater contamination would not be a problem. At a depth of 8,600 feet, the detonation should be well below the water table, below any aquifers, and in rock having very low permeability. Provided there are no fissures (created or natural) allowing fluid communication with these near-surface mobile-water features, radioactivity would not spread by aqueous transport.

It will be necessary to ensure isolation of the chimney from near-surface waters by very careful cementing of well casings in the upper strata. Absence of explosion produced radioactivity in wells and aquifers would be confirmed by postshot measurement.

Following the explosion and construction of the power plant, care must be exercised in plant operation to ensure that the steam and condensate remain isolated from near-surface water features.

Radiological Safety: Radiological safety consists of protecting people against radiation exposure and documenting radiologic exposure if any should occur.

Operation and maintenance of radiological monitoring systems would be carried out by distributing film badges, protective equipment, and radiation-detection and measuring equipment. Radiological safety coverage would be provided from the time a nuclear explosive is received at the site, through plant operations with the degree of coverage at any time dictated by the potential seriousness of any emergency situation and the probability of its occurrence. The radiological safety effort would also include:

(1) Conducting an environmental sampling and monitoring program in the off-site area prior to and following the detonation,

(2) Maintaining a record of fluctuations in the background radioactivity,

(3) Providing information to state public health agencies and enlisting their cooperation and participation,

(4) Being prepared to effect appropriate protective measures should an unplanned release of radioactivity occur.

Seismic Considerations: A consideration in any underground nuclear explosion is the possibility of damage to manmade or natural structures. The possibility of damage arises from ground motion resulting from stress waves traveling through the earth from the detonation point. Ground motion is described by the acceleration, velocity, and displacement of the earth's surface at the point of interest. Primarily these vary with the surface distance, the depth of burst, the yield of the explosive, the geology of the region, the medium in which the explosion occurs, and the type of earth or rock on which a structure rests.

Figure 4.9 which shows the variation of these surface-movement parameters with distance was scaled from measurements on many tests at the Nevada Test Site and is typical of the effect expected for the movement of an alluvial surface at distances beyond a few miles from a detonation in a deep hard-rock medium, assuming a 1,000-kiloton explosion at a depth of 8,500 feet. Similar curves have been developed for other yields and depths of burial.

Structure damage will depend in part on the frequencies of the disturbing wave (a function of soil type, explosive yield, depth of explosion, etc.) and on the natural vibration-response frequencies of the individual structures involved. More sophisticated predictive methods for taking these factors into account are available and can be used once the detailed information on proposed locations, explosive yields, and geology are known. The use of small explosives to seismically calibrate the area and experimentally measure natural frequencies would aid these predictions.

In addition to prediction of structural response, a "condition survey" of facilities would be conducted and protective measures for those facilities that might be adversely affected by ground motions from the nuclear detonation would be recommended. This survey would include the investigation of foundation conditions, structural response, and natural ground-vibration levels for structures in the general vicinity.

A pre- and postshot survey would be made of mines, slopes, potential rock falls and oil and gas wells in the area that might be affected by the detonation. The preshot survey includes examination and documentation of the conditions of mines, wells, and associated surface facilities which might be affected and, if warranted, development of recommendations to install shoring or other measures required to prevent or reduce the damage potential.

FIGURE 4.9: EFFECTS ON ALLUVIAL SURFACE

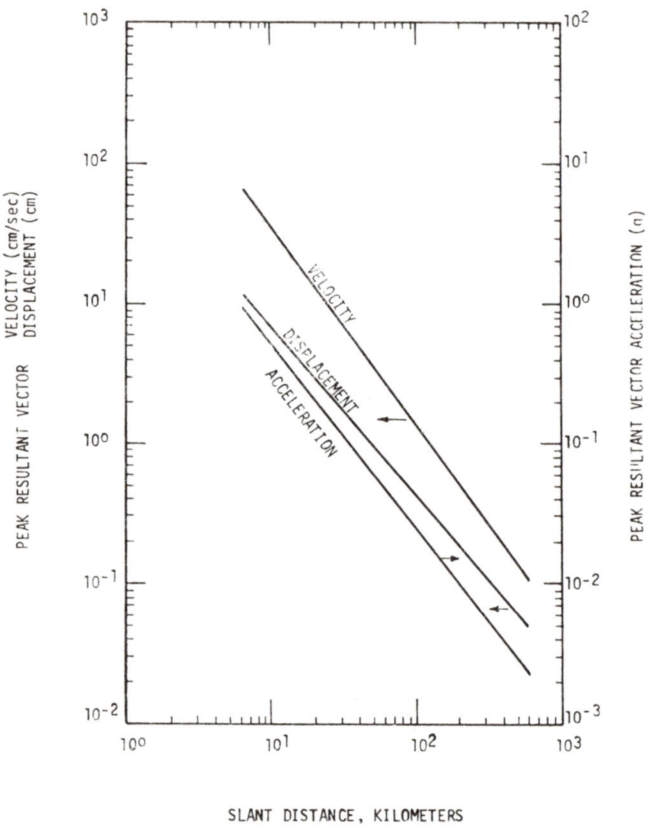

Source: PNE-1550

Aftershocks: Experience from several years of testing indicates that underground nuclear explosions may affect the local seismicity. The type and intensity of the effects are dependent upon two major factors: the energy yield of the explosion and the state of stress and tectonic condition of the region. The largest contained nuclear explosions have caused small aftershocks within the immediate vicinity of the explosion–shocks apart from the usually recognizable signals resulting from the collapse of the explosion cavity. The largest of these aftershocks measured to date have been about 100 times or more smaller than the signal from the explosion. It is believed that the origin of these signals arises from readjustment of the surrounding ground to the transient perturbation of the explosion. The distance from the source of these signals has been a maximum of about 15 kilometers for an explosion in the low-megaton range. Natural seismic events are similar to large underground explosions in that the main shock is generally followed by numerous aftershocks of less magnitude.

The question is whether a "natural" damaging earthquake could be produced as a result of an underground explosion, either in the immediate vicinity of the explosion or at some distant area. This has not happened on any previous underground explosion and it is unlikely that it could ever happen, especially in a tectonically stable area. In a tectonically unstable area (which may be the situation for a geothermal site) or where local stresses may have accumulated to a level at which a natural earthquake is imminent, there is a low, nonnegligible, probability that a seismic event might occur prematurely.

Volcanic Stimulation: Surface extrusion of molten rock resulting from an underground nuclear explosion appears to be a remote possibility. Theoretically:

(1) The rock melted by the explosion might rise through the overlying fractured rock to reach the surface.

(2) Pre-existing magma residing in a natural reservoir at a greater depth might be mobilized and erupt through an established vent or through fractures opened by the explosion.

The likelihood of such occurrences will be considered separately. Data on previous tests in similar rock show that the amount of molten rock produced is about 1,000 metric tons per kiloton of yield. A 1,000-kiloton device would produce about 400,000 cubic meters of molten rock. Even though this volume of molten rock is large, past experience in over 300 underground tests has shown that this material is completely confined in the immediate vicinity of the explosion if the explosive is buried deep enough and if no natural fractures extending to the surface intersect the explosion region. The 3-kilometer depth used in this study is sufficient to completely contain any explosion considered in this study. The second condition, that in which a natural pocket of molten magma below the site of detonation would be mobilized by the shock, may be possible.

At least three factors are to be considered. First, the shock might open fractures or old vents through which the magma could rise hydrostatically. Second, the compressional shock might increase the fluidity of the liquid, as it does in certain thixotropic materials, and cause it to be injected into fractures, just as natural earthquakes often cause eruptions of wet sand and clay. And finally, the shock could cause a sudden nucleation of gases dissolved in the magma, so that bubbles form and cause the liquid to froth, just as shaking a carbonated beverage causes it to foam and overflow.

The prospects of such effects triggering a volcanic eruption will depend on the proximity of the magma body and the availability of open fractures or older vents through which the magma can rise to the surface. If reasonable precautions are observed, the risk of a surface eruption triggered by the explosion can be greatly reduced or completely eliminated.

Hydrothermal Explosions: A hydrothermal explosion differs from a volcanic eruption in that no magma is involved. They occur in regions where a water circulation system can convect heat from some depth to the surface. Some of the water near the surface may be very close to the temperature at which it will flash to steam. Also, pressures greater than hydrostatic might be stored up due to "self sealing" resulting from mineral deposit and rock alteration.

If a large thickness of material is at the flash point, an instability, such as pressure release at the surface, could possibly trigger the whole section to flash to steam and produce a hydrothermal explosion. Such explosions are infrequent. There are many hot springs areas in the world; however, few have produced hydrothermal explosions. Conditions for this occurrence must therefore be very special.

The Plowshare geothermal application being considered here is not intended for use where a natural water circulation system is present. Neither is the system to be operated in a manner that would require large volumes of water to be maintained at their flash points. Possibly, if there were hot springs in the vicinity of a Plowshare geothermal site near their flash point, the seismic signal might act as a triggering mechanism to cause a hydrothermal explosion. The probability of this occurrence and the potential hazard it could represent would be evaluated for an actual site.

Ecological Considerations: An ecological investigation of the proposed Plowshare site would be made which would include assessing the effects of the nuclear detonation and the construction and operation of the power plant on living organisms of the region and on various environmental factors. Because construction methods and materials for a geothermal power plant will be typical of those used in contemporary construction, no new ecological effects are anticipated. However, as in any construction project, certain precautions would be taken to protect the ecology as much as possible.

A cooling pond poses potential problems during the operational phase as it affects the level and movement of groundwater. Studies on the effect of percolation from the cooling pond will be required to predict the effect of groundwater levels and the movement of thermal water.

Present local and regional use of groundwater will have to be determined. Predictions of future use will also need to be made, since establishment of the geothermal plant and regional development are likely to intensify local land use for irrigation, manufacturing, and potable-water supplies. Makeup water for the cooling pond will be required to replace that lost by percolation and evaporation, and these requirements must be estimated. Additionally, disposal of replacement water for the cooling pond will be required if salinity levels are to be maintained at an acceptable level for plant operations. Failure to do so is likely to cause adverse biological effects.

Two aspects of the nuclear detonation and plant operation might be considered new—ground motion from the detonation and the possibility of release of radioactivity during the detonation or during plant operation. However, these aspects are not really new as potential and actual effects on the environment from these sources have been analyzed for many years in many areas.

A classical environmental study that was performed is the "Environment of the Cape Thompson Region, Alaska," conducted in 1959 to 1961 under the sponsorship of the AEC's Plowshare program. Such detailed study would not be conducted for each site, however, sufficient ecological data would be collected to enable analyses of possible effects on the environment and to provide a basis to develop means to minimize or eliminate adverse environmental effects.

Public Acceptance

To be commercially implemented, the Plowshare geothermal program must have the understanding and acceptance of the public. These must be sought early, by fully informing the local population and public officials of the risk, goals and benefits of the program.

Briefings, public meetings, exhibits and distribution of factual literature should be used to give all segments of the public the opportunity to voice their support of or opposition to the project and seek answers to their questions concerning the project. The predicted effects of the explosion and plant operation and the possibility of accident situations with effects greater than expected should be fully and candidly explained. The public should also be informed of the precautions taken to avoid unexpected occurrences and the contingency plans to assure public safety even in case of accident.

The safe and practical use of Plowshare technology is the joint responsibility of both the user industry and public agencies. The public exercises its control through the local, state and federal government. Projects embodying this technology would be subject to the review of local, state, and federal agencies. It is encumbent upon the user industry to perform those studies necessary to demonstrate that such projects may be conducted without undue risk to the public and to the environment.

HEAT EXTRACTION

The estimated power that can be produced from the heat available solely from the chimney is too little to be of interest for a 200-megawatt power plant, even if very large explosives could be used. Consequently, it is necessary to remove additional heat from the fractured region surrounding the chimney. This might be done by sequentially firing the explosives in an array so that their fracture zones overlap. The success of this scheme depends on being able to flow water or steam through the chimney and the fractured region in order to extract the heat. The removal of sufficient heat depends primarily on the resistance to fluid flow in the fractured region and on whether the flow can be distributed to a large enough volume of the fractured region.

Both the flow of water or steam in the fractured region and the transfer of heat to the fluid can be analyzed in order to determine if the heat can be removed from the rock in a way that meets the steamflow requirements of the power plant. It is not meaningful to perform detailed flow calculations in the fractured region because so little is known about the characteristics of the fractures. However, some indication of the fluid losses, flow regimes, and heat transfer can be obtained for simplified geometries. The pressure required to flow the fluid from the pipe to the chimney, or between chimneys, can be estimated. The pressure drop is expected to be acceptable if the average permeability of the formation is greater than roughly 10 millidarcies. The estimated flow velocities and Reynolds number are small enough that the flow can be considered slow and smooth. Analysis of the heat flow indicates that it is sufficient to extract the heat from the source rock in both the chimney and the fractured region, but it is not known how much the heat removal is affected by fluid flow in more permeable regions at the expense of fluid flow in less permeable regions.

Fluid Losses and Permeability

The resistance of a medium to flow depends on its permeability and is measured by the fluid pressure required to maintain the flow. The pressure available to drive the flow is estimated to be about 250 atmospheres, assuming a 3-kilometer-deep water-filled pipe to the fractured region and a 50-atmosphere pressure in the steam-exhaust line to the turbine. The pressure required to force the flow through the chimney region is much less than this, as can be estimated from Darcy's law. In the fractured region, flow might be introduced from a vertical hole alongside the chimney or from an adjacent chimney. Viewed from above:

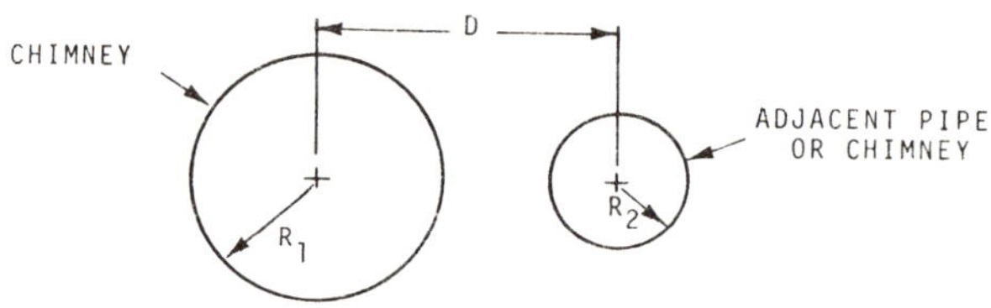

The pressure required to maintain a starting flow rate of m = 10^6 kilograms per hour, assuming R_1/R_2 = 200, is shown for various spacings and permeabilities in Table 4.6. The available head of about 250 atmospheres is exceeded if the medium's average permeability is less than roughly 10 millidarcies.

TABLE 4.6: REQUIRED PRESSURE FOR FLOW THROUGH THE FRACTURED REGION BETWEEN A PIPE AND AN ADJACENT CHIMNEY WITH R_1/R_2 = 200

Fractured-region permeability (millidarcies)	Required pressure (in atmosphere) for spacing (D/R_1) of:			
	2	4	6	8
1000	6.60	8.25	9.13	9.76
100	66.0	82.5	91.3	97.6
10	660	825	914	976

Source: PNE-1550

The results for two adjacent chimneys of equal size (R_1 = R_2) for the same condition is shown in Table 5.7 set forth on the following page.

TABLE 4.7: REQUIRED PRESSURE FOR FLOW THROUGH THE FRACTURED REGION BETWEEN TWO ADJACENT CHIMNEYS ($R_1 = R_2$)

Fractured-region permeability (millidarcies)	Required pressure (in atmosphere) for a spacing (D/R_1) of:				
	2	3	4	6	8
1000	0	1.97	2.70	3.63	4.40
100	0	19.7	27.0	36.3	44.0
10	0	197	270	363	440

Source: PNE-1550

The available 250-atmosphere head needed to drive the flow will be inadequate if the region permeability is about 10 millidarcies or less and if the spacing is greater than about D = 4R. More pipes and more chimneys to share the flow rate will permit correspondingly less permeability and/or greater spacing.

The permeability of the fractured region is not well known. It will change from a large value near the chimney to a relatively small value characteristic of the formation near the edge of the fractured region. The constant permeability used in the necessary calculations should consequently be interpreted as an average value. The above results indicate that the useful portion of the fractured region will be a zone surrounding the chimney in which the average permeability is about 10 millidarcies or more.

Certain fluid paths will find less resistance to flow than others because of the variation in permeability in the fractured region. Consequently, flow may be channeled into more permeable paths at the expense of flow in less permeable paths and reduce the volume of the fractured region from which heat can be extracted. Although it is theoretically possible, it is not meaningful to perform detailed calculations of flow channeling at this time because so little is known about the actual characteristics of fractured regions.

Flow Regimes

Fluid flow in the chimney and the fractured region will be characterized by its velocity and Reynolds number. The flow velocities and Reynolds number are small enough in both the chimney and the fractured region to allow the flow to be considered slow, viscous, and nonturbulent.

The mass flow rate of steam at the start of operations will be about 10^6 kilograms per hour, and the density of steam in the cavity will be 30.6 kilograms per cubic meter. The flow velocity necessary to produce this mass flow rate, assuming a representative chimney of circular cross section with radius of 60 meters and 20% porosity, is 0.4 centimeter per second.

At the start of operations, the kinematic viscosity of the steam is 0.0069 square centimeter per second. If L is the width of the flow channel in the chimney expressed in centimeters, the Reynolds number of the flow is 58 L.

At the end of cavity life when the final operating conditions of the turbine have been reached, the mass flow rate will increase to 1.64×10^6 kilograms per hour, the cavity-steam density will drop to 6.66 kilograms per cubic meter, and the kinematic viscosity will be 0.02 square centimeter per second. At these conditions, the velocity will be 109 meters per hour and the Reynolds number will be 128 L.

In the fractured region surrounding the chimney, the cross-sectional area of flow will be about four times that of the chimney, but the porosity will decrease to about 1.0%. Consequently, the flow velocity and Reynolds number calculated for the chimney will be about five times greater for the fractured zone. Thus, as stated above, the flow in both the chimney and the fractured zone is slow, viscous, and nonturbulent.

Heat Transfer from the Chimney

An estimate of the time required to transfer heat from the chimney rubble is obtained by considering the heat transfer from a sphere of radius that is considered to represent a piece of rubble immersed in a bath of constant temperature. The following heat transfer times are calculated:

Radius (centimeters)	Time
10	6.4 hours
50	160 hours
100	26.6 days
500	1.82 years

All of the chimney rubble is expected to have a radius less than about 100 centimeters, and about 80% will have a radius of less than 50 centimeters. The above calculation shows that ample time is available to transfer heat from the chimney rubble to the steam. A similar analysis can be made on other rubble shapes with the same qualitative results.

Heat Removal from the Fractured Region

Heat removal from the fractured region in a shot array is difficult to calculate. However, a great deal of insight is gained by analyzing the heat transfer to a fluid flowing in a single fissure that traverses a medium of infinite extent.

Calculations show that most of the heat contained in a region about 0.5 meter on both sides of a fissure and about 1 meter along a fissure having the above characteristics can be removed in about 1 year. Thus, it appears that intersecting cracks with a spacing of about one crack of millimeter size per meter would be sufficient to transfer all the heat from the fractured region.

Unfortunately, so little is known about the cracking induced in the fractured region that it is impossible to know the distance from the explosion point at which this crack frequency is reached, nor is there data on crack widths in the fractured zone.

PLANT OPERATION

Steam Quality

The purity of the steam produced in the fractured region will have a major effect on the design, cost, and operation of a Plowshare geothermal plant. Both the quantity and the type of contaminants carried up by the steam may affect the turbine-condenser design, the operating efficiency of the plant, and even safety requirements.

Although a number of assumptions must be made as to the chemical compositions formed, their deposition in the cavity, and their reaction with the water-steam, calculations have been made of the radioactivity present in the steam. Some estimates have also been made of the silica and salt content of the steam. A survey has been made of noncondensable gases and of the corrosive elements in natural geothermal steam.

Radioactivity Produced by the Explosive: The amount of radioactivity produced depends on the yield of the explosive and the fission/fusion ratio. In this section the radioactivity is estimated for the largest yield included in the ground rules and for the limit of either all fission or all fusion. Table 4.8 lists the radioactive nuclides consisting if fission products and activities induced in the soil, that are likely to be present in significant amounts for a 1,000-kiloton explosive 180 days after detonation, and that may have some volatile behavior under the assumed operating conditions.

These nuclides are the ones most likely to be able to form a species sufficiently volatile that a significant fraction of the nuclide might exist in the vapor phase with the steam. Included in Table 4.8 are the half-lives for each nuclide, the kilocuries of each expected to be present 180 days after a 1,000-kiloton explosion, and the corresponding gamma radiation from each nuclide. The latter quantity is needed for calculation of any gamma-radiation dose rate.

It should be noted that the induced activities indicated in Table 4.8 are for an explosive with no neutron shielding. Calculations show a factor of about 10 reduction in soil activation for every 15 centimeters of boric-acid neutron shielding. The induced activities also depend on soil composition (the calculations in Table 4.8 performed for Hardhat granite) and details of the explosive design. It would be reasonable to expect that the explosive could be shielded with downhole reaming and liquid shielding material.

TABLE 4.8: POTENTIALLY VOLATILE RADIOACTIVE NUCLIDES

		Fission		Fusion (3 kt of Fission)	
Nuclide	Half-life	kilocuries	MeV/sec	kilocuries	MeV/sec
^{85}Kr	10.7 years	19	1.41×10^{12}	0.057	4.2×10^{9}
^{103}Ru	40 days	1150	2.1×10^{16}	3.45	6.3×10^{13}
^{106}Ru	1 year	1000	3.37×10^{15}	3.0	1×10^{13}
^{125}Sb	2.7 years	60	9.6×10^{14}	0.18	2.9×10^{12}

(continued)

TABLE 4.8: (continued)

		Fission		Fusion (3 kt of Fission)	
Nuclide	Half-life	kilo-curies	MeV/sec	kilo-curies	MeV/sec
^{127m}Te	109 days	90	6.7×10^{11}	0.27	2.0×10^{9}
^{137}Cs	30 years	180	3.8×10^{15}	0.54	1.14×10^{13}
Induced activity in soil[a]					
^{3}H	12.3 years	220	--	20,290[b]	--
^{22}Na	2.6 years	--	--	0.6	4.9×10^{13}
^{32}P	14.3 days	2	--	2.5	--
^{35}S	88 days	29	--	40.0	--
^{37}Ar	35 days	70	--	200	--
^{134}Cs	2 years	14	7.8×10^{14}	18.3	1.0×10^{15}

[a]The amounts of these activities are derived by a direct linear scaling of amounts produced by 50 kiloton fission and fusion devices calculated by by J. Green and R. Lessler. Lawrence Radiation Laboratory, private communication.

[b]To this 290 kilocurie of soil-induced activity has been added the residual tritium from the thermonuclear fuel. Assuming 2 grams of residual tritium per kiloton of fusion yield, this is an additional 20,000 kilocurie.

Source: PNE-1550

Disposal of Radioactive Noncondensable Gases: Noncondensable gases present in the steam will collect in the condenser. In natural steam fields noncondensable gases are eliminated from the condenser by venting them to the atmosphere with exhaust fans. For the Plowshare geothermal application, the gases could be collected and stored underground provided that the amount of noncondensable gases present in the formation is small enough for easy storage. The amount of noncondensable gases present in the Geysers of Larderello fields would be too great for storage without processing and separating out the radioactive gases.

It might be possible to use Freon-exchange equipment to separate Kr (the most bothersome gas), which could be compressed and stored, or to compress and recycle the gases to the cavity. Fusion explosives, if used, would result in several orders of magnitude less krypton, but much more argon. They would also produce much more tritium. Once the tritium exchanged with the hydrogen in the water and/or steam, it would be most difficult to remove and adequate measures to insure its safe containment would be needed.

Radioactivity Expected in the Steam: The radioactive contaminants from either fission or fusion nuclear explosives that might be volatile in steam after 180 days are the fission products ^{85}Kr, ^{103}Ru, ^{106}Ru, ^{106}Rh, ^{125}Sb, ^{127m}Te, and ^{137}Cs. The induced activities are ^{3}H, ^{22}Na, ^{32}P, ^{35}S, ^{37}Ar, and ^{134}Cs. Significant gamma emitters have been underlined. For 1,000 kilotons of yield, the amount of gamma radiation after 180 days for the above species is 3.0×10^{16} MeV/sec for an all-fission device and 1.1×10^{15} MeV/sec for a fusion having 3 kilotons of fission. Although the fusion device shows a lower level of gamma radiation for volatile species, it has a very high level of beta activity in the form of tritium.

Actual volatilities of the various gamma emitting species using a thermodynamic model will be used to estimate the gamma activity of the steam since the assumption of complete volatility of these species appears to be unnecessarily extreme. In order to apply thermodynamic calculations to the problem, certain assumptions need to be made about processes and reactions occurring during formation of the explosion cavity.

At short times after the nuclear event, virtually all of the vaporized rock and explosive materials will have condensed (except for steam and permanent gases). The refractory radioactive materials will have condensed, as well as some portion of the volatile radioactive materials. The last of the vaporized rock constituents to condense would be residual amounts of the alkali metals and silica. This late condensate would be richer in volatile radioactive materials than the main condensate and would be expected to be physically displaced from it, as for example by having migrated to the chimney region before condensing.

Sodium and potassium are the main alkalies present in the rock and determine the conditions for the late-time condensation. Their principal vapor species are believed to be NaOH(g) and KOH(g) under the above conditions. The condensed forms are probably carbonates and silicates that can be assumed for purposes of thermodynamic calculation to be $Na_2Si_2O_5$, $NaAlSi_3O_8$, K_2CO_3, $K_2Si_4O_9$, and $KAlSi_3O_8$. Whether an alkali condenses as a carbonate or as a silicate will depend upon the partial pressure of CO_2 in the gas and upon kinetic factors.

Condensed NaOH and KOH will not form because of their high vapor pressures under these conditions. Based on available thermodynamic data, the alkali hydroxide vapor pressure should exceed 10^{-3} atmospheres. The vapor pressure of volatile silicon hydroxide species under these conditions also exceeds 10^{-3} atmospheres so that direct gas phase reactions can also lead to silicate formation. Thus, either gas-surface reactions or gas-gas reactions can occur, which will lead to late-time condensation of alkali silicates and carbonates at temperatures of about 1300°K or less. Assuming that the vapor phase contains 10^{-3} atmospheres each of NaOH(g) and KOH(g) in a volume of 10^6 cubic meters means that about 2×10^4 gram-atoms of alkali will condense at temperatures below about 1300°K (that is, at temperatures below the freezing point of the rock).

All of the potentially volatile species listed in Table 4.9 with the exception of permanent gases, are proposed to be associated with the late-time condensation of alkali silicates and carbonates. It is further assumed that this condensation occurs over a time scale of seconds to minutes. Because of the late-time nature of this condensate (as compared to the main puddle glass), it will probably extend into the chimney region and coat rock surfaces.

It may be concluded from Table 4.9 that if cavity conditions are reducing and contain some residual CO_2, the ruthenium volatilities will be suppressed and the cesium volatilities will be limited to that of CsCl. At start-up of the power plant, the cavity gas will contain 1.4×10^{15} MeV/sec as the maximum gamma activity, representing all of the ^{125}Sb and 10% of the ^{137}Cs and ^{134}Cs.

TABLE 4.9: RELATIVE VOLATILITIES IN STEAM AFTER 180 DAYS OF POTENTIALLY VOLATILE GAMMA EMITTERS

Nuclide	Half-life	γ-activity,[a] MeV/sec	gram-atoms of Nuclide, 180 days	Volatility in steam, 80 atm, 623°K
^{85}Kr	10.7 years	0.14×10^{13}	0.57	Permanent gas
^{103}Ru	40 days	2040×10^{13}	0.35	Oxidizing - high Reducing - low
$^{106}Ru(^{106}Rh)$	1.0 year	337×10^{13}	2.8	Oxidizing - high Reducing - low
^{125}Sb	2.7 years	96×10^{13}	0.45	Probably high
^{127m}Te	109 days	0.07×10^{13}	0.075	High
^{137}Cs	30 years	374×10^{13}	15.1	With CO_2 - low Without CO_2 - high (30%)
^{134}Cs[b]	2 years	78×10^{13}	0.08	With Chloride - high (10%)
		Total 2955×10^{13}		

[a]The mean gamma energy is about 0.5 MeV.

[b]Cesium-134 is an activation product - no neutron shielding is assumed.

Source: PNE-1550

With the maximum activity above, the gamma intensity at the surface of an 18-inch-diameter pipe can be estimated, assuming an infinite-cylinder geometry and no deposition of radioactivity on the pipe walls. This results in a surface dose rate of about 25 mr/hr.

The maximum calculated activity in the steam is such that it would be inadvisable to use a barometric condenser in which the steam condensate is mixed with the cooling-pond water. When experimental data on the steam radioactivity becomes available, it will be useful to re-evaluate this condensing method, for it has technical and economic advantages. In the interim, however, the reference design of the plant in this study will use a surface condenser.

The "order of magnitude" nature of the volatility calculations used in Table 4.9 should be emphasized. The primary objective has been to make reasonable estimates of the concentration of volatile radioactive species in the steam. As experimental information becomes available, these conclusions may change. ^{137}Cs and ^{125}Sb may be less volatile than anticipated, or at least less available in the chimney region for volatilization to occur. Other activities that have been considered refractory here may actually be more significant. It is clear that these complex chemical interactions must be studied in the laboratory and the field before they can be defined accurately.

Deposition of Radioactive Solids: The potential problems due to silica deposition on the turbine and in the condenser may be extremely important. The deposition of salts and silica (and associated radioactivity) on turbine blades is dependent upon the impurities in steam and the steam pressure. Salt deposits on steam-turbine blades are water-soluble, and with proper operational procedures can be washed off with water. The most important of these salts are NaCl, NaOH, sodium silicates, Na_2CO_3 and sometimes Na_2SO_4. Deposits that are insoluble in water are most commonly silicic acid, amorphous silica, and the various crystalline forms of silica. Deposits of iron oxides and $CaCO_3$ that are insoluble in water may also occur.

The maximum permissible silica concentration in steam necessary to reduce silica deposits to a negligible level at 80 atmospheres of steam pressure at the turbine inlet is about 0.05 ppm. The total dissolved solid content (salts plus silica) of steam used in normal turbine operations should not exceed 0.05 to 0.2 ppm. Such purity of steam is difficult to achieve even in the best boiler systems.

An estimate of the silica concentration in steam for the geothermal application may be obtained from Heitman's work, in which silica gel, silicic acid, or quartz was equilibrated with high-pressure steam (H. Heitmann, *Glastechn. Ber.* 38, page 41, 1965). Results on silica gel are presented as smoothed data in Figure 4.10.

FIGURE 4.10: TEMPERATURE AND PRESSURE DEPENDENCE OF SILICA CONCENTRATIONS

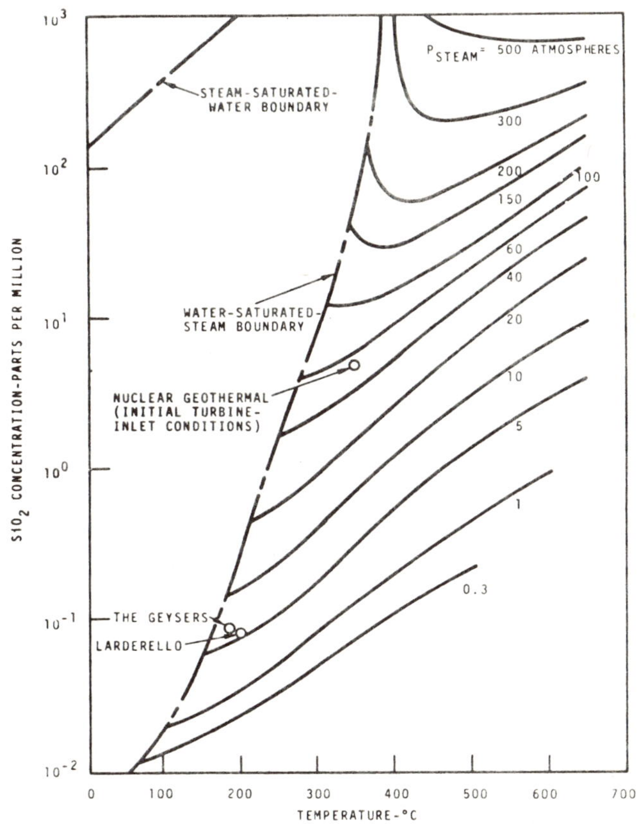

Source: PNE-1550

Silica gel and the various silicic acids will interconvert according to temperature and steam-pressure conditions. Hence, no distinction has been made among them in the figure. Quartz shows a lower solubility in water than does silica gel by a factor ranging from 10 at 100°C to 2 at 350°C. In steam, the volatilized silica concentrations are lower for quartz than for silica gel, but the differences are usually small. The maximum difference amounts to a factor of 2 at about 350°C and 30 to 50 atmospheres of steam pressure.

From Figure 4.10, the silica concentration in steam is found to be 9 ppm at 623°K and 80 atmospheres of steam pressure. It is, therefore, estimated to be in the range of 5 to 9 ppm for the underground cavity condition since either quartz or the more active forms of silica may be present. It becomes apparent that the silica content of 80 atmospheres of steam from the underground cavity is too high by about 2 orders of magnitude for normal turbine operations, and that severe deposition will occur on the turbine blades if the steam is fed directly into the turbine. The situation is further complicated by the presence of radioactivity in the steam, for ^{125}Sb, ^{137}Cs, and ^{134}Cs may coprecipitate on the turbine blades along with the silica and various salts.

Natural steam fields operate with "dirty" steam but at lower temperatures and pressures, as is shown in Figure 4.10. Their turbine cycles are correspondingly less efficient, but this inefficiency is counterbalanced by the low cost of natural steam.

Corrosion and Scaling: This section summarizes some of the corrosion and scaling problems by water and steam that can be anticipated in a nuclear geothermal power plant. Much of the information on corrosion and scaling is of necessity based upon natural-geothermal-well experience. Caution needs to be exercised in extrapolating this information to the nuclear case because of the different steam conditions and geology expected from the nuclear wells.

A comparison of conditions for natural geothermal wells with the proposed nuclear wells shows that natural wells have source waters at temperatures in the range of 200° to 300°C, as compared with about 350°C for the proposed nuclear geothermal rock temperatures. Natural wells typically use pipes 20 to 25 centimeters in diameter at depths of 0.15 to 0.6 kilometer and produce steam at pressures of 3 to 12 atmosphere and 160° to 200°C at the well head. In the nuclear case, the wells are proposed to have a pipe 1.3 to 1.8 meters in diameter at a depth of about 3 kilometers. The well head temperatures and pressures are expected to be about 350°C and 80 atmospheres at start-up and about 150°C and 4.5 atmospheres at shut-down of the plant.

In order to compare the corrosion and scaling phenomena for the natural geothermal and nuclear cases, an assessment must be made of the impurities present in the rocks and source waters of the wells. It may then be possible to derive some conclusions concerning the condensable and noncondensable impurities picked up by the steam and the chemical effects of these impurities on the steam-turbine system.

Chemical analyses of geothermal well waters from Iceland and New Zealand and surface waters from Chile are shown in Table 4.10 and geothermal steam analyses from several sources are shown in Table 4.11.

TABLE 4.10: CHEMICAL ANALYSES OF GEOTHERMAL WATERS FROM SEVERAL SOURCES

	Hengill, Iceland	Krysuvik, Iceland	Wairakei, New Zealand	Waiotapu, New Zealand	Tatio, Chile	Reykjavik, Iceland
Maximum Source Temperature °C	230	230	265	295	40 to 90[a]	88[a]
Rock Type	Basaltic lava	Basaltic lava	Ignimbrites and pumiceous breccias	Ignimbrites and pumiceous breccias	Volcanic origin	Volcanic origin
Total Solids (ppm)	913	2030	∼5000	∼3500	--	285
Na^+	174	500	1285	825	--	159
K^+	10	68	201	122	--	1.4
NH_4^{+b}	0.1	0.0	0.2	1.0	--	0.0
Mg^{++}	0.0	0.5	2.6	3.5	--	0.3
Ca^{++}	2.8	8.7	12	9	7-28	1.9
H_3BO_3	4.9	9.7	160	85	60-1050	0.3
Al^{+++}	0.4	0.1	--	--	3-19	0.1
Fe^{+++}	0.1	0.0	--	--	3-27	0.0
SiO_2	283	425	602	470	11-36	126
Cl^-	152	735	2178	1380	--	30
HCO_3^{-b}	24	0	71	145	--	--
$CO_3^{=b}$	57	50	35	106	--	43
$SO_4^=$	72	67	34	52	--	17
Total sulfide as H_2S^b (ppm)	5.2	7.0	2.4	11	--	0.2

[a]Surface waters.

[b]Analyses for total ammonia, total carbon dioxide, and total H_2S will vary according to the method used to separate water from steam. The values determined by chemical analyses are lower than the actual concentrations in the underground water.

Source: PNE-1550

Except for the Showa-shinzan Volcano, analyses were taken from various reports in the 1961 "Proceedings of the United Nations Conference on New Sources of Energy." The Showa-shinzan Volcano analysis was taken from K.B. Kruskopf, *Introduction to Geochemistry,* McGraw-Hill Book Co., page 461, 1967, New York. A number of minor constituents have not been included in the tables. The rock types represented are primarily igneous in origin and should be comparable to those expected in the nuclear application.

In Table 4.10 analyses for solutes that can form gaseous products are not representative of the actual solute concentrations in the underground well waters because of losses incurred during separation of the geothermal waters from steam. During the separation, volatile solutes tend to concentrate in the steam phase, whereas nonvolatile solutes tend to concentrate in the water. As an example, the actual CO_2 content of Wairakei, New Zealand well waters is about 200 to 600 ppm in the underground condition, and for Waiotapu, New Zealand it is about 800 to 2,500 ppm, as compared with the much lower values for HCO_3^- and $CO_3^=$ listed in Table 4.10.

TABLE 4.11: CHEMICAL ANALYSES OF NONCONDENSABLE GASES IN GEOTHERMAL STEAM

	Hengill, Iceland	Hveragerdi, Iceland	Krysuvik, Iceland	Wairakei, New Zealand	Waiotapu, New Zealand
Maximum source temperature °C	230	230	230	265	295
Well-head temperature °C	∿160	∿160	∿160	∿195	--
Noncondensable gas content of total discharge (wt%)	0.3	0.1	1.3	0.01 to 0.5	0.07 to 0.2
CO_2 (mole %)	84.6	78.5	83.9	93.0	90.0
H_2 (mole %)	2.1	1.1	5.4	0.8	1.5
H_2S (mole %)	4.9	17.2	9.6	3.8	7.8
CH_4 (mole %)	0.0	—	0.1	0.8	0.3
NH_3 (mole %)	--	—	--	∿0.2	∿0.2
N_2 (mole %)	8.4	3.2	1.0	--	--

	Larderello, Italy (1870)	Larderello, Italy (1960)	The Geysers, California	Showa-shinzan Volcano, Japan	Showa-shinzan Volcano, Japan
Maximum source temperature °C	--	245	>205	--	--
Well-head temperature °C	--	180-200	175	194	328
Noncondensable gas content of total discharge (wt%)	—	4.5	0.7	0.6	2.2
CO_2 (mole %)	90.5	92.4	69.3	69.4	84.9
H_2 (mole %)	2.0	1.4	12.7	12.4	6.6
H_2S (mole %)	4.2	2.5	3.0	3.9	1.0
CH_4 (mole %)	1.4	1.0	11.8	0.1	0.1
NH_3 (mole %)	--	1.7	1.6	0.0	0.0
N_2 (mole %)	1.9	0.6	1.6	9.1	5.2
Residuals (mole %)	--	0.4[a]	--	5.1[b]	2.2[b]

[a] Primarily H_3BO_3.

[b] Primarily HCl, with lesser amounts of HF and SO_2.

Source: PNE-1550

Tables 4.10 and 4.11 show that substances known to cause corrosion and scaling are present in significant amounts. Similar substances are to be expected for the nuclear case because of the similarity of rock types. Whether or not corrosion and scaling reactions actually occur will depend upon the specific conditions in each case. Chlorides, NH_3, CO_2, and H_2S are known to be corrosive to metals in certain instances. Steam and hot water are corrosive to all silicate materials, such as grouting in pipe joints, reinforced concrete pipes or tanks, silicate filter beds, etc. All such materials should either be avoided, or protected with suitable coatings.

Scaling (or deposition) may occur in pipes, turbines, or condensers as a consequence of volatile impurities in the steam, such as the chlorides, hydroxides, carbonates, sulfates, and silicates of the alkalies, calcium carbonate, and silica.

Noncondensable impurities in the steam, such as CO_2, H_2S, H_2, CH_4, etc., may not necessarily be corrosive, but will need to be exhausted from the turbine discharge in order to have an efficient turbine operation.

A limited amount of information is available on the corrosion resistance of common alloys and other materials to geothermal fluids. The alloys most resistant to corrosion are found to be titanium, austenitic stainless steels, and chromium plated steels. Epoxy resins are found to be resistant to corrosion at temperatures up to 100°C. Tests have been performed on alloy coupons with water at 50° to 240°C and with steam at 100° to 200°C for periods ranging up to 150 days.

Although a wide variety of corrosion mechanisms have been identified, the most serious in alloys is probably that of stress corrosion. It is known that stress corrosion can be induced by trace amounts of either chlorides or sulfides and can lead to formation of fissures and fractures in a variety of alloys that are used in the construction of turbines. A minimum concentration and temperature requirement for chloride stress corrosion is believed to be 5 ppm of chlorides at 50°C.

The presence of oxygen accelerates the rate of stress corrosion. There is no limiting stress below which cracking will not occur. Sulfide induced stress cracking can occur in the presence of H_2S at temperatures up to at least 190°C. H_2S also causes a phenomenon known as hydrogen infusion, which can lead to the embrittlement, blistering, and fracture of stainless steels.

Other corrosion mechanisms that should be mentioned are chemical corrosion and mechanical erosion. Chemical corrosion can occur from the presence of NH_3, H_2S, CO_2, and chlorides in geothermal steam. In special circumstances, such as in early experiences at The Geysers, mercury may also be present as a corrosive agent. Generally, the corrosive action of these chemicals is enhanced by the presence of air. Chemical corrosion has been minimized or avoided in most natural geothermal power plants by the proper selection of alloys and by avoiding aeration. At the higher steam pressures and higher temperatures proposed for the nuclear application, however, chemical corrosion may be a more serious problem.

The problem of mechanical erosion is encountered when wet steam at high velocities impinges on the turbine blades. Such impingement can lead to a combined corrosion-erosion process that can seriously damage the turbine blades. A certain amount of success has been achieved in erosion resistance through the use of 13 chromium-stainless steel alloys.

The experiences gained on corrosion and scaling in natural geothermal power plants make it apparent that similar problems are likely to be encountered in the nuclear case. These problems can be explored more fully by performing exposure tests on alloy coupons in geothermal steam and water under the proposed nuclear application conditions. It is important in such tests that the compositions of the steam and the water be representative of the particular geothermal formation. Steam taken directly from a nuclear well would be most desirable, or lacking an actual well, it may be possible to infer the impurities to be expected in the steam from analyses of core samples from the formation. It seems likely, as has been the case for natural geothermal wells, that suitable alloys and other construction materials for turbine systems can be found through such screening tests.

Steam Cleanup: If the geothermal steam is to be used directly in turbine operations at high pressures, it is likely that it will need purification to avoid inefficient turbine operation and excessive radioactive buildup. A solid phase scrubber such as limestone or dolomite might prove effective in removing silica. A caustic (NaOH) steam wash might also prove effective, but the operation is time consuming and requires considerable care to avoid turbine damage.

Another solution is the addition of a heat exchanger in which water would be boiled and the steam from the fractured region condensed. While this additional equipment would decrease the thermal efficiency of the process, it would concentrate the scaling and radioactivity. The economic tradeoff in using this approach lies in the ability to use more conventional equipment in the rest of the plant. If scaling in the heat exchangers proves severe, two could be provided or chemical or mechanical descalers could be added.

If the steam is cleaned, the contaminants will be present in the scrubbing systems. The disposal of radioactivity and contaminants present in the scrubbing system could possibly be carried out by re-injection into the fractured region if a fluid scrubbing system is used. The necessity for steam scrubbing, the methods of disposal of the contaminants scrubbed from the steam and the cost of this process are areas needing further investigation.

System Thermodynamics

The system can be characterized thermodynamically by the conditions in the fractured region, the losses incurred in transporting the heat from the fractured region to the turbine, and the conditions required at the turbine throttle. This section describes the turbine with its required throttle condition and the various transport systems with their associated thermodynamic losses.

Turbine Performance: The turbine for this plant has an unusual and diverse set of conditions to meet. The heat source produces steam of varying quality and chemical and gaseous content, as well as radioactive contaminants. Although it is desirable to maintain constant power output over the life of the plant, the wide variation in steam pressure and temperature make the achievement of that objective difficult. The design of the turbine is far from the normal current-day central-station practice, resembling more nearly that used for combat ships in World War II.

Figure 4.11 is a schematic of the turbine showing two cylinders, that is, a high pressure and a low pressure cylinder (the actual machine would have to have two low pressure cylinders to accommodate the volume of steam which must be passed). After being partially expanded in the high pressure cylinder, the steam enters a moisture separator where it is stripped of most of its moisture content before entering the low pressure cylinder. No reheater will be used after moisture separation because the cost of energy is too low to justify the added capital expenditure.

FIGURE 4.11: PLOWSHARE GEOTHERMAL TURBINE

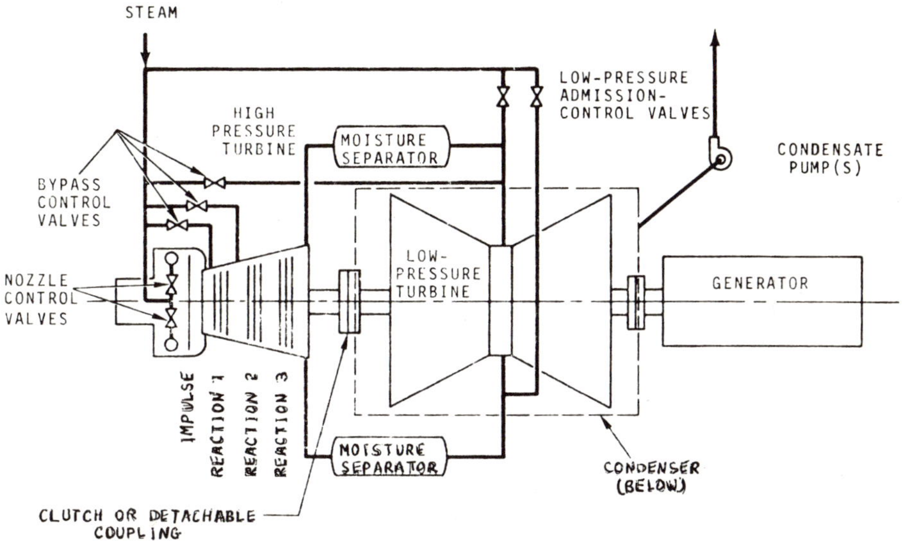

Source: PNE-1550

The high pressure cylinder is built with a partial admission first stage and by-pass steam admission valves and steam chests at the first and second reaction groups. Initially, it is desired to admit the steam at conditions of 800 psia and 650°F (see Figure 4.12) to the impulse stage through one half of its frontal area. As the geothermal heat source cools, feedwater flow is regulated so as to supply steam of declining pressure and temperature down to 425 psia saturated.

At the turbine, admission of the increasing volumetric flow rate of steam is accommodated by increasing the arc of admission in steps until it is open the full 360 degrees. Beyond that point, the steam begins to by-pass the impulse stage and proceeds until by-passing of the first reaction group begins as the pressure and temperature of the steam continues to fall to 225 psia and 400°F. Admission is now shifted directly to the low pressure cylinder completely by-passing the high pressure cylinder and moisture separators. The high pressure cylinder is uncoupled to eliminate windage losses and possible damage.

The feedwater is controlled so as to provide steam at about 85 psia and 400°F. As the temperature continues to fall, superheat is reduced and pressure increased somewhat to 90 psia in order to obtain increased steam flow. From this point onward, pressure and temperature fall as well as power output until the end point is reached at 67 psia, 300°F, and 150 Mw.

FIGURE 4.12: PERFORMANCE OF THE PLOWSHARE GEOTHERMAL TURBINE

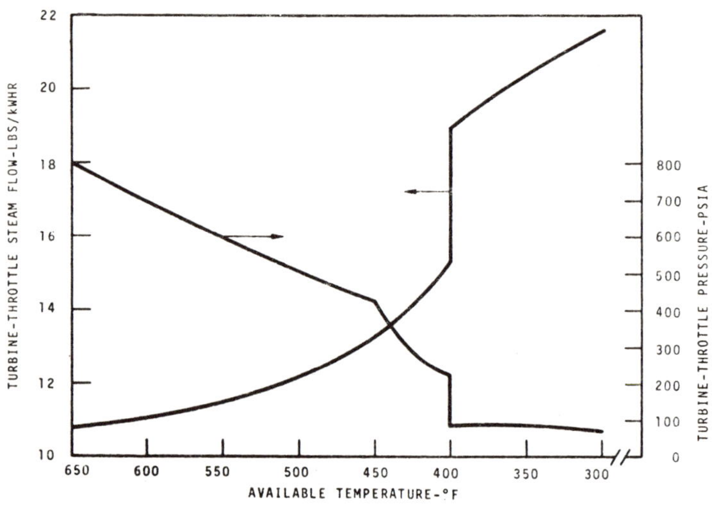

Source: PNE-1550

The plant will not employ feedwater heaters for the same reason that steam reheaters will be omitted; that is, the cost of energy is too low to justify the added capital expenditure. The steam flow rate for 200 Mw begins at about 2,200,000 pounds per hour and increases to 3,600,000 pounds per hour at the end of the cycle at a load of 150 Mw. Unlike a normal steam plant where only half the throttle steam is exhausted to the condenser (the other half is extracted for regenerative feedwater heating), this turbine passes all of the steam through the last stages and into the condenser. This has the chief effect of requiring twice the turbine exhaust annulus area.

The geothermal plant is assumed to be located in a remote area where land is inexpensive and structures, such as cooling towers, are more costly than would normally be the case, indicating the use of a cooling pond as a heat sink. For purposes of this study, a 4-inch-HgA condenser pressure was assumed. In an actual design, the condenser pressure would be optimized, taking into consideration an economic balance between the effects on power production, condenser cost, cooling pond size (cost) and water requirements which would be critical in this concept.

Figure 4.13 is typical of the effect which condenser pressure has on turbine-generator output for a fixed throttle pressure. The fact that this concept employs a sliding throttle pressure complicates the calculation somewhat. The actual effect of condenser pressure is expected to be more severe than that shown in Figure 4.13.

FIGURE 4.13: TURBINE OUTPUT VERSUS EXHAUST PRESSURE

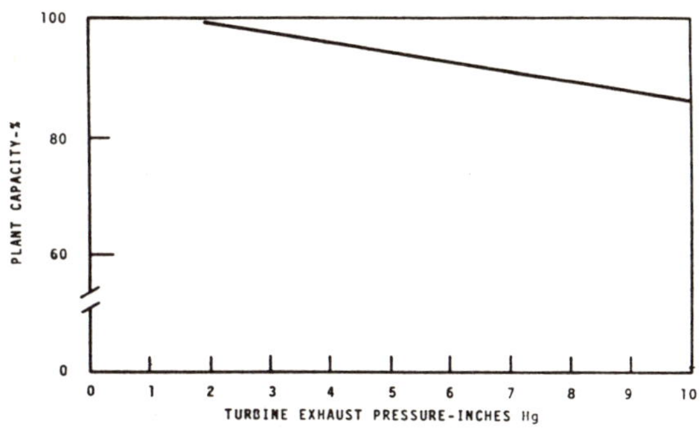

FIGURE 4.14: COOLING POND REQUIREMENTS

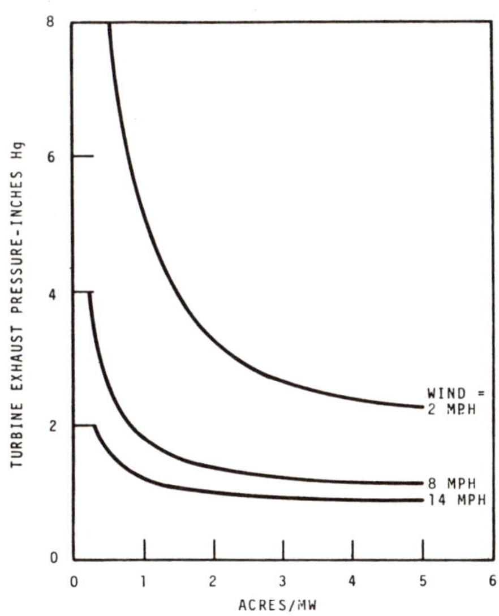

Source: PNE-1550

Figure 4.14 is a typical set of curves of turbine-exhaust (condenser) pressure versus cooling pond size with wind speed as a parameter. Not shown but included in the calculations are effects of relative humidity, dry bulb ambient temperature, cloud cover, pondwater temperature, condenser-discharge temperature, and cooling water flow rate. The pond size recommended for the present is 1 acre/Mw or 200 acres with an average depth of 4 feet. The expected water makeup requirement is 72 inches per year or 2,250 acre-feet per year. This figure does not include other plant requirements such as the initial inventory for the steam wells or makeup.

Pressurized and Boiling Water Cycles: Two methods were studied for transporting the heat from the fractured geothermal rock to the turbine: a pressurized water cycle (PWC) and a boiling water cycle (BWC). In the pressurized water cycle, the cavity is completely flooded with water and maintained at a high pressure so that no boiling can occur. In the primary loop, the water is mechanically pumped down to and through the fractured region up to a heat exchanger (boiler), and back down to the fractured region. In the secondary loop, the steam is produced in the boiler and then expanded through the turbine.

In the boiling water cycle, a steam-water interface is maintained in the fractured region. Feedwater is introduced at the bottom of the fractured region and is boiled to steam as it rises through it. The steam is piped directly to the turbine and the condensate returned to the fractured region. In this concept, the fractured region would be slowly filled with water during the life of the system.

The PWC and BWC have a number of similarities. The principle losses in both systems are caused by the flow friction in the piping and fractured region. The varying steam conditions require that any design take account of the most severe conditions at both beginning and end of life; for example, any piping system must be designed to withstand the high pressures at beginning of life and to handle the high mass flow rates at the end of life. The major considerations when comparing the PWC and the BWC are scale formation and associated radioactivity and the pumping, piping, and heat-exchanger costs.

The PWC uses a boiler to produce relatively clean steam for the turbine and condenser. Hence, a conventional turbine and condenser could be used. The boiler, however, would require large allowances in size and the use of special materials to compensate for the heavy scale formation which would occur. Also, since the throttle conditions degrade with time, the heat transferred in the condenser is two and one-half times as much at the end of the cycle as at the beginning, again increasing the size of the boiler.

The pumping power required by the PWC is, of course, dependent on the size of pipes used. Any system would have to be optimized for a balance between pumping power, pipe size, and boiler size. For the plant considered in this study, the pumping power was limited to about 10% of the plant output at reasonable pipe sizes.

The BWC diminishes the amount of scale and radioactivity brought to the surface by volatile gases and small amounts of other materials carried by or dissolved in the steam. The impurities which reach the surface in the BWC must be passed on through the turbine and surface condenser and returned with the condensate to the fractured region. The possible corrosive and scaling effects of the geothermal steam on the turbine are of primary importance.

Some means of providing fairly clean steam to the turbine throttle is necessary. As with the boiler in the PWC, the condenser in the BWC must be designed with allowances for the additional heat-transfer capability at end of life. Also, the geothermal steam may have a relatively high, noncondensable gas content. A mere 1% by weight of gas would require 50% greater surface area than the conventional condenser. In addition, stainless-steel construction and a large size allowance for fouling would be necessary to permit reasonable operating periods between tube cleaning operations.

In the BWC the static head in the feedwater line provides the system pressure. Since the principal thermodynamic losses in the BWC are a function of the losses in the piping, the end of cavity life is reached when the saturation pressure of the steam at the cavity exit temperature equals the pressure drop in the pipe plus the throttle pressure.

Preliminary analysis of the PWC and the BWC showed that both were technically feasible. The PWC required primary circulation pumps and a steam generator; on the other hand, this concept could utilize a smaller exhaust line, and operate with a clean secondary system which would allow use of a low cost barometric condenser. The advantages of one system over the other on an economic basis are not clear but if the transfer of radioactivity and silica to the surface heat exchanger proved to be severe the BWC would reduce this transfer by many orders of magnitude. This is due to the lower solubility of all materials in the steam phase. It was primarily on this basis that the BWC was selected for detailed analysis.

Analysis of System Losses: In designing the transport system for the BWC, the thermodynamic degradation of the steam in the pipes must be balanced against the cost of larger pipes. Numerous piping schemes are available. The simplest system would be one pipe with a wall thickness large enough to withstand the beginning of life pressure and with a diameter sufficient to carry the required end of life mass flow rate without excessive pressure drop. An alternate to this system would be a small diameter high pressure line for beginning of life conditions and a large diameter low pressure line for end of life use.

In order to perform detailed parametric analyses of various piping systems for a boiling water cycle, a computer code was written to calculate pressures and temperatures required at various points in the system for given throttle conditions (pressure, enthalpy, mass flow rate) and pipe sizes. The pressure drops in the pipes were calculated using the well known Moody relationship with a fraction factor of 0.02. The pressure drop across the heat extraction volume is calculated using Darcy's Law with a cylindrical geometry and a permeability of 0.5 darcy. Calculations were done for the range of parameters shown in Table 4.12.

The effect of pipe diameter on the amount of energy that can be extracted from the fractured region is shown in Figure 4.15. These curves were calculated for identical throttle conditions and pipe lengths. One hundred percent energy extraction would be realized if the final average fractured source rock temperature were the same as the final throttle temperature. Of course, with the system losses described above, 100% extraction could never be achieved. These curves make it evident that for a given exhaust pipe diameter there is a maximum energy which can be extracted. This maximum is insensitive to surface pipe diameter, i.e., with a 3-foot exhaust line, the maximum occurs somewhere above a 4.5-foot surface line. A set of curves similar to these can be obtained for each different set of throttle conditions and pipe lengths. These curves provide the information necessary for choosing a piping system that will yield minimum energy costs.

TABLE 4.12: PARAMETER CALCULATIONS

Parameters	Values
Throttle conditions	
Pressure	2300 - 50 psi
Enthalpy	1335 - 1170 Btu/lb
Mass flow rate	2-4 × 10^6 lbs/hr
Surface pipe	
Length	2000 - 31,650 ft
Diameter	1.5 - 8 ft
Vertical pipe	
Length	8600 - 9200 ft
Diameter	2-6 ft

FIGURE 4.15: FRACTION OF ENERGY EXTRACTED VERSUS SURFACE PIPE DIAMETER

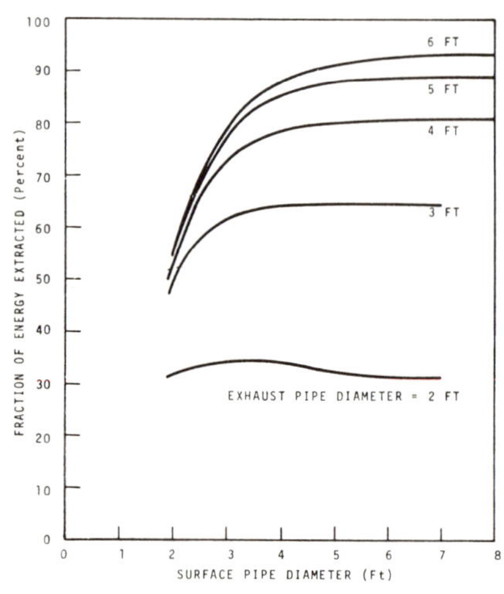

Source: PNE-1550

The relative energy extracted from the cavity during the various stages of the turbine operation is shown in Figure 4.16. These two curves were calculated for identical throttle conditions and pipe lengths. The 1.5-foot curve in Figure 4.16 stops at a throttle pressure of 370 psi with 33.5% of energy extracted. Physically, this means that for the given mass flow rates and pipe lengths used to make the calculations, one cannot obtain a throttle pressure less than 320 psi because the pressure drop due to friction in the 1.5-foot pipe is too large. (Note that as the throttle pressure decreases, the mass flow rate must increase as shown in Figure 4.12).

By using a 4.5-foot pipe, the final throttle pressure of 70 psi can be reached. The vertical portion of the 4.5-foot curve represents the coastdown phase of plant operation. During this coastdown operation, approximately 15% of the energy extracted from the fractured region is used.

FIGURE 4.16: FRACTION OF ENERGY EXTRACTED AS A FUNCTION OF THROTTLE PRESSURE

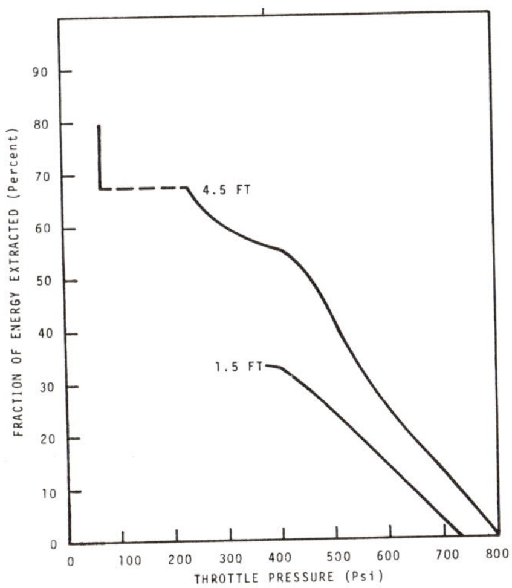

Source: PNE-1550

Losses with Preshot Design: One of the main advantages of the preshot design is that the turbine-generator plant can be moved in close to the cavity exhaust hole, thus minimizing the thermodynamic losses and pipe costs. For the reference plant, a distance of 2,000 feet was felt to be adequate.

The total energy extracted from a given array during its lifetime can be estimated by the end-of-life bulk fluid temperature at the geothermal source. Turbine operation is terminated when the throttle steam conditions are 70 psi and 303°F. The end-of-life source rock temperature is as shown in Figure 4.17. If it is assumed that the bulk fluid temperature is directly proportional to the average fractured source rock temperature, the lower temperatures in Figure 4.17 represent more energy extracted from the cavity. As expected, the larger the exhaust and surface lines, the lower the cavity final temperature, and thus the more heat extracted.

FIGURE 4.17: BULK FLUID TEMPERATURE FOR VARIOUS PIPING SYSTEMS

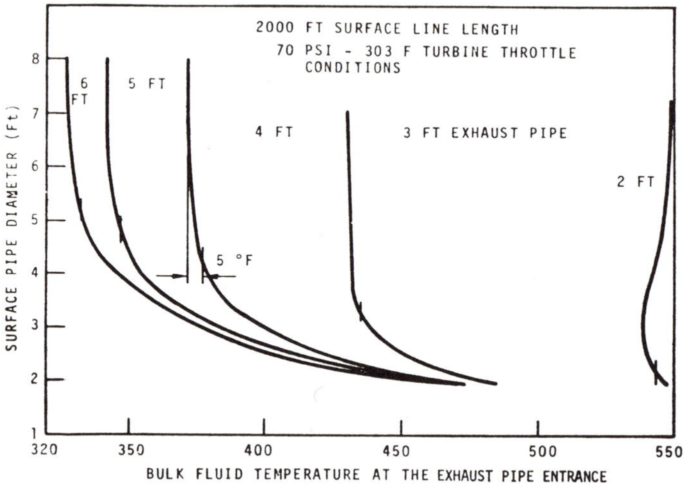

Source: PNE-1550

The pipe line losses are the difference between the fluid temperature at the exhaust line entrance and the turbine throttle. In general, this is made up of two components: the losses in the surface line and the losses in the exhaust pipe. Therefore, for a given exhaust pipe diameter, increasing the size of the surface pipe beyond the point where its losses are negligible does not result in lower source temperatures (Figure 4.17). The curves are, therefore, asymptotic to some source temperature which represents the losses due to the exhaust pipe only.

Rather than optimize the surface pipe diameter for each exhaust pipe size, it was found that reasonable relationships could be developed by selecting a surface pipe that resulted in a source temperature within 5°F of the asymptotic source temperature for each exhaust pipe. This results in the relationship shown in Figure 4.18.

Each of the piping systems in Figure 4.18 results in a specific final source temperature. This, in turn, implies each extracts different amounts of energy from the geothermal heat source. If, as a reference, it is assumed that one unit of energy is extracted in taking the source rock from 662°F to 303°F, the relative energy extracted by the various piping systems can be obtained from Figures 4.17 and 4.18. The results are shown in Figure 4.19.

FIGURE 4.18: OPTIMAL SURFACE PIPE DIAMETER AS A FUNCTION OF EXHAUST PIPE DIAMETER

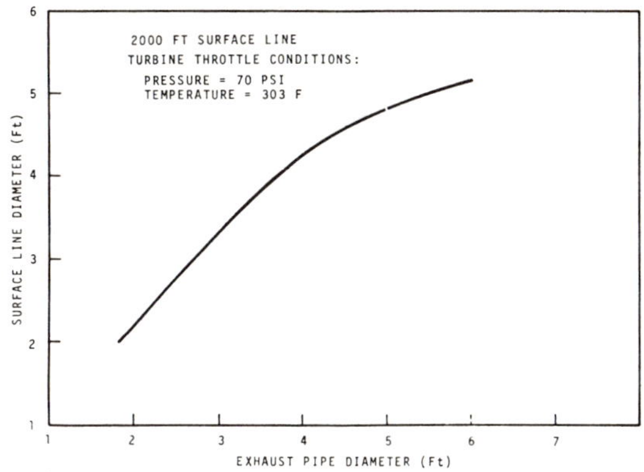

FIGURE 4.19: RELATIVE ENERGY EXTRACTED AS A FUNCTION OF EXHAUST PIPE DIAMETER

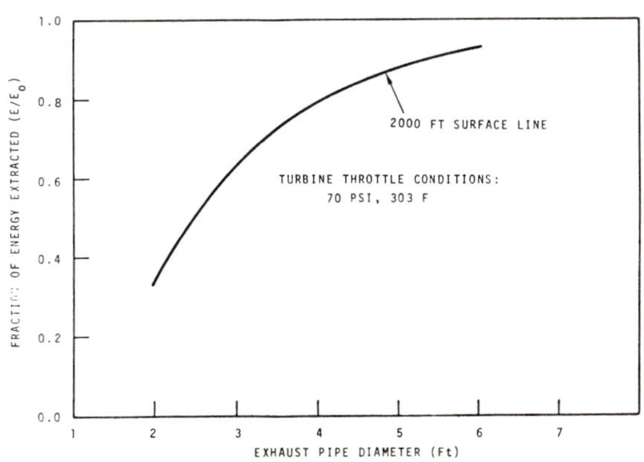

Source: PNE-1550

Figure 4.19, then, represents the performance of the various piping systems. This, combined with an energy base for the preshot array and piping systems and other field development costs, forms a basis for selecting the piping system for the reference design.

Losses with Aseismic Plant Design: The thermodynamic performance of the power conversion system for the hardened plant design is the same as for the Reference Plant, but the pipe line losses for the last twenty years of plant life are unique to the remote location of the seismically hardened plant. During the first 10-year period of plant operation, the heat is transported 2,000 feet in a surface pipe, and the losses would be similar to those discussed for the reference preshot design. The thermodynamic performance of several combinations of exhaust (vertical) and surface pipe diameters that transport the steam from the cavity to a plant located 4 miles from the exhaust hole are next considered.

To assist in selecting a combination of exhaust and surface pipe diameters, the bulk fluid temperature at the source outlet is estimated for end of life turbine throttle conditions. From this estimated temperature can be determined the relative energy extracted from a given field by various piping systems. Some deviation from the methods used for the preshot design are necessary.

In the former analysis, the diameter of the surface pipe was arbitrarily increased until the bulk fluid temperature at the source exit was within 5°F of the temperature that would be achieved by using an exhaust pipe only. This resulted in a relationship between surface pipe diameter and exhaust pipe diameter that simplified the ensuing calculations. With the cost of 4 miles of large surface pipe involved, it was felt necessary to optimize the surface pipe diameter for each one considered. Figure 4.20 shows the relative energy extracted by the various piping systems considered in the optimization calculations for the potshot alternate design. This, together with an energy base (E_0) and the cost of the various piping systems and other field development costs, forms a basis for selecting a system.

FIGURE 4.20: RELATIVE ENERGY EXTRACTED

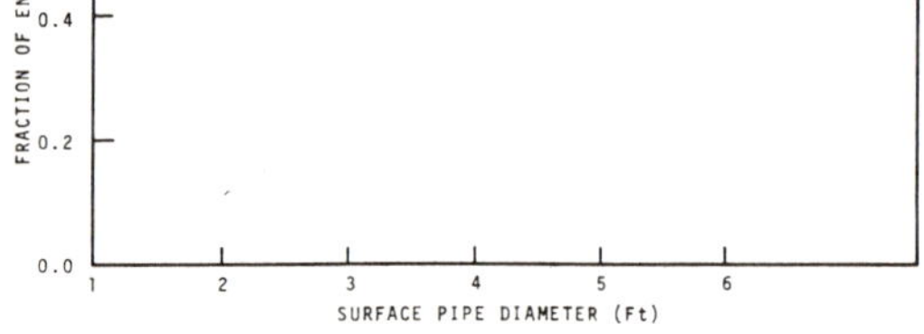

Source: PNE-1550

Hypothetical Venting Accident

An analysis was made of a hypothetical vent accident. It was assumed that the plant was located 4 miles from the geothermal field and that a rupture occurred in the middle of the surface steam line. After the leak was detected (6 seconds) by a pressure sensing device, quick acting valves at both ends of the pipe closed the system down in another 15 seconds. The rupture was assumed to be of the same diameter as the pipe and pointed upward. Calculations were made for both the high pressure (18-inch diameter) and low pressure (54 inches) pipes. Although there are differences in the numbers for the two calculations, the conclusions are the same.

Table 4.13 summarizes the various conditions associated with this accident. It is assumed that at the end of the flow of the steam through the rupture the conditions inside the pipe are at 1 atmosphere and 100°C. The rupture is on the top side of the pipe and the steam flow is directed upwards.

TABLE 4.13: STEAM PIPE RUPTURE CONDITIONS

Pipe length = 4 miles
Shutdown time = 15 sec
Wind speed = 3 m/sec

Parameter	18 in. Diameter Pipe
Inlet temperature	349°C
Inlet pressure	73.5 atmos.
Turbine pressure	50.0 atmos.
Average temperature	349°C
Average pressure	61 atmos.
Average sound speed in steam (γ = 1.28)	606 m/sec
Time required to sense pressure drop 2 miles from rupture	6.0 sec
Total shut down time	21.0 sec
Volume in 4 mi of pipe	1060 m^3
Average density in 4 mile pipe	24,600 g/m^3
Steam flow rate at 160 MWe	2.18×10^5 g/sec
Density of steam at 1 atmos. and 100°C	597 g/m^3
Mass flow before shut down	0.46×10^7 g
Mass removed in lowering pipe to 1 atmos. (assuming no condensation)	2.54×10^7 g
TOTAL	3.00×10^7 g
Initial release rate to atmosphere at sonic speed (330 m/sec)	1.3×10^6 g/sec
Total vent time assuming constant flow	22 sec
Initial heat flux rate (Cp = 0.48, Te = 15°C)	2.1×10^8 cal/sec
Top of effluent at end of vent time	190 m
Height at transition distance if continuous plume	1370 m
Stabilization height used	600 m
Stabilization diameter used	60 m
Volume	1.7×10^6 m^3

Source: PNE-1550

There are two types of meteorological problems associated with this type of venting: (1) the trajectory of the plume centerline, and (2) the diffusion of material about this centerline. The effluent plume height was calculated at the end of vent time, but this value was judged to be too low since the plume would continue to rise after the end of venting. Likewise, the value calculated for equilibrium height for a continuous point source (at the initial heat and momentum flux rates) was judged to be high since the initial fluxes would not be maintained through the entire release. Intermediate heights of 600 m and 1,200 m were chosen for the 18 inch and 54 inch pipes, respectively. Lower heights would be expected if the rupture were horizontal or downward.

At six months postshot many of the nuclides have decayed to insignificant amounts. Of those left only a few are thought to be soluble. These are listed in Table 4.14. It is assumed that they are uniformly mixed throughout the water prior to the accident and that the noncondensable gases (^{85}Kr and ^{37}Ar) have been disposed of prior to the accident. The resulting water concentrations are also given in Table 4.14.

TABLE 4.14: SIGNIFICANT RADIOACTIVITY AVAILABLE IN STEAM AT SIX MONTHS POSTSHOT

	Fission Explosive		Fusion Explosive	
Nuclide	Production kCi	Concentration in Steam pCi/g	Production kCi	Concentration in Steam pCi/g
H^3	220	4.4(5)*	20,290	4.1(7)
Sb^{125}	60	1.2(5)	0.18	3.6(2)
Te^{127m}	90	1.8(5)	0.27	5.4(2)
Cs^{134}	14	2.8(3)	18	3.6(3)
Cs^{137}	180	3.6(4)	0.54	1.1(2)

Note: It is assumed that 100% of the production is uniformly mixed through the total water volume (5×10^5 m^3) except for the Cs^{134} and Cs^{137}. For these two nuclides 10% of the production is assumed to go into solution.

*4.4(5) = 4.4×10^5 pCi/g

Source: PNE-1550

The distance from centerline to edge of a turbulent jet is about 10% of the distance from the nozzle. This results in a cone shaped initial cloud volume. For simplicity, this volume was assumed to be a cylinder with a radius of 5% of the height, and it was also assumed that the vented activity was uniformly mixed throughout this volume. This will make the ground level concentrations conservative for two reasons: (1) the rate of activity release is the greatest at the time of rupture, therefore, proportionally more activity would be at the higher altitudes and (2) this procedure increases the integrated concentrations, over cloud passage, by giving the cloud more width at ground level than would be expected.

It should be noted that any depletion by condensation and precipitation has been neglected, even though it is expected that the vent would be visible unless atmospheric conditions are very hot and dry. This is also conservative as it maximizes downward airborne concentrations.

The radioactivity released is given in Table 4.15. This is calculated from the water concentration data given in Table 4.14 and the total water release from Table 4.13. This assumes a rupture at six months postshot right after plant startup but after having disposed of the noncondensable radioactive gases.

TABLE 4.15: RADIOACTIVITY RELEASE

18 in. Pipe Rupture

Nuclide	Fission pCi	Fusion pCi
H^3	1.3 (13)*	1.2 (15)
Sb^{125}	3.5 (12)	1.1 (10)
Te^{127m}	5.5 (12)	1.6 (10)
Cs^{134}	8.0 (10)	1.1 (11)
Cs^{137}	1.1 (12)	3.3 (9)

*1.3 (13) = 1.3×10^{13} pCi

Source: PNE-1550

As the cylindrical cloud of effluent moved downwind at 3 m/sec, it was allowed to diffuse horizontally at rates dictated by the predictions of atmospheric turbulence by similarity theory. Values of turbulent dissipation were used which are consistent with the above values of wind speed and the height of cloud center above ground. Vertical diffusion above the initial cloud top was neglected.

Table 4.16 presents the calculated ground level concentrations as a function of distance for a rupture in the 18-inch steam pipe. In the case of the fission explosive, the ^{127m}Te and ^{137}Cs nuclides just barely exceed their MPC at 10 kilometers. The ^3H from the fusion explosive is an order of magnitude above its MPC at 10 kilometers. And yet the MPC is based on an annual exposure whereas this vent cloud will pass any particular point in a matter of minutes. This point is amply illustrated in Table 4.17 which gives the integrated concentrations over the period of cloud passage and compares this to the MPC's multiplied by the number of seconds in a year. In this case, for the worst situation, ^3H from a fusion explosive, the exposure at 100 meters downwind is only about 0.25% of the integrated MPC.

Even if all atmospheric entrainment and plume rise had been neglected, so that we had a spherical cloud of steam whose initial volume was only given by the adiabatic expansion of the vented material to atmospheric pressure and which traveled downwind at ground level, the numbers in Tables 4.16 and 4.17 would only be one or two orders of magnitude higher close to the leak and would approach the values in those tables at distances of 10 km and further. Although the model used here is conservative compared to a more rigorous one, the total exposures calculated are orders of magnitude less than the annual exposures on which the MPC tables are based.

TABLE 4.16: DOWNWIND AIR CONCENTRATION FROM VENTING 6 MONTHS POSTSHOT, 1,000-KILOTON EXPLOSIVE

Downwind	Relative Concentration	Fission Explosive H^3	Fission Explosive Sb^{125}	Fission Explosive Te^{127m}	Fission Explosive Cs^{134}	Fission Explosive Cs^{137}	Fusion Explosive H^3	Fusion Explosive Sb^{125}	Fusion Explosive Te^{127m}	Fusion Explosive Cs^{134}	Fusion Explosive Cs^{137}
m	$1/m^3$	pCi/m^3	pCi/m^3	pCi/m^3	pCi/m^3	pCi/m^3	pCi/m^3	pCi/m^3	pCi/m^3	pCi/m^3	pCi/m^3
100	5.9(-7)	7.7(6)	2.1(6)	3.3(6)	4.7(4)	6.5(5)	7.1(8)	6.5(3)	9.4(3)	6.5(4)	1.9(3)
300	5.8(-7)	7.5(6)	2.0(6)	3.2(6)	4.6(4)	6.4(5)	7.0(8)	6.4(3)	9.3(3)	6.4(4)	1.9(3)
500	5.7(-7)	7.4(6)	2.0(6)	3.1(6)	4.6(4)	6.3(5)	6.9(8)	6.3(3)	9.1(3)	6.3(4)	1.9(3)
1,000	4.5(-7)	5.9(6)	1.6(6)	2.5(6)	3.6(4)	5.0(5)	5.4(8)	5.0(3)	7.2(3)	5.0(4)	1.5(3)
3,000	6.5(-8)	8.4(5)	2.3(5)	3.6(5)	5.2(3)	7.2(4)	7.8(7)	7.2(2)	1.0(3)	7.2(3)	2.1(2)
5,000	1.5(-8)	2.0(5)	5.3(4)	8.2(4)	1.2(3)	1.7(4)	1.8(7)	1.7(2)	2.4(2)	1.7(3)	5.0(1)
10,000	2.0(-9)	2.6(4)	7.0(3)	1.1(4)	1.6(2)	2.2(3)	2.4(6)	2.2(1)	3.2(1)	2.2(2)	6.6(0)
MPC in air - pCi/m^3		2 (5)	2 (4)	5 (3)	1 (3)	2 (3)	2 (5)	2 (4)	5 (3)	1 (3)	2 (3)

TABLE 4.17: DOWNWIND INTEGRATED AIR CONCENTRATION FROM VENTING

Downwind	Integrated Relative Concentration	Fission Device H^3	Fission Device Sb^{125}	Fission Device Te^{127m}	Fission Device Cs^{134}	Fission Device Cs^{137}	Fusion Device H^3	Fusion Device Sb^{125}	Fusion Device Te^{127m}	Fusion Device Cs^{134}	Fusion Device Cs^{137}
m	$\frac{sec}{m^3}$	$\frac{pCi\ sec}{m^3}$	$\frac{pCi\ sec}{m^3}$	$\frac{pCi\ sec}{m^3}$	$\frac{pCi\ sec}{m^3}$	$\frac{pCi\ sec}{m^3}$	$\frac{pCi\ sec}{m^3}$	$\frac{pCi\ sec}{m^3}$	$\frac{pCi\ sec}{m^3}$	$\frac{pCi\ sec}{m^3}$	$\frac{pCi\ sec}{m^3}$
100	1.2(-5)	1.6(8)	4.2(7)	6.6(7)	9.6(5)	1.3(7)	1.4(10)	1.3(5)	1.9(5)	1.3(6)	4.0(4)
300	1.2(-5)	1.6(8)	4.2(7)	6.6(7)	9.6(5)	1.3(7)	1.4(10)	1.3(5)	1.9(5)	1.3(6)	4.0(4)
500	1.2(-5)	1.6(8)	4.2(7)	6.6(7)	9.6(5)	1.3(7)	1.4(10)	1.3(5)	1.9(5)	1.3(6)	4.0(4)
1,000	1.0(-5)	1.3(8)	3.5(7)	5.5(7)	8.0(5)	1.1(7)	1.2(10)	1.1(5)	1.6(5)	1.1(6)	3.3(4)
3,000	3.9(-6)	5.1(7)	1.4(7)	2.1(7)	3.1(5)	4.3(6)	4.7(9)	4.3(4)	6.2(4)	4.3(5)	1.3(4)
5,000	1.9(-6)	2.5(7)]	6.7(6)	1.0(7)	1.5(5)	2.1(6)	2.3(9)	2.1(4)	3.0(4)	2.1(5)	6.3(3)
10,000	6.8(-7)	8.8(6)	2.4(6)	3.7(6)	5.4(4)	7.5(5)	8.2(8)	7.5(3)	1.1(4)	7.5(4)	2.2(3)
Integrated MPC - $\frac{pCi\ sec}{m^3}$		6 (12)	6 (11)	1.5(11)	3 (10)	6 (10)	6 (12)	6 (11)	1.5(11)	3 (10)	6 (10)

Source: PNE-1550

Seismic Design Considerations

For the postshot design it is necessary to arrive at a reasonable estimate of the costs associated with the construction of the power plant and its equipment to withstand the shocks from the detonations subsequent to construction. This requires a knowledge of ground response from deep underground detonations which in turn depends on the yield of the explosive, depth of burial and the physical properties of the site geology.

From an aseismic (shock isolation or hardening) design viewpoint, the most important aspect of the anticipated ground motion is the breakaway (spall) of the upper earth strata, separating it from lower strata, as the nuclear explosion-produced shock wave interacts with the earth's free surface. When the spalled mass free falls back and collides with the lower earth strata (denoted as slap down) it causes an important secondary source of ground motion.

In general terms, the observed seismic disturbance within the slap-down region is characterized by the first arrival shock pulse (compression wave) followed by a sharp spall closure pulse which may be much larger than the first arrival shock. Outside the spall zone the seismic disturbance is made up of many oscillations with greatly reduced peak values and the noticeable absence of a spall closure pulse. In this far out region, somewhere beyond 80,000 feet, shear, reflected, and surface waves become more important so that the recorded ground motion is similar to an earthquake-produced disturbance.

Somewhere between the sharp slap-down zone and the far out zone, a transition region exists where the first arrival is the peak pulse amplitude. Thus, as range increases, not only does the amplitude of the ground motion decrease, but also the time dependent character of the signal is drastically changed. Figure 4.21 illustrates these typical differences for a hypothetical 300-kiloton, contained, nuclear explosion.

FIGURE 4.21: COMPARISON OF TYPICAL GROUND MOTION HISTORIES

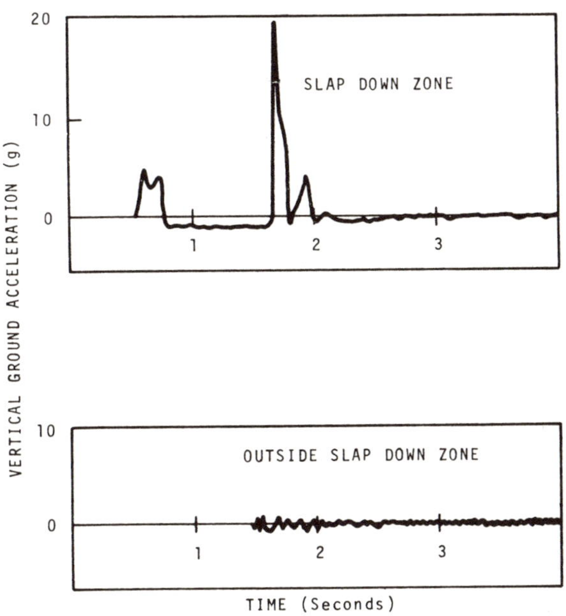

Source: PNE-1550

The upper acceleration trace might be typical of an observation taken 1,000 to 2,000 feet from surface ground zero (SGZ). The initial shock pulse is evident as are, in this case, two spall closure pulses which arrive about 1 second after the first shock arrival. The lower curve reflects the increased travel time to an imaginary station located several miles from SGZ. Absence of slap down and reduction in amplitude is evident.

Owing to the newness of any geothermal site and detonation media, a standard geometric deviation of three has been suggested to apply to the initial ground motion predictions. A linear plot of predicted peak acceleration is shown in Figure 4.22 for three device yields and serves to illustrate the influence of range on intensity of ground motion. It is readily apparent that with sufficient range the anticipated peak ground acceleration will be comparable to that expected in a conventional earthquake seismic zone. However, for reduced ranges the predicted ground acceleration is seen to increase sharply.

FIGURE 4.22: PREDICTED PEAK GROUND ACCELERATION

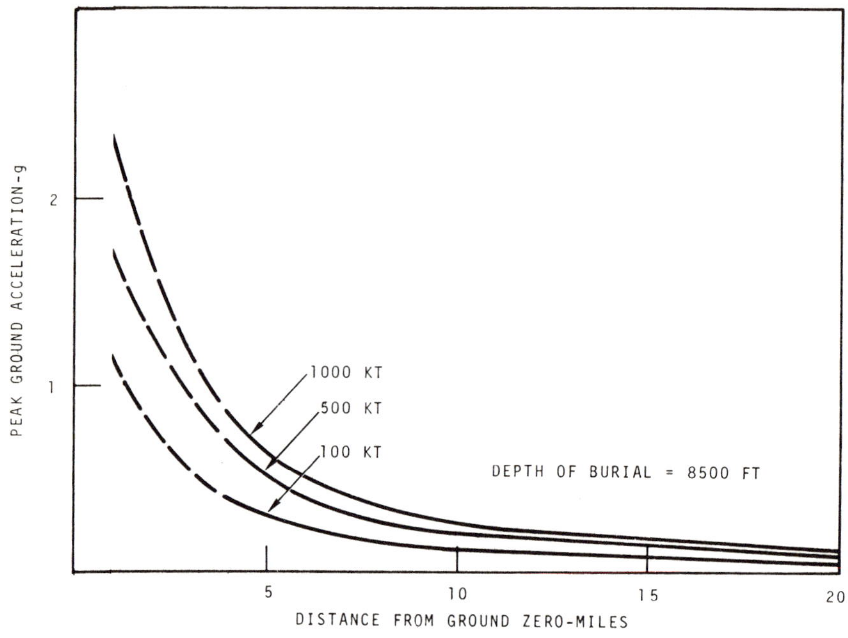

Source: PNE-1550

Although the curves represent average values taken from many field observations, these observations were taken no closer to SGZ than the edge of the slap-down region. Thus, the data should not be used to predict ground motion closer than that range. For close-in locations where accelerations exceed 1 gram and slap down plays an important role, far fewer field observations are available upon which to base reliable predictions of the ground motion. Limited data has been reported and several papers have investigated techniques for improving or establishing close-in ground motion predictions although much more needs to be done.

At present only very limited data are available upon which to predict close-in motion within the spall zone. Of the limited data published to date, none of it was obtained for the same combination of explosive yields and burial depths anticipated to be used in a geothermal Plowshare application, although both large yield and deep depth measurements have separately been obtained.

It should be emphasized that the currently available ground motion predictions provide only resultant peak vectors of acceleration, velocity, and displacement. Orthogonal components, one vertical and two horizontal, are not defined except by the general observations of recorded ground motion records. These indicate that, as a general rule, very little horizontal components exist close to SGZ; the motion is nearly all vertical. At great distances, well removed from the slap-down region, the horizontal components predominate. Near the edge of the slap-down zone, the three components may be more nearly equal. In addition, the current prediction techniques do not provide a means to assess the influence on ground motion of interaction of fractured zones of multiple shots.

If, as in the barge design, the plant is located at the end of an 8-mile canal, it would receive essentially the same seismic intensity, for a 1,000-kiloton device, as would be expected from an earthquake with a Zone 3 seismic region. Smaller devices would require correspondingly shorter canals. Thus, with the barge located at the extreme end of the canal, the ground motion would be considerably reduced and the disturbances would pass without incident. At the completion of the firings, the barge would be towed back along the canal and connected again to the main downhole steam lines.

The operating concept for the floating power plant is as follows: the construction of the barge with the deck-mounted turbine-generator-condenser assembly would proceed concurrently with excavation, grading, soil compaction and installation of an impermeable membrane canal liner. Once completed, however, the barge would be secured at the one end of the canal awaiting the detonation of the first series of shots at the other end. It might be of some advantage to construct a temporary dike just ahead of the floating assembly to prevent loss of water in case part of the canal or cooling pond close in to SGZ were destroyed during the detonations. After necessary canal or downhole steam pipe repairs, the floating barge would be towed to the cooling pond, secured within a stilling basin provided by a surrounding concrete wall and connected to the steam lines.

It is anticipated that the heat available after each series of shots would last for approximately 10 years. Therefore, sometime during the 9th year the plant would be shut down and the barge-mounted plant would be towed to the far end of the canal for a second series of detonations. Afterward, the barge would again be towed back to the cooling pond and connected to the original steam lines to utilize the remaining energy. During the 10th year, the power plant would be moved and attached to the second 10-year heat supply. The same procedure would be followed for a third and final series of detonations. The study concluded that under certain conditions power could be produced for from 5 to 7 mills/Kwh.

According to D.H. Stewart of Batelle, Pacific Northwest Laboratories in *The Status of Plowshare Geothermal Power,* BNWL-SA-4079, October 1971, for several reasons these numbers may prove conservative. Depletion allowance and terrain which would permit the mobile plant concept to be used could reduce these costs by 0.4 mill/Kwh and 0.8 mill/Kwh, respectively. It was announced after the study was completed that the large diameter holes (24") required for the large device emplacement can now be much smaller, around 8" to 9". This would also effect a major reduction in field development costs and power costs of approximately 1.25 mills/Kwh.

Key problems are the location of a geothermal site in an area which large nuclear devices could be detonated (>200 KT), a site with sufficient ground or surface water available to operate the power plant condenser, and a site reasonably near a large electrical market. Public opposition could develop and should be anticipated, but it is hoped that the relatively nonpolluting aspects of geothermal energy could offset any public opposition.

DIXON PATENT

In the process of *R.P. Dixon; U.S. Patent 3,640,336; February 8, 1972; assigned to the U.S. Atomic Energy Commission* a nuclear explosive is detonated underground within a zone above hot igneous rock. The zone for the explosive is selected far enough below ground to be nonventing and far enough above the igneous rock so that the ambient temperature is too low to interfere with the functioning of the nuclear explosive. The nuclear explosive must form a chimney which extends into a zone of rock having a temperature greater than 200°C.

Water is then introduced into the chimney, producing steam when it comes into contact with the stratum of igneous rock. The steam produced is withdrawn and is used as a source of energy. If the steam generated in the first stage after the nuclear detonation contains an excessive amount of radioactivity, it may be collected at the surface and segregated for safe disposal until a subsequent portion of the steam has a radioactivity level which is acceptable for the intended use.

PLOWSHARE STIMULATION OF NATURAL GEOTHERMAL RESOURCES

The following information is from *Survey of Expert Geological Opinion on Feasibility of Plowshare Stimulation of Natural Geothermal Systems,* BNWL-B-134, September 1971, by D.H. Stewart and J.B. Burnham of Batelle, Pacific Northwest Laboratories.

There are two major differences between the proposed stimulation study and the Plowshare geothermal feasibility study. In the dry geothermal case, large nuclear devices are required (>200 KT) and no natural underground water system is involved. In the natural stimulation case, the device sizes required are small (1 to 5 KT) and a natural underground water system is present. It is expected that nuclear stimulation could provide some or all of the following advantages to natural geothermal systems.

(1) Provide an increased well bore diameter effect. This effect is analogous to the natural gas stimulation experiments where a nuclear device can make a non-economic well economically productive by increasing the flow rate and total gas recovery through the extensive fracturing it produces.

(2) Provide a void volume for flashing water to steam. Several good producing wells have been abandoned when after 12 to 18 months of operation the steam flow was choked off by the deposition of minerals from the water as it flashed to steam at the well bore. The proposed void volume might provide adequate space to make the well productive for 10 to 20 years. It may also be possible to change a wet well to dry steam production.

(3) Provide a void and fractured region to permit reinjection of the condensed water and salts produced from the formation. In many areas (i.e., Salton Sea), it will be necessary to return all fluids into the formation from which they were produced as they contain too much mineral concentration for surface disposal. Reinjection could also offer the advantage of providing continuing underground water replenishment in areas where water inflow is limited.

The principal problems raised in Plowshare Stimulation which require analysis are: the fate of the radioactivity in the geothermal water and whether this could present any hazard, and whether a nuclear explosion would be likely to destroy the natural system by shifting a key fault or some other such mechanism.

It should be noted that an attempt to stimulate natural geothermal systems by hydro-fracturing has been attempted successfully in some cases. This success provides a reason for anticipating nuclear stimulation would prove even more successful. It should be noted that the permeability in the Geysers and Larderello geothermal fields is secondary permeability, that is, permeability due to fractures. Nuclear detonations are likely to provide a link between the different fractures and thus enhance the permeability, the key factor for a commercial production.

Discussions were held with the following geothermal experts as to their opinions on the desirability of Plowshare stimulation of natural geothermal systems and whether it could be of interest to them for their particular system.

Iceland:
- Dr. G. Palmason - Department of Natural Heat
- Mr. J. Zoega - Reykjavik Municipal District Heating Service
- Mr. K. Ragnars - Department of Natural Heat

Italy:
Dr. Pietro Aron - ENEL - Rome
Dr. P. Cataldi - ENEL - Rome
Mr. Gian Carlo Ferra - Larderello
Mr. Pier Domenico Burgassi - Larderello
Dr. Mario Farrelli - Geologist, Pisa
Kay-Han Gurani - Eng. from Turkey

New Zealand:
Mr. J.H. Smith - Ministry of Works, Wellington
Mr. W. Basil Stillwell - Project Engineer, Wairakei
Mr. J. Giardin - Plant Manager, Wairakei

Japan:
Dr. Hisayoshi Nakamura - Geological Survey of Japan
Dr. Konosuki Sato - The Dawa Mining Company
Professor Hideo Iga - Japan Geothermal Energy Association
Mr. Yoshimi Hara - Chief Engineer, Otake Geothermal Plant

Hawaii:
Dr. Walter Arnell - Center for Engineering Research
Dr. John W. Shupe - Dean, School of Engineering, University of Hawaii
Dr. Richard Fand - Head of Mechanical Engineering Department, University of Hawaii

Geysers, California:
Dr. Carel Otte - Union Oil Co.
Mr. Tony Chastain - Geysers
Mr. Al Bruce - PG&E

As their comments were so similar, their opinions are summarized below.

(1) All expressed concern as to the fate and level of the radioactivity produced. (This is a function of the type and size of the nuclear device and the nature of the natural system.) Assuming this problem was successfully resolved, all believed that stimulation could aid in the economic recovery of geothermal steam.

(2) No one expressed concern about the destruction of the natural system. All believed that the additional fracturing and void volume created by the device would be very likely to improve steam and water recovery rates.

(3) All have considered at one time or another reinjection of the water produced from the field, as the limitations of natural systems appear to be the water inflow and not the available heat. Reinjection has been tried to a limited degree in the Geysers area of California and no loss in production from adjacent wells was noted. Several felt that a small cavity could aid in reinjection.

(4) Nearly all felt the information developed in the study and in an experimental program would be of value to their geothermal plans and development.

(5) One expressed an interest in a test in his geothermal field. Several expressed interest in trying Plowshare techniques once they were developed.

THREE EXPERIMENTAL STUDIES

SCALE FORMATION IN SIMULATED GEOTHERMAL BRINE

An experimental study of scale formation from a simulated geothermal brine and the effect of the scale on heat transfer is described by P.H. Harley in ORNL-TM-4137, June, 1973, *SiO_2 Scale Formation in Simulated Geothermal Brine* from which the following material is derived.

The Office of Saline Water, U.S. Department of the Interior, is preparing to design a pilot plant for investigating the production of fresh water by evaporation of geothermal brine using the upflow vertical tube evaporator (VTE) process. As part of that work, the Oak Ridge National Laboratory was asked to provide preliminary information on the tendency of silica normally dissolved in the brine to deposit on heated surfaces of the evaporator tubes.

Since ORNL had just completed an investigation of the heat transfer performance of a single tube, upflow VTE evaporator using nonscaling brine solutions in the temperature range of interest, i.e., 250° to 390°F, a loop was available which could be easily modified for tests to observe the scale formation tendencies of the silicate-containing brine.

Description of the Experiment

The high temperature single tube VTE loop which was built to measure heat transfer performance with synthetic geothermal brines was converted from a circulating system to a once-through system for these scaling tests. The loop modification consisted of the addition of three tanks for mixing and supplying the brine feed solution, a deaerator for removing oxygen from the feed, an acid addition system for converting the sodium silicate in the feed to silicic acid after the brine was heated to the operating temperature and a letdown system for cooling and discharging brine after it had passed through the evaporator. Lighted viewing ports were installed after the brine heater at the inlet to the evaporator and in the discharge line from the evaporator so that visual observation for precipitate could be made during loop operation. The flowsheet of the system for these tests is shown in Figure 5.1.

FIGURE 5.1: GEOTHERMAL BRINE LOOP SCALING TEST FLOWSHEET

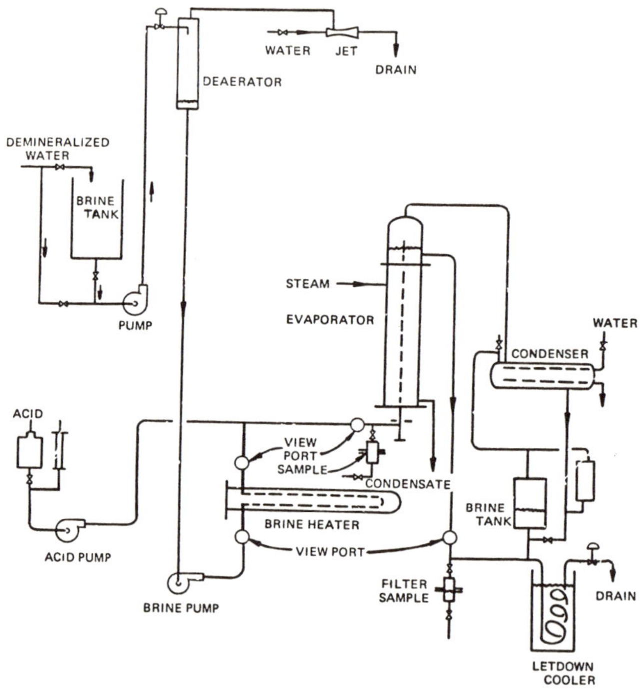

Source: ORNL-TM-4137

The synthetic brine was prepared in the three 300 gal stainless steel (SS) tanks. Either the feed brine or demineralized water could be pumped to the vacuum deaerator through a ⅜" plastic tubing supply line and level control valve which maintained a level in the bottom of the deaerator. The deaerator consisted of 8 ft of 4" glass pipe packed with 1" ceramic saddles. A vacuum of ∿29" of Hg was maintained by a Schutte and Koerting water jet ejector.

The deaerator liquid was fed by gravity to the brine feed pump located 38 ft below the deaerator. This maintained a suction head for the brine pump of about 5 ft of water. Liquid flow was controlled by varying the frequency to the brine pump motor and was adjusted to produce about 0.5 gpm flow rate.

The cold feed was pumped through a flow indicating orifice, brought to the desired temperature by passing through a steam-heated brine heater and fed into the inlet (bottom) of the evaporator tube through an adjustable orifice which controlled the flashdown rate on entering. The tube used for the test was a 1" OD double-fluted, 90–10 Cu-Ni tube. It was heated on the outside by the condensation of building steam which was fed in the 6" SS jacket surrounding the evaporator tube. The liquid left the upper head of the evaporator and flowed by gravity to a level control tank and a letdown system.

The vapor formed in the tube by evaporation passed through a stainless steel wire mesh entrainment separator into the condenser. The condensate passed through the level control tank and to the letdown system where it was mixed with the discharge brine.

When silica runs were made 22 weight percent hydrochloric acid was pumped into the brine at a point in the piping between the brine heater and the evaporator tube inlet. Sufficient acid was pumped to convert the sodium silicate and sodium carbonate contained in the feed to silicic acid, and sodium chloride. The CO_2 formed was kept in solution by overpressure until the liquid flashed into the evaporator where the CO_2 came out of solution and was then vented from the condenser.

In tests in which calcium and magnesium salts were included in the feed solution, they were added as chlorides to the acid being used in order to prevent scale deposition on the brine heater.

The letdown system consisted of 30 ft of ⅜" OD, 316 stainless steel tubing coiled in a 30 gal cooling water drum. The brine was cooled by process water to below 100°F and was then discharged through a level control valve to a large holding tank or directly to the drain.

Viewing ports consisting of 1" diameter by ½" thick Pyrex windows were located at the brine heater inlet and exit line and the evaporator inlet and discharge line and were used to observe the formation of precipitates in the system.

Sample lines and electrically heated sample bombs were located at the inlet and exit of the evaporator. The sample bombs contained a screen supported filter paper with a 5 to 10 micron pore size for collecting solids at the operating temperature.

Operation

All modifications to the original system were made with materials already on hand or available in plant stores. Initial operation of the system using water proved satisfactory except for the deaerator.

The original deaerator was fabricated from 8 ft of 2" glass pipe packed with 1" ceramic saddles. This apparently was too small a tube-to-packing diameter ratio since the minimum oxygen concentration obtained was only 200 ppb. Heating the feed to 100° to 110°F to reduce the oxygen partial pressure did not improve the performance of the uninsulated deaerator so the original unit was replaced by a 4" diameter column of the same height and packing size. The oxygen concentration in the feed water following the modification was reduced to about

50 ppb which was considered satisfactory. Later tests with brine indicated 20 ppb of oxygen. The only other major operational problem occurred during the tests involving acid injection.

Although corrosion of stainless steel tubing by the hydrochloric acid was expected, it was not anticipated that it would be so rapid as to interfere with the tests. Thus, the acid addition system was built using 316 SS tubing and supply tank. The Lapp micro pulsafeeder acid pump was equipped with a Carpenter 20 SS head and Hastelloy C check valves.

During early tests using HCl, gas formation from corrosion in the pump suction line was sufficient to gas bind the small volume pump. To minimize this gas evolution the supply tank was replaced by a glass carboy and the suction line by polyethylene tubing. In addition, it was necessary to line the pump head with an epoxy coating where possible. Even then, adequate pumping was obtained only for a period of several hours until the flow rate would decrease from gas still being generated in the small passages between the pump check valves and the main pump diaphragm cavity. Pumping capacity was improved by a periodic 10 to 20 minute water flush.

During hot operation severe corrosion at the acid line penetration to the main loop caused a shutdown on two occasions. The first occurred in a stainless steel acid line, and the second with a titanium tube which was used as a replacement. Finally, ½" diameter tantalum tubing was successfully used.

Test Program

The composition of the simulated brine used in these tests was the same as that used by Envirogenics in tests at Wrightsville Beach, N.C. The composition in g/l is as follows:

$NaCl$	13.7
KCl	1.84
$LiCl$	0.0893
$NaBr$	0.0937
H_3BO_3	0.0562
Na_2CO_3	1.126
Na_2SO_4	0.014
$Na_2SiO_3 \cdot 9H_2O$	3.786
$CaCl_2$	0.590
$MgCl_2 \cdot 6H_2O$	0.105
HCl	To give pH of 7

In order to qualitatively define the composition of the scale deposits encountered, the brine tests were divided into three phases: (1) Operation with calcium and magnesium but no silicate. (2) Operation with silicate in the feed but no calcium or magnesium. (3) Operation with all components.

Phase 1: In this series of tests, the stoichiometric amount of HCl was added to convert the carbonate to chloride and water, with the release of carbon dioxide. In runs at 270°F no precipitate was observed upstream of the evaporator, while at 310°F a slight precipitate formed, and at 360°F a rather heavy precipitate was observed. During this latter series of tests the stainless steel acid line corrosion occurring at the penetration to the main loop may have reduced the amount of acid available to react with the carbonate and contributed to the observed precipitation, which was probably $CaCO_3$.

Phase 2: Early operation during the silica-forming runs was hampered by acid leaks and poorly mixed brine feed. Operation was improved by installing a tantalum tube at the acid line penetration and obtaining an agitator to improve the mixing of the chemicals. (Earlier mixing had been by recirculation.)

Visual observations during runs at 360°F and 310°F inlet brine temperatures showed a clear solution at the evaporator inlet. At the evaporator discharge after the brine had flashed down 5° to 20°F and the CO_2 was stripped, the brine had a cloudy or hazy appearance in the viewport. Then after the discharge brine was mixed with condensate from the condenser, cooled in the letdown system, samples were usually quite clear suggesting either that precipitate remaining in the liquid at the discharge from the evaporator was scaling the letdown cooler or that it was redissolving.

The longest run, 13 hours at a brine outlet temperature of 290°F, was terminated due to a restriction in the flow through the letdown cooler to a point where 0.5 gpm would not pass with the 43 psig driving force available. This was accompanied by an increase in the letdown temperature to about 140°F.

When operating with a normal letdown stream with a pH of 7, isolating the vapor condensate (as was required to measure the condensing rate) resulted in an increase in the letdown stream pH to about 8 to 8.5. This indicates that the condensate itself was acidic, probably resulting from dissolved CO_2.

Phase 3: This consisted of one run of about two hours duration. During this period, there was a rapid deposition containing Mg and Ca. The run was made with a 360°F inlet brine temperature in order to avoid the erratic regulation of the steam to the brine heater at lower steam pressures.

When the acid flow was first started there was a considerable amount of precipitate at the evaporator inlet but this cleared as the acid-brine stream reached normal strength.

As the acid pump capacity decreased, caused by gas accumulation in the acid pump, solids reappeared at the inlet viewport. Increasing the acid pumping rate back to normal eliminated the inlet precipitation. During the entire test precipitate in the exit viewport was so heavy that no light would pass and the letdown samples contained a black flocculant precipitate which is believed to have been primarily magnesium hydroxide. This test was terminated by acid pumping problems and not continued, as scale formation had been demonstrated.

Heat Transfer Results

For all tests the liquid flow was 0.5 gpm. Scale formation tests were made with inlet brine and steam at 360°F and 310°F with flashdown of 20°F and 5°F into the tube.

During preoperational tests on water, heat transfer coefficients were measured to provide a reference. In addition, during Phase 1, while investigating precipitate formation from $CaCl_2$ and $MgCl_2$ additions at the three proposed operating temperatures, more heat transfer measurements were made to insure that the tube was not fouled by the precipitate formed.

Most of the effort was concentrated in Phase 2, operation with silica but without calcium and magnesium present. Although there was considerable difficulty with the acid addition system during the earlier runs, the tests followed a pattern similar to the later runs.

Figures 5.2, 5.3 and 5.4 are plots of overall heat transfer coefficients versus silica formation time (acid pumping time). Figure 5.2 shows the results obtained at 20° ΔT with steam and inlet brine at 360°F and 310°F. It was quite noticeable that the heat transfer performance recovered during overnight operation on demineralized water or following a shutdown and restart using water, a process which required about 4 hours. On one occasion a permanent drop in performance occurred when a hot shutdown was required.

FIGURE 5.2: EFFECT OF SiO_2 SCALE ON HEAT TRANSFER-20°F ΔT

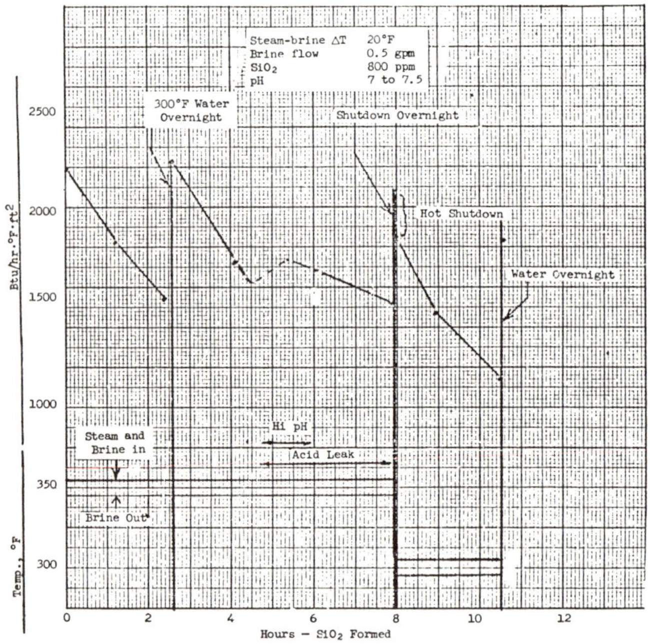

Source: ORNL-TM-4137

FIGURE 5.3: EFFECT OF SiO_2 SCALE ON HEAT TRANSFER-5°F ΔT

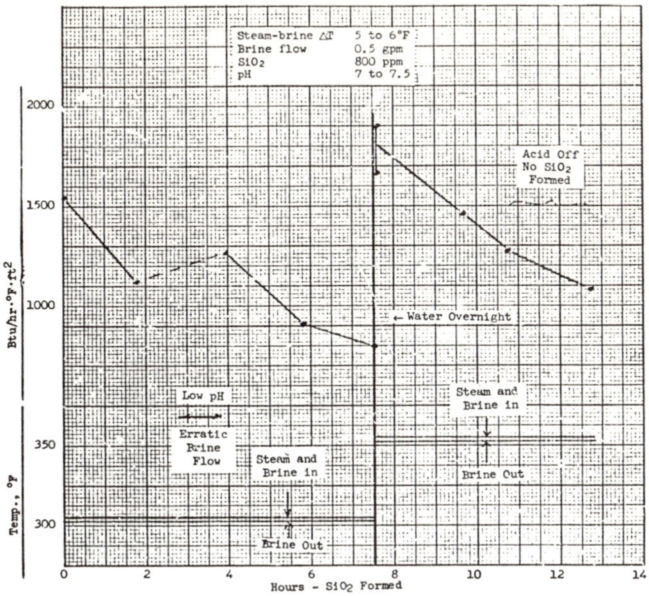

FIGURE 5.4: HEAT TRANSFER OF SIMULATED GEOTHERMAL BRINE

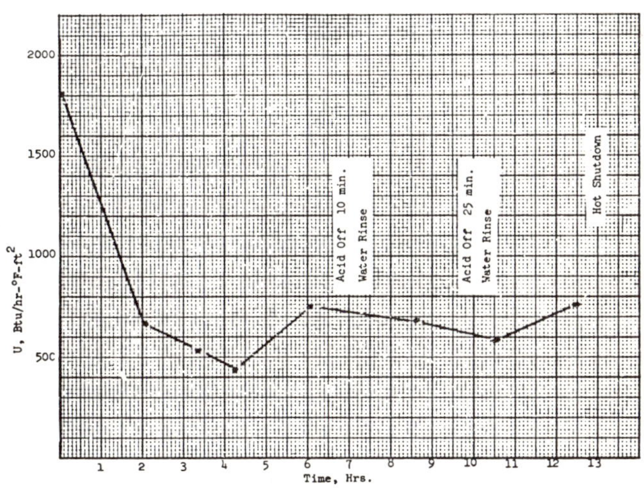

Steam In - 310°F
Brine In - 310°F
Brine Out - 290°F

Brine Flow - 250 lb/hr
No Ca or Mg in Brine

Source: ORNL-TM-4137

Runs were made at 5° ΔT to determine the effect of ΔT on scale formation. Results of these tests are shown in Figure 5.3. Although the rate of scale formation was lower than for 20° ΔT, it was still quite severe. During the 5° ΔT tests there was less precision and larger variation in heat balance data. This resulted from slower collection rates for the steam and vapor condensate in the measuring tanks which permitted the system temperatures to drift more than during tests at higher ΔTs. In addition, after scale had formed, the measured condensing rate was reduced to the same magnitude as the system heat losses, resulting in an additional loss in precision.

Most of the scaling tests involved periods of continuous operation of less than 8 hours. One long run was made to obtain a good scale sample for analysis and to determine if an equilibrium amount of scale deposition would be established. A 16 hour run was planned. However scale deposition caused sufficient flow restriction in the letdown cooler after 13 hours of the run to force an early shutdown. Figure 5.4 shows the heat transfer results obtained during this test which was made at 310°F brine inlet with a 20°F flashdown. The loop was shut down without rinsing, drained, and allowed to cool under an inert atmosphere.

The discharge end of the evaporator was opened to obtain a scale sample. A visual examination indicated that the scale was uniformly distributed on the inner surface of the tube and not significantly concentrated in the bottom of the grooves between flutes.

After removing a scale sample, the loop was reclosed and started up on water. After the 1½ hours required to heat to the 310° to 290°F operating condition, the overall coefficient had recovered to a value of 907 Btu/hr/ft²/°F compared to an equivalent coefficient of ~500 at the time of shutdown. During the next 5 hours the coefficient varied between 953 and 988. After overnight operation on water the coefficient rose to 1,217 Btu/hr/°F/ft² which is comparable to 1,300 to 1,360 Btu/hr/°F/ft² obtained during the reference water runs. The silica also apparently dissolved from the letdown system as no difficulty was observed with level control.

Phase 3 operation consisted of one short run made at 360°F inlet steam and brine with a 20° ΔT. Plans had been to make this run at 310°F but loss of good steam regulation to the brine heater after making the zero acid run forced the test to 350°F to obtain better steam regulation. The zero acid run yielded a coefficient at 2,100. After 1½ hours operation with acid addition containing $MgCl_2$ and $CaCl_2$, the heat transfer coefficient had fallen to 696.

During this run the acid pump lost capacity more rapidly than usual and needed flushing so instead of continuing operation on brine, the loop was switched to water and flushed for two hours and then shut down. During the two hours on water two sets of data yielded coefficients of 1,280 and 1,261 Btu/hr/ft²/°F (compared to 1,670 for a clean tube).

The loop was started up and operated for an additional 24 hours to test for removal of scale. Two more sets of data were taken at 360°F and 20° ΔT with coefficients of 1,356 and 1,298 Btu/hr/°F/ft².

The loop was shut down and cooled to obtain a sample of the remaining scale. When examined the inside of the tube was clean except for a dark purplish oxide film. However, there was evidence of leakage past the upper tube to tube sheet O-ring and a red, heavy scale on the outside of the tube which indicates that failure of the coefficient to recover completely may have been due to fouling of the outside surface or possibly a combination of the oxidation fouling of both inside and outside surfaces of the tube.

Samples

Only a few samples were taken during this series of tests. During the early part of Phase 2 (silica forming tests), samples were qualitatively analyzed and showed considerable iron present which was originally thought to be from generalized loop corrosion. After failure of the acid line-loop penetration, it was concluded that the dissolved iron had come from that location.

Brine samples containing suspended solids analyzed by Analytical Chemistry Division of ORNL confirmed the silica concentration in the feed and indicated that the precipitate was amorphous (no crystal pattern on x-ray analysis).

A filtered precipitate sample from the evaporator discharge bomb and a scale sample removed from the discharge end of the evaporator tube following the long (13 hour) run were analyzed by semiquantitative spectrographic analysis and by x-ray diffraction. The significant anions found by spectrographic analysis are shown in Table 5.1.

TABLE 5.1

Sample	Filtered Solids Weight Percent	Tube Scale Weight Percent
Fe	4	1
Cu	0.2	0.3
Ca	0.3	0.02
K	2	0.6
Na	10	5
Si	Most	Most
Ti	0.2	0.3

Source: ORNL-TM-4137

The x-ray pattern showed a halo where the strongest SiO_2 line would be found indicating that the SiO_2 was amorphous. The filter sample indicated the presence of NaCl and Fe_3O_4 while the only compound other than SiO_2 found in the scale sample was a trace of Cu_2O.

Pressure Drop

Pressure drop measurements were made for all runs. An improved computer program for calculating pressure drop was developed based on the model presented by Standiford which included terms for frictional losses, static head and acceleration losses (F.C. Standiford, *Relative Economics of Upflow Versus Downflow VTE Evaporators and Development Program for Upflow VTE Evaporator,* Report 517, April 12, 1972, pp 23 to 25). However, the present work covers a very narrow flow range (225 to 260 lb/hr). The discrepancies noted between measured values at similar operating conditions could be accounted for by an erratic flow rate by the fact that the pressure taps (in the fluted tube) were not truly static taps or that there was a drift of the d/P cell.

In most cases for water runs the measured pressure drop was about 140% of the calculated value, but for brine runs the measured value was only 44 to 80% of the calculated value. Since static head accounted for more than 90% of the calculated drop, a different type of flow apparently existed during brine operation which reduces the effect of static head. This may have been an increase in foam. Reynolds number calculations place both water and brine tests in the turbulent-turbulent two phase flow regime.

Results and Conclusions

Using a simulated brine composition which contained about 800 ppm of SiO_2, the following results were observed during operation of the high temperature VTE upflow evaporator.

(1) Scaling of the evaporator occurred within a few hours after startup. The heat transfer coefficient dropped to as low as one-third of the initial value in tests at exit brine temperatures of 355° to 290°F.

(2) Scale formation occurred at steam-exit brine vapor ΔT's from 5° to 20°F, with the rate of scale formation increasing with increasing ΔT.

(3) Further cooling of the discharge brine after being remixed with the vapor condensate resulted in scaling of the letdown cooling coil. A 13 hour run was terminated when level control was lost due to flow restriction through the letdown cooler.

(4) Scale formed with and without calcium and magnesium present.

(5) The silica which precipitated was amorphous and the scale appeared to reach an equilibrium thickness in a few hours.

(6) The scale redissolved when the loop was operated with demineralized water overnight.

Based upon these results, the conclusion is reached that silica scale will form on heated surfaces of geothermal brine evaporators for brines similar to this synthetic solution.

WAIRAKEI WELL ANALYSIS

The following analysis of water from well 20 of the Wairakei, New Zealand geothermal field is given in PB 219,376, 1973, *ENTHALP, A Computer Program for the Calculation of Aquifer Chemistry in Hot-Water Geothermal Systems* by A.H. Truesdell.

The aquifer fluid flows toward and into the well along a gradient of decreasing pressure resulting from the lower density of the two-phase steam-water mixture in the well. The decrease in pressure causes steam separation and a decrease in temperature. These processes continue up the well and into the separator, which was operated at a pressure of 16.3 bars absolute and a temperature of 202°C. With this decrease of pressure and temperature, the fluid is now 10.1% steam.

The gases originally dissolved in the water have largely exsolved into the gas phase (experimental distribution coefficients for CO_2 and H_2S greatly favor the gas phase, particularly near 200°C). At this point, the steam is separated and analyzed for CO_2 and H_2S (Table 5.2). The steam and water leave the separator in separate lines, and the water enters the silencer, which is actually a second separator operating at atmospheric pressure and (at Wairakei) 99°C.

Here an additional 20% of the original water is flashed to steam, and the remaining water flows out the weir box where it is sampled. The sample in a capped polyethylene bottle is cooled by conduction and taken to the laboratory where it is analyzed and the pH is measured. The CO_2 and H_2S dissolved in the water after separation are not analyzed. This quantity of gas is assumed negligible relative to that which partitions into the gas phase.

The physical and chemical properties of the water as analyzed are given in Table 5.2. The steam and water analyses were made by the staff of the Chemistry Division, DSIR, New Zealand.

TABLE 5.2: ANALYSIS OF WATER AND STEAM FROM WELL 20, WAIRAKEI, NEW ZEALAND

Temperature of steam separation	202°C
Temperature of water separation	202°and 99°C
CO_2 in high pressure steam	142 mmols/100 mols H_2O
H_2S in high pressure steam	4.5 mmols/100 mols H_2O
Water pH measured at 20°C	8.0

	Ppm*		Ppm*
Li	14.8	Br	5.4
Na	1,260	I	0.4
K	228	SO_4	35
Rb	3.0	B	26
Cs	2.5	HCO_3	7.7
Ca	25	CO_3	0
Mg	0	SiO_2	650
F	8.8	NH_3	0.3
Cl	2,164	H_2S	0

*Dissolved constituents in the water as analyzed at 20°C

Source: PB 219,376

PLOWSHARE GEOTHERMAL STEAM CHEMISTRY

A two month exploratory study is described in BNWL-1614, July 1971, *Plow-*

share Geothermal Steam Chemistry, Introductory Studies by L.A. Charlot, W.C. Morgan and I.M. Warner of Battelle, Pacific Northwest Laboratories. The material below is derived from this study.

The purpose of the study was determination of the kinds and concentrations of radioactive and nonradioactive contaminants likely to be brought to the surface by the export steam.

These objectives included: design and construction of a system to determine the steam and water distribution coefficients of rock constituents, development of analytical procedures for trace concentrations of solids in water and methods for sampling steam and water phases equilibrated with Piledriver rock samples. In the following sections the test apparatus, test material, mode of operation, and analytical methods are described.

Experimental

Test Apparatus: The system constructed to determine distribution of selected chemical species is shown in a schematic drawing in Figure 5.5. It consists of three basic sections, (1) the electrically heated 5 liter Inconel autoclave contained the water, rock samples and a demister for separating water droplets from the exported steam, (2) the constant temperature ⅜" OD Inconel steam riser maintained at a temperature of 3° to 5°C above the autoclave temperature by encasing it with a heating unit and asbestos covering, and (3) the ¼" OD air cooled Inconel steam condensate leg used for condensing steam and returning it in liquid form to the autoclave.

As shown in Figure 5.5, provisions were made for necessary monitoring operations including: sampling ports, pressure gauge, and thermocouples for measuring temperature at various points in the apparatus. Both water and steam sampling ports included a water cooled condenser and two regulating valves for a controlled sample withdrawal rate. A water pump was used to replenish the system after each sampling operation. For monitoring pressure, a gauge designed to operate at a maximum pressure of 5,000 psi was employed. To ensure safe performance of the system, a rupture disc, designed to operate at a maximum pressure of 3,500 psi was placed between the autoclave and the constant temperature steam riser. Eight thermocouples were used for monitoring the temperature of the system.

Thermocouple **1** was used for monitoring the temperature of the water phase, thermocouple **2** placed just above the water level monitored the temperature of the steam, thermocouples **3** and **4** monitored the temperature of the steam at the bottom and top of the steam riser section, thermocouple **5** monitored the steam condensate temperature at the end of the steam condensate leg, thermocouples **6** and **7** monitored the temperature of inlet and outlet air on the steam condenser and thermocouple **8** monitored the wall temperature of the steam riser. A demister device was placed in the autoclave above the boiling water surface to minimize water droplet entrainment in the steam riser section.

Test Material: About 3 kg of granitic rock were obtained from Lawrence Radiation Laboratory. The rock samples are small sections of drill core taken from the lower hemisphere portion of the Piledriver cavity. They represent rock from the cracked or crushed zone of the blast area and into the cavity zone proper.

FIGURE 5.5: DISTRIBUTION COEFFICIENT MEASUREMENT APPARATUS

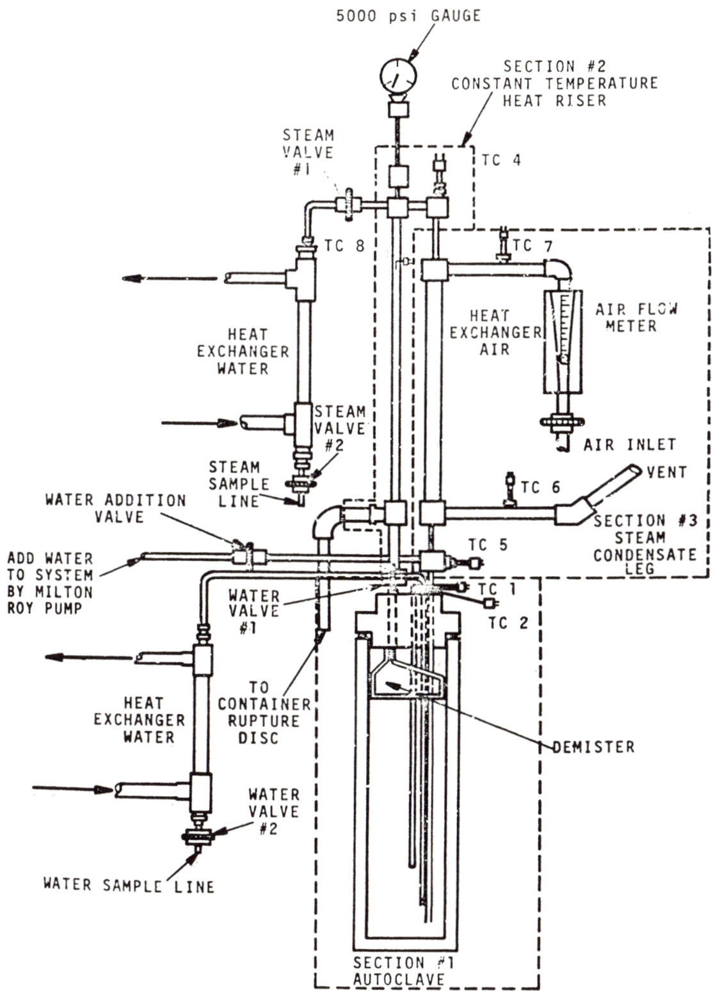

Source: BNWL-1614

Core sections of both melted (fused) and nonmelted rock were placed in the test vessel in the as-received condition. No attempt was made to wash or clean the rock. No attempt was made to increase the rock surface area by crushing or grinding.

Before testing a representative fragment of the dark, fused rock and a fragment of nonfused rock were counted for the major gamma emitting isotopes fixed on or within the rock. Both fragments showed the presence of the same major isotopes, but, as expected, the fused fragment showed orders of magnitude greater activity than did the nonfused fragment. The following gamma emitting isotopes were found: ^{60}Co, ^{106}Ru, ^{125}Sb, ^{134}Cs, ^{137}Cs, ^{144}Ce, ^{144}Pr, ^{155}Eu, ^{154}Eu, ^{152}Eu.

Operating Procedure: During a test run, the autoclave was filled with 1,800 ml of water, sealed and electrically heated to the desired operating temperature. The system was allowed to equilibrate at temperature for 18 hours before sampling. The constant temperature steam riser was turned on two hours before sampling and adjusted to maintain the temperature of thermocouple **8** at 2° to 5°C above that of thermocouple **1**. Air flow through the steam return leg was adjusted to assure essentially complete condensation but little or no supercooling of the condensed water. The condensation rate of steam in the steam condensate leg was calculated. Observed water condensation rates were 350, 350 and 250 grams H_2O/min at 650°F (343°C), 550°F (288°C) and 450°F (232°C).

The sampling procedure used was the following: The cold water flow to the sample cooler was turned on. Valve #**2** on the sample line was closed. Valve #**1** on the sample line was slowly opened and the polyethylene sample bottle was placed under the sample line and valve #**2** was adjusted for a flow of ~10 ml/min. Sample 1, 50 ml, was withdrawn and marked for discard. Samples 2 through 5, 50 ml each, were retained for analysis. On completion of sampling, valve #**2** and then valve #**1** were closed and valve #**2** was reopened to drain the sample line.

Water samples were always taken before steam samples. Steam and water sampling procedures were essentially the same, except that the water sampling line was disconnected at valve #**1** and the cooling line flushed with cleaning solutions to guard against loss of materials through deposition in the line. The flushing sequence for the disconnected sampling system consisted of 20 ml of 20% HCl, 20 ml demineralized water, 20 ml of 20% NH_4OH and 20 ml of demineralized water. Analysis of the flush solutions indicated only minor deposition took place during cooling of the water samples.

The water supply in the autoclave was replenished after both water and steam samples had been obtained. Replenishment was accomplished by pumping a volume of water equivalent to the sampling withdrawals through the water addition valve to the autoclave. When the desired amount of water had been added, system conditions were reestablished and the system allowed to equilibrate at the selected temperature before sampling again.

Methods of Analysis: Analytical methods were required for trace element analysis of water and steam equilibrated with radioactive (very low level) rock samples. Analytical methods appropriate to elements selected for analysis are listed, along with what are considered the lower detection limits. The basis of the element selection is the literature data on granitic rock composition and the anticipated fission products.

Factors which must be considered in selecting the analytical methods include: low concentration of elements of concern in the rock, radioactive decay of short lived and intermediate half life radioisotopes, e.g., $t_{1/2} \leqslant 6$ months, and finally interference of one element with analytical methods otherwise useful for other constituents.

The distribution coefficients of radionuclides are clearly of major importance in Plowshare-geothermal steam systems. Both direct gamma analysis of radioactive species and of augmented samples (50 ppm spike + neutron activation) were tried.

In the determination of nonradioactive species, modifications of established analytical methods, were employed. Atomic absorption, flame emission spectroscopy, emission spectroscopy and colorimetric methods of analysis were used. Listed in Table 5.3 are the elements determined, method of analysis and lower detection limits. The colorimetric methods used and their references are as follows:

Chrome azurol S	P. Pakalns, "Spectrophotometric Determination of Aluminum with Chrome Azurol S," *Analytica Chemica Acta,* Vol 32, pp 57-63, 1965
Eriochrome Cyanine R	J.C. Langford, *Flamephotometer and Colorimetric Procedures,* BNWL-CC-849, Pacific Northwest Laboratory, Richland, Washington, September, 1966
Azure C	L. Pasztor and J.D. Bode, "Thionin Derivatives in the Extraction and Direct Photometric Determination of Boron," *Analytica Chemica Acta,* Vol 24, pp 467-473, 1961
Ferrozine	L.L. Stookey, "Ferrozine–A New Spectrophotometric Reagent for Iron," *Analytical Chemistry,* Vol 42, Number 7, pp 779-781, 1970
Eriochrome Black T	A.E. Harvey, J.M. Komarmy and G.M. Wyatt, "Colorimetric Determination of Magnesium with Eriochrome Black T," *Analytical Chemistry,* Vol 25, Number 3, pp 498-500, 1963
Molybdenum Blue	O.A. Kenyon and H.A. Berwick, "Photometric Determination of Silica in Alkalies," *Analytical Chemistry,* Vol 25, Number 1, pp 145-148, 1953

Analytical results are reported as simple averages of results obtained from one or more analytical procedures. Determinations were made for all of the listed elements. In cases where water samples contained less than 100 ppb of the element, no steam analyses were made, except where gamma spectroscopy techniques were applicable.

TABLE 5.3: ANALYTICAL METHODS

Element	Method of Analysis	Detection Limit
Aluminum	Chrome azurol S (colorimetric)	3 ppb (10 cm cells)
	Atomic Absorption Spectroscopy	0.3 ppm
	Eriochrome Cyanine R (colorimetric)	10 ppb (1 cm cell)
Antimony	Gamma Spectroscopy	20 ppb to 10^{-6} ppb*

(continued)

TABLE 5.3: (continued)

Element	Method of Analysis	Detection Limit
Boron	Azure C (colorimetric)	20 ppb (1 cm cell)
	Atomic Absorption Spectroscopy	0.5 ppm
Calcium	Flame Spectroscopy	10 ppb
	Atomic Absorption Spectroscopy	10 ppb
Cerium	Gamma Spectroscopy	20 ppb to 10^{-6} ppb*
Cesium	Gamma Spectroscopy	20 ppb to 10^{-6} ppb*
Cobalt	Atomic Absorption Spectroscopy	10 ppb
	Activation Analyses	20 ppb to 10^{-6} ppb*
Copper	Flame Spectroscopy	25 ppb
	Atomic Absorption Spectroscopy	10 ppb
Iron	Flame Spectroscopy	200 ppb
	Atomic Absorption Spectroscopy	25 ppb
	Ferrozine (colorimetric)	2 ppb (10 cm cell)
Magnesium	Flame Spectroscopy	100 ppb
	Atomic Absorption Spectroscopy	1 ppb
	Eriochrome Black T (colorimetric)	10 ppb (1 cm cell)
Manganese	Flame Spectroscopy	20 ppb
	Atomic Absorption Spectroscopy	1 ppb
Potassium	Flame Spectroscopy	1 ppb
Ruthenium	Gamma Spectroscopy	20 ppb to 10^{-6} ppb*
Silica	Molybdenum Blue (colorimetric)	3 ppb (10 cm cell)
	Atomic Absorption Spectroscopy	0.5 ppm
Sodium	Flame Spectroscopy	0.2 ppb
Strontium	Atomic Absorption Spectroscopy	1 ppb
	Flame Spectroscopy	20 ppb

*Sensitivity of gamma spectroscopy determination varies widely, depending on interferences from other radionuclides present, total gamma signal of sample, activation time, counting time, half life, etc.

Source: BNWL-1614

Results and Discussion

Water and Steam Leached Contaminants: Practically all of the previous investigations of the distribution coefficients have been conducted on simple systems, that is a single compound dissolved in pure water. An apparent distribution coefficient, K'_d has been obtained from the total concentrations of an element in the steam-condensate (C_s) and water (C_w) phases that is roughly dependent on the ratio C_s/C_w. A discussion of the determination of K'_d is found in BNWL-1614 pages 13 to 14 and 25 to 28.

The chemistry of a mixed oxide-hydroxide system one obtains under these conditions is quite complex, the ionic state of any element in solution being dependent on the ionic states and solubilities of all other compounds in the solution.

Based on this hypothesis, and bearing in mind that SiO_2 is relatively soluble and the major constituent of granite, it seems reasonable to assume the the K'_d of silica would be least affected by changes in the ionic state or solubilities of all other constituents, as is observed from the results.

Figures 5.6 and 5.7 show the results as a function of saturated steam temperature. Preliminary investigations of the leaching of the rock samples at 650°F (343°C) indicated that the radionucludes of prime importance, such as Eu, Ru, Ce, and Co, were not leached from the samples in sufficient quantity to be directly detectable by gamma spectroscopy. Moreover, the radioactive species which did leach, such as Sb and Cs, were present in the water in such minute quantities that they were barely above detection level in the steam samples.

FIGURE 5.6: APPARENT DISTRIBUTION COEFFICIENTS FOR CHLORIDE ION AND ALKALI METALS

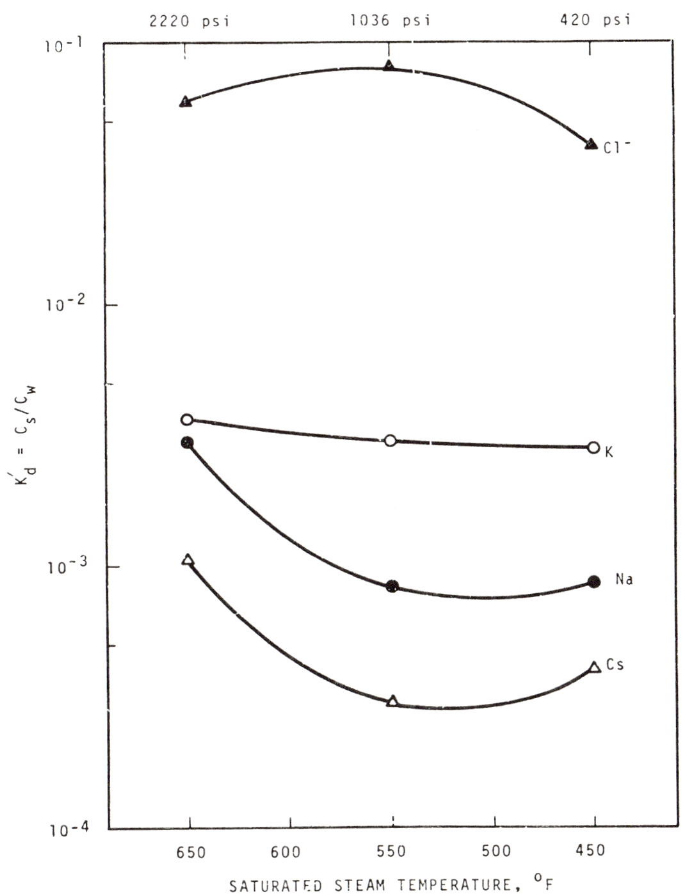

Source: BNWL-1614

FIGURE 5.7: APPARENT DISTRIBUTION COEFFICIENTS FOR Al, B, Sb AND SiO_2

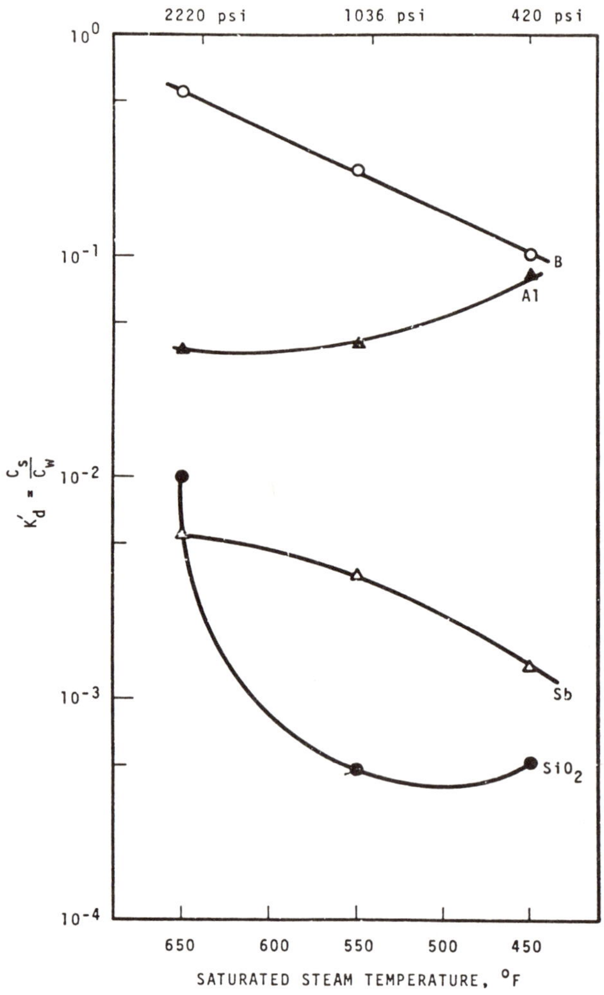

Source: BNWL-1614

It was not clear whether the low concentration of these radionuclides were due to (1) radioactive decay due to sample age (~6 years), (2) a slow leaching process, or (3) the insolubility of compounds containing the radionuclides. If the low concentrations of the radioactive species were due to processes (1) or (2), isotopic dilution followed by neutron activation would make possible the determination of distribution coefficients for the desired radioactive species. Therefore, 50 ppm of the elements Cs, Sb, Ru, Co, and Ce were added in the form of

TABLE 5.4: LIQUID PHASE CONCENTRATION AND APPARENT STEAM-WATER DISTRIBUTION COEFFICIENTS

	650°F (343°C)		550°F (288°C)		450°F (232°C)	
	Concentration (L)	Distrib. Coeff.	Concentration (L)	Distrib. Coeff.	Concentration (L)	Distrib. Coeff.
Silicon (SiO_2)	1000 ppm	10^{-2}	800	5×10^{-4}	600	5×10^{-4}
Sodium	20	3×10^{-3}	50	8×10^{-4}	60	9×10^{-4}
Potassium	10	4×10^{-3}	20	2×10^{-3}	30	2×10^{-3}
Chloride	12	6×10^{-2}	14	8×10^{-2}	20	4×10^{-2}
Aluminum	5	4×10^{-2}	6	4×10^{-2}	8	8×10^{-2}
Antimony*	2	6×10^{-3}	1	4×10^{-3}	1	10^{-3}
Boron	1	6×10^{-1}	0.7	3×10^{-1}	0.8	2×10^{-1}
Cesium*	0.08	10^{-3}	0.1	3×10^{-4}	0.1	4×10^{-4}
Iron	0.1	-	0.05	-	0.04	-
Cobalt*	0.05	-	0.03	-	0.03	-
Manganese	0.02	-	0.01	-	0.01	-
Calcium	<0.01	-	<0.01	-	<0.01	-
Copper	<0.005	-	<0.005	-	<0.005	-
Strontium	<0.005	-	<0.005	-	<0.005	-
Magnesium	<0.002	-	<0.002	-	<0.002	-
Cerium*	<0.02	-	<0.02	-	<0.02	-
Ruthenium*	<0.02	-	<0.02	-	<0.02	-
Europium*	<0.02	-	<0.02	-	<0.02	-

* Experiment spiked with 50 ppm of specified element as oxide or nitrate

Source: BNWL-1614

oxide or nitrate compounds. In a separate series of runs, 50 ppm Eu was added as the oxide. Separation of the Eu from the other isotopes was required to avoid anticipated counting interferences arising from the complex Eu gamma spectra.

Results obtained from the activation analysis indicated that essentially the same species dominated the solution, i.e., small amounts of Sb, ~2 ppm maximum and Cs, 0.1 ppm maximum. Small amounts of Co were also found in the water and steam phase samples, however, Co was at a concentration level which might be relatable to Inconel or stainless steel corrosion product. Other elements added, Ru, Ce, and Eu were not present in detectable quantities. It was concluded, therefore, that the low concentrations of these elements in steam and water is attributable to the low solubility of the chemical compounds in which they are present.

Calculated distribution coefficients are given in Table 5.4 and Figures 5.6 and 5.7. Distribution coefficients in a simple system are expected to remain concentration and temperature independent near the critical temperature. Due to the complex chemical nature of the system, results do not follow such a simple relationship. Some species analyzed exhibited an inverse water solubility. This was evident in the case of sodium.

The result is not very surprising since it is known that sodium silicate does exhibit an inverse solubility. It is not expected that the results of Figures 5.6 and 5.7 can be generally applied to rocks of differing chemistry. Furthermore, it is not expected that all possible troublesome species have been defined. However, it is expected that if rock samples of similar composition were allowed to equilibrate at the temperatures investigated the constituents of major interest will follow a relationship similar to that shown in Figures 5.6 and 5.7.

Autoclave Deposit Analysis: Upon conclusion of the rock leaching studies, the autoclave walls were observed to be coated with a hard, flaky material extending above the water level. This material was resistant to concentrated hydrochloric acid, nitric acid, aqua regia, and 50% caustic solution even when heated to ~90°C. The only effective method of dissolving the species was a 3 to 5 hour exposure to 48% hydrofluoric acid solution. To identify this autoclave deposit, chemical and radiochemical analyses were performed. Results are indicated in Table 5.5 and Table 5.6.

TABLE 5.5: RADIOCHEMICAL COMPARISON OF AUTOCLAVE DEPOSITS WITH ROCK AND WATER SAMPLES

	Ratio	
Substance	Cs^{137}/Cs^{134}	Sb^{125}/Cs^{134}
Water Sample	2.7	3.6
Rock Sample	1.4	0.53
Autoclave Flakes	2.6	3.0

Source: BNWL-1614

TABLE 5.6: FLAKE COMPOSITION COMPARED TO OTHER SAMPLES EMISSION SPECTRA

Element	650°F (343°C) H_2O Sample	650°F (343°C) Steam Sample	Rock	Flakes
Ag	0.0004 ppm	-	-	0.001%
Al	4.8	0.09 ppm	7.4%	2.0
B	1.3	0.7	0.85	1.1
Ca	0.2	<0.02	5	0.5
Cr	0.02	0.002	-	0.02
Cu	0.008	0.0008	0.0005	0.01
Fe	0.1	0.08	2	0.2
Mg	0.008	0.004	0.02	0.005
Na	20	0.1	1.3	0.42
Sb	1.7	0.009	-	0.2
Si (as SiO_2)	1000	15	65	90
V	0.08	-	0.02	0.05
Ru	<0.02	<0.02	?	Trace

Source: BNWL-1614

The results in Table 5.5 suggest that the formation process for these flakes was primarily one of concentration by evaporation and possibly spattering just above the water level. This is evident from the fact that the radioactive species present had approximately the same elemental ratio as that in the water sample.

Conclusions

During this study a system was constructed and used to investigate the leaching and steam transport of rock constituents over a temperature range of interest for geothermal steam power plant operation. Suitable techniques have been selected for the analysis of radioactive and nonradioactive materials leached from nuclear blast affected rock and transported in the steam phase. Steam-water distribution ratios of these compounds have been measured at temperatures of 650°F (343°C), 550°F (288°C), and 450°F (232°C).

The results indicate that the chemistry of the rock-water system is quite complex. It appears that both the total solubility of an element and the ratios of the ionic species containing that element may be influenced by solubilities and ionic states of other rock constituents as well as by pH and oxygen ion concentrations. These complex interactions result in apparent steam-water distribution coefficients which, in general, do not follow the linear relationship observed for single solute system. Several radioactive products, such as the isotopes of Eu, Ru, and Ce, were found to have such low solubilities, under the conditions of this investigation, that they should present no problems for future geothermal steam plants. For other isotopes, such as ^{137}Cs and ^{125}Sb solubilities are sufficiently high that their transport in geothermal steam could be a problem.

Silica is the principal dissolved constituent of the rock, representing 80% or more of the water-soluble material. Of the species studied, silica solubility characteristics are, predictably, most like those observed in simple, one solute systems.

A STUDY OF GEOTHERMAL RESOURCES OF CALIFORNIA

PB 212,753, May 1972, *Geothermal Resources in California—Potentials and Problems* from which the following material is derived attempts to describe the nature of the geothermal resource, its extent, and the technology by which it can be exploited. Initial research and writing of this report to the Assembly General Research Committee of the California Legislature was done by M. Goldsmith and E. Rolph. The findings and conclusions were formulated by the Panel on Energy Planning and Programs.

NATURE OF THE RESOURCE

Improved understanding of the basic forces at work within the earth is giving greater insight into the sources of geothermal energy. From earliest recorded times men have noticed at various places on the surface of the earth manifestations of heat coming from below. These manifestations have included geysers, flows of hot water from springs, or steam issuing from fissures in the earth. At some of these locations men have drilled wells in an effort to tap the resource for useful purposes. In a now increasing number of cases these efforts have met with success and useful quantities of hot steam or hot water are being obtained from the wells.

What is the source of the geothermal heat? It is known that, in general, as one penetrates more deeply into the earth, the temperature of the earth's materials increase. These temperatures are believed to reach very high values in the central part of the earth's interior. The source of this heat is presently thought to be the decay of radioactive elements over long periods of time, as well as frictional (tidal) forces.

Studies have been made to see if a working fluid, such as water, could be heated by passing it down a hole drilled into warm regions of the earth, there to be heated and returned to the surface. The heat would then be used, e.g., for the production of electricity. The earth's crust is made up of gigantic plates, of continental size, which move in relation to one another and have fissures between them. The hotter material from the interior of the earth rises in circulation patterns along the joints between these plates. Along uplifted areas, the earth's crust is thinner and it is fractured. Thus in these regions, the heat flow from the interior to the surface is substantially higher than average. This means that basement rocks in those areas are apt to be hotter than would be expected for materials at that depth in other parts of the world.

One of these great rift areas is located in the eastern part of the Pacific Ocean and passes up the Gulf of California. In fact, the separation of Baja California from mainland Mexico is attributed to this rift separation between the continental plates. The continuation of this ocean rise is marked, in part at least, by the San Andreas fault zone. Many thermal anomolies are known to exist along this zone.

If water should flow underground through this heated zone, it will in turn become heated. In some areas, there is connection between the surface and these underground reservoirs and we note surface manifestations (e.g., geysers, hot springs) of the heated body below. It is in these areas that men have explored for geothermal resources. Until fairly recently it was commonly felt that these manifestations were to be found only along fracture zones and the bodies of water beneath the surface were localized in nature.

However, better understanding of the geologic conditions, and the recognition that some of the geothermal fields extended over areas much broader than localized fracture zones, has caused geologists to revise their theories about the nature of the geothermal reservoirs. Without entering into the details of the scientific facts or arguments, it suffices to say that it is now believed that in many cases geothermal reservoirs may extend over considerable areas and have substantial volume.

In the early days of oil exploration, men usually drilled where surface manifestations were present, but elaborate and detailed theories and practices enable men to find oil even under the bed of the sea. In the same way, geothermal resource exploration is in its early stages; thus attention is concentrated upon areas having surface manifestations. However, with the knowledge that has been gained in modern geology, progress is being made rapidly in devising methods to reveal the presence of geothermal resources where the surface is innocent of its appearance.

When wells are drilled to tap the geothermal resource, one of several conditions is found. In some cases, the product is dry steam unaccompanied by liquid water. This is the case at one of the earliest of the geothermal steam fields, that of Lardarello in Italy. It is also the case of The Geysers steam field in northern California, which is currently being used to produce electricity by the Pacific Gas and Electric Company. In this case, the subterranean waters are flashed into steam by the hot rocks beneath the surface of the ground, forming a large steam reservoir, often at elevated pressure. This steam will then pass up the well and may be put to useful purpose at the surface.

In other cases, the geothermal wells encounter a reservoir of extremely hot water.

Because of the depth and configuration of these reservoirs, the waters are usually found at high pressure. But the well, now open to the surface of the earth, can be used to relieve the pressure on part of the water reservoir. When the pressure is reduced on the very hot water, part of it is evaporated into steam. The steam, in evaporating and expanding, will try to rush up the well pipe and in so doing will entrain a substantial amount of the water. In this case, a flow of mixed steam and water will emerge from the well head, still at a very high temperature although somewhat cooled because of the boiling and expansion process that occurs during the flow through the well.

The action in appearance is very similar to that seen in a coffee percolator, where the bubbles of steam drive the flow of water before them. At the wellhead, the steam and water may be separated by a simple centrifugal device, sometimes called a cyclone separator. The mixture passes into the separator, which produces water from one pipe, and steam from another. Because the steam and the water are in thermodynamic equilibrium, i.e., both are at the same temperature and pressure, this steam is called saturated. This class of geothermal resource has been found in New Zealand, in Mexico, in the Imperial Valley and other places.

The nature of the water in the subterranean reservoir differs however. In New Zealand the waters have a very low mineral content. There the steam is used to produce electricity, and the water is simply allowed to flow into the sea. New Zealand is an area of abundant rainfall and the geothermal waters are of small value. The geothermal waters found in the Imperial Valley near the Salton Sea, for example, (at the area called Buttes) are of the opposite extreme. Their dissolved mineral content is extremely high, up to 30% by weight.

This should be compared to sea water whose concentration of dissolved minerals is about 3.3%. Because of the high mineral content, the steam that is obtained from these wells is corrosive and has proven very difficult to use in electric generating equipment, in small-scale experiments. However, the prime purpose of the Buttes development was to obtain chemicals from the highly saline brines. This remains one of the possible objectives of geothermal resources development, that of obtaining chemicals from the geothermal waters as a prime product or as a by-product from other processes.

Still a different type of water is found at the Cerro Prieto region in northern Baja California. There the waters in the underground reservoir are neither fresh nor highly concentrated in salts. The geothermal water at Cerro Prieto contains approximately 2½% dissolved minerals. This is far too saline for use as potable water or for irrigation purposes, but it is reasonable enough in its properties that the steam obtained from the wells is capable of being used in electric generating equipment. At this time the Mexican government is engaged in the construction of a 75 Mw electric power plant to use the steam obtained from the Cerro Prieto field.

In the general case, the water that also flows from the separator at the wellhead is hot, as previously noted. While the steam might be used to generate electricity or for other purposes, the water is not without value. In some cases, of course, it would simply be wasted to the ocean or reinjected into the ground for disposal. As noted, it might also be exploited for its mineral content. Serious consideration has been given to running such water through a desalting plant in order to rid it of its dissolved minerals, thus producing fresh water suitable for agricultural, municipal, or industrial purposes. This is particularly true in the Imperial Valley and other naturally arid areas of the Western United States. It has not yet been found economically desirable to produce fresh water via the desalting process from sea water in California. But because the geothermal waters are already heated, it may be possible to desalt them more cheaply than sea water and thus they might form a valuable resource.

One thing is certain however. The separated water, if not used, considering its chemical content, cannot be permitted to enter surface streams of shallow aquifers where it might mix with other water supplies. Either a conveyance system must be provided to transport it to the ocean (if this is found permissible) or it must be injected deep into the ground, probably near the level of the producing zone.

In summary then, the nature of the geothermal resource manifests itself in two forms: one, dry steam, and the other very hot water in the ground. In the latter case, the water could be pumped from the well without permitting the pressure to drop, thereby preventing formation of steam. On the other hand, the pressure can be reduced permitting the steam to flash from the water and the mixture to flow from the well where it may then be separated. The case of the underground water reservoir may be further subdivided into fresh, highly mineralized, and intermediately brackish types. This division reflects the utility of the resource and the technical problems to be expected in its use.

EXTENT OF THE RESOURCE

Several reports have recently been issued that delineate the known geothermal resource areas of the State of California. The data have been accumulated by the State and Federal agencies and are published in various forms. The known geothermal resource areas delineated by the United States Geological Survey have been published in the Federal Register and were reprinted by the State of California in a special issue of *Geothermal Hotline* published in July, 1971.

The total California acreage so delineated amounts to over one million acres, but it should not be assumed that all of these acres embody proven resources. In the report, *Economic Potential of Geothermal Resources in California,* issued by the Geothermal Resources Board of the State in January 1971, a listing of known geothermal areas in California was included.

The State document divided the areas into categories such as those producing power, significant discoveries, potentially significant, and other areas. Other areas include, for example, the various hot springs locations in the State. Potentially significant areas include those where it is strongly suspected that significant geothermal resources exist but where detailed exploration has not taken place. Without extensive exploration it is not possible to make any estimates of the potential values of these areas. Three areas within the State have received most recent attention and will be discussed below. The first of these is The Geysers area which presently is being exploited for electric power. The second is the Imperial County area and the third is the Mono Lake-Casa Diabo region.

The Geysers

At The Geysers, a group of steam-producing companies, headed by the Union Oil Company, drill for and produce steam which is sold to the Pacific Gas & Electric Company. The utility uses the steam to produce electricity and reimburses the steam producers on the basis of kwh of electricity produced. The first generating unit, Geysers No. 1, went on line September 1960. It is able to produce 12 Mw on a continuous basis. This original experimental commercial unit was successful, and through the 1960's PG&E increased its producing capacity until in 1968 a capacity of 83 Mw was achieved. Four separate generating units were used. In 1971, 5 and 6 went on line with 55 Mw each. PG&E has announced its intention to increase capacity at The Geysers by 110 Mw each year until 600 Mw is reached.

An additional 8 geothermal generating units of 53 Mw capacity each are listed as coming into service through 1979 in *Environmental Report—Unit 1 and 2, Mendocino Power Plant,* PG&E, September 1971. It is assumed that they would be in the general vicinity of the Geysers field. Thus, in 1979 PG&E intends that 1,000 Mw, approximately the equivalent of a single modern nuclear power plant unit, would be installed in its system. Statements of PG&E and oil company personnel indicate that the capacity of The Geysers field is at least 1,000 Mw.

It is reported that the area of the known steam reservoir is greater than 10 square miles. Further exploration may indicate that even this estimate is conservative. At the present time the steam wells are spaced at approximately one per forty acres (recent wells average about 8 Mw electrical production each) and some estimates have been made that a spacing of one well for twenty acres may be possible. Production at The Geysers consists of dry, slightly superheated steam, without flow of water from the wells. The steam produced is collected and passed to the steam generating stations where it is condensed after passing through the turbine. The resulting water is evaporated in the cooling towers and only a small residual fraction is reinjected in the earth. Still in question is the total amount of time that production may be maintained from The Geysers field.

Depletion is estimated at 5% per year wellhead pressure drop. Depletion is presumed to occur because the inflow of water to the steam field is less than the withdrawal. For fiscal purposes, the PG&E Company is amortizing their investment over a period of 35 years.

Imperial Valley

The second area in California that has received intensive attention is the Imperial Valley. Near the Southeast shore of the Salton Sea, surface manifestations such as fumeroles and mudpots had been noted. In 1927, a well was drilled in that area but it was found that the steam encountered was insufficient to produce power on a commercial basis. However, carbon dioxide gases from the field were used for dry ice production for many years, until flooded out by the increasing level of the Salton Sea. The Imperial Valley has also been explored for oil and in 1957, an exploratory oil well was drilled nearby and encountered hot brine at depths near 5,000 feet. While no oil or gas was discovered, interest was reborn in the expoitation of geothermal resources. Subsequently, additional wells were drilled in the general area known as the Buttes field; a number of these wells were intended for production of minerals from the highly saline brine.

In this area the wells encounter very hot water at depth, which then flashes into a mixture of steam and water as it flows up the well pipe. The percentage of minerals in the water amounts to as much as 25 to 30% by weight. A detailed account of the drilling problems involved in this area is in "Drilling Production and Disposal Technology in the Salton Sea Geothermal Area, Imperial County, Calif." by Carel Otte of the Union Oil Company, in Compendium of Papers, Geothermal Hearing, October 1970, The Resources Agency, California, and a summary of the mineral production situation is given in "Testimony Before Geothermal Resources Board", by H.W. Diamond, Compendium of Papers, Geothermal Hearing, October 1970, The Resources Agency, California, by the Morton Salt Company.

In spite of the fact that many millions of dollars were invested in well drilling and production equipment, changes in the price structure of chemicals such as potash prevented the successful economic exploitation of the mineral resources. Experiments at producing electric power using steam from these wells were unsuccessful due to the highly corrosive nature of the mineralized brine. Because of this history, further exploration for geothermal resources in the Imperial Valley was inhibited for a number of years. While large quantities of water were obviously present, and heated to high temperatures, the mineral content of the water seemed an insurmountable obstacle to successful commercial exploitation.

Mexican research and development in the Cerro Prieto steam field have served to stimulate further intensive exploration of the geothermal potential of the Imperial Valley area lying away from the Salton Sea, between that body and the Mexican border. Much of this investigation has been carried out by Dr. Robert Rex and his group at the University of California, Riverside.

In "Phase I—The Imperial Valley Project", by R.W. Rex, Institute of Geophysics and Planetary Physics, University of California, Riverside, October 1968, Rex points out that the Buttes area has two separate types of geothermal brines beneath it, the deeper being the very hot, hypersaline brine, while a cooler, less saline brine lies above it. Review of geological and well drilling data indicates that the contact line between these brines dips to the South.

Thus, at what would be useful producing depths of perhaps 5,000 feet in the southerly reaches of the Imperial Valley, Rex expects that the less saline brines, similar to those found in the Cerro Prieto field, would be found. The depth of basement in the Imperial Valley is very great, perhaps 20,000 feet, and the basin is considered to be filled with porous, sedimentary formations.

In *Investigation of Geothermal Resources in the Imperial Valley and Their Potential Value for Desalination of Water and Electricity Production,* by Robert W. Rex, University of California, Riverside, June 1, 1970, Rex has estimated the total volume of water contained in the Imperial Valley basin and his estimates indicated that over a billion acre-feet of water are contained in the reservoirs using what he feels to be conservative assumptions. Less conservative assumptions lead to estimates ranging up to nearly 5 billion acre feet of water. The origin of this water is considered to be meteoric (originating from rainfall), coming from the Colorado River system.

However, simply having a body of water in the ground is not sufficient to provide an exploitable geothermal resource. The water body must be heated, at least in some places. The UCR and the Bureau of Reclamation exploration programs in the area have been directed at the identification of thermal anomalies, that is, areas where the heat flow from the interior of the earth is substantially higher than normal. Modern geophysical exploration methods have been used to conduct these investigations.

Gravity measurements, resistivity measurements, as well as shallow drill holes, have been used to produce plots of thermal anomalies. The current status of these investigations have been summarized in *Geothermal Resource Investigations, Imperial Valley, California, Status Report,* April 1971, U.S. Bureau of Reclamation. Geophysical surveys have been made of the Imperial Valley area and the heat flows, in the form of thermal gradients, have been measured or estimated.

The methodology that has been used by Rex to determine the total potential of the geothermal resource is straightforward. He has assumed that any area having a temperature gradient of greater than 6° to 8°F per hundred foot depth represents an area of useful economic potential. This value is chosen on the consideration that for an economic operation, water approximately 600°F or greater should be found at depths of less than 8,000 feet, which represents a limit observed at Buttes and Cerro Prieto for commercial production.

The measurements indicate that perhaps 100,000 acres of land lie within these high heat flow anomaly areas. Rex's calculations have assumed that 30 acres per producing well are required. This is contrasted to the Mexican practice where substantially closer well spacing has been employed in the well clusters. If it is assumed that each well is capable of producing 10 Mw of electricity, one is quickly led to the conclusion that approximately 30,000 Mw of electrical potential exists in the Imperial Valley. The 10 Mw value corresponds to approximately 200,000 lb/hr of well flow steam and is in excess of that experienced in the Mexican wells where somewhat smaller well diameters are employed as compared to those that would be anticipated for U.S. use. However, formation permeability may be limiting, and these estimates may be optimistic.

Mono Lake

Geothermal manifestations have also been noted in the Mono Lake-Long Valley-Casa Diablo area of the Eastern Sierra Nevada regions of California. Shallow wells have been drilled in the Casa Diablo area and hot water was discovered. No commercial exploitation has been carried out. At least one deep well has been drilled in the Mono Lake area by the Southern California Edison Company and its cooperating associates. The well drilled in September of 1971 was found to be both dry and cold. Additional holes will be drilled in the immediate future. The potential of this area cannot be assessed at this time.

TECHNOLOGY OF EXPLOITATION

The means by which the geothermal resources are utilized will depend first on the nature of the resource, that is, whether the fluids obtained from the ground are dry steam or a mixture of steam and water. Likewise, the relative values placed on electric power or distilled water will determine which product is to be emphasized and what methods will be used to produce them. Because all of these processes are thermodynamic in nature, it is required that some form of cooling be utilized. Again, the means of cooling will depend on products desired and on the availability of cooling media, such as local water. This section describes the equipment that is necessary and available for use in geothermal processes.

Dry Steam

Dry steam could be utilized to produce either electric power or to act as the heat source of a desalting plant. The desalting plant would be fed by local waters or sea water. Known dry steam resources in California, however, occur in regions where water is not in short supply. Therefore, it is not considered worthwhile to use geothermal dry steam for the production of water, although the technology is straightforward.

In this case, ordinary desalting plants, as constructed for use with the available feed waters, would simpy use the geothermal steam rather than steam generated from another source, such as an oil-fired boiler or low pressure exhaust steam from an electric power plant.

The means of using dry geothermal steam to generate electric power is likewise straightforward. The steam is usually available at the wellhead at low pressure as compared to that usually used in fossil or nuclear-powered electric plants. The steam itself seems to require no special treatment except filtering to be sure that no abrasive particles are entrained in the steam and pass through the turbine. This is usually accomplished with rather simple particle collectors.

Because of the low pressure of the feed steam, the turbines themselves differ in design from standard modern power plant practice. For a given power output, the throttling valves, for example, are much larger in size. Likewise, high pressure stages in the turbine are absent. For a given power output, the turbine using geothermal steam is substantially larger than one designed to use high pressure steam; however, the general appearance of the geothermal turbine is that of the low pressure sections of conventional turbines. The largest parts of the turbines are the lowest pressure stages and the exhaust steam piping. Because geothermal steam is low pressure, a great deal more steam is required to produce electricity as compared to a conventional plant.

Therefore, for a given output, the exit sections of the turbine may be substantially larger than for conventional practice. This is also influenced by the design back-pressure of the turbine. We should not then expect that the maximum output of geothermal turbines will approach the maximum production of turbines used in conventional plants. While the largest steam turbines presently in use today approximate 1,000 Mw in size, the largest geothermal turbines presently in use are 55 Mw in size. The steam is expanded through the turbine thereby turning a conventional electric generator. Several means of disposal of the steam passing out of the turbine can be envisioned. The simplest would be simply to discharge into the atmosphere at atmospheric pressure. However, in conventional fossil and nuclear power plants the steam is invariably discharged to a condenser at a pressure very much lower than atmospheric. While condensers are expensive, power plant cost-effectiveness analysis always indicates that it is better to use the condenser, thus extracting greater energy from the turbine, rather than to waste the steam to atmosphere.

The same is true for analyses performed on geothermal plants. Therefore the steam will be discharged to a condenser, of which there are several types. A condenser is nothing more than a heat exchanger which condenses the steam into water, which may then be pumped to atmospheric pressure and disposed of, or in conventional plants, returned to the boiler. In the process of condensing the steam, a great deal of heat is released. This heat must be transferred to the atmosphere or some cooling medium. In many cases ocean water, lake water, or water from rivers is passed through the condenser and used to remove the heat. In other cases, a recirculating supply of cooling water passes through the condenser and then to a cooling tower where the heat is rejected to the atmosphere.

Aspects of cooling that are pertinent to the geothermal power and water cycle will be discussed in a later part of this section. Thus, it can be seen that the production of electric power from dry geothermal steam is a straighforward engineering problem and has been met successfully, for example, at The Geysers in Northern California as well as other locations in the world.

Power Production from Geothermal Water

If the geothermal resource consists of water at high pressure and temperature deep in the ground, relief of that pressure through the well pipe will cause some of the water to flash into steam and start to drive up the pipe in the manner of a percolator. The steam and entrained water will therefore flow from the wellhead as a mixture and can then be separated into the steam and water components. This separation process will be caused to occur at some elevated pressure.

The steam so produced may then be transported through pipes to a power plant using conventional steam turbine equipment. Here again, the steam pressure is low as compared to conventional fossil and nuclear steam power equipment. This geothermal steam is then expanded through the turbine and into a condenser as previously discussed.

The water that was separated from the steam may be flashed again at a lower pressure producing some additional steam at a pressure intermediate between the first separation and the pressure found in the condenser. This additional steam may be introduced into a low-pressure section of the power turbine, thus increasing the total power output. The residual geothermal water may then be used for desalting purposes, process heating, chemical production or disposed of otherwise.

Desalination of Geothermal Water

The product of geothermal wells can be used to produce pure water. (If, of course, the geothermal waters are themselves pure, no further treatment is required. It is not expected that this will usually be the case in California.) For discussion, let us assume that the water that has come from the geothermal well, mixed with steam, has been separated into its steam and water components.

In the case most often considered, the steam is drawn off to operate a power plant. The water remaining is then introduced into a desalting plant. Desalting is accomplished in one of several ways, among which are membrane processes, distillation processes, and others. Processes such as reverse osmosis do not employ heat as a driving mechanism, and thus the heat that is available in our geothermal brines is of no advantage. Thus, it is usually considered that one of the distillation processes would be used for the desalting of geothermal water.

The distillation processes involve the evaporation of part of the feed water and the subsequent condensation of the vapors into pure distilled water. The remainder of the feed stock is either further evaporated and distilled or returned to waste. This is the process commonly associated with the desalting of sea water.

The geothermal situation is unique in that the feed waters are already heated. One of the major operating expenses of a conventional desalting plant is supplying the heat to the water. Because this heat is expensive, conventional plants are carefully optimized to maximize the use of the heat. The water is heated and then as it evaporates and is cooled, it passes to subsequent evaporation sections. The coolant used for the condensing of vapors is actually feed water at a lower temperature. Thus the reject heat of one distillation section is used for the heating of another.

In the geothermal case, some of this regenerative effect will not be required and thus will lead to simpler plant designs. In conventional desalting practice careful trade-offs are made between the cost of energy used and the capital costs of the equipment. Because higher capital costs are required for greater efficiency, these two factors will combine to display an optimum operating condition and design. The same factors apply in the geothermal case. That is, the cost of producing the geothermal waters with their concomitant heat will be balanced against the greater capital cost of using the resource more efficiently. It is expected, however, in the geothermal case that the optimum point will occur for plants of lesser efficiency, although this point has yet to be clearly established.

In the final sections of the geothermal distillation plant, some heat must be rejected either to the atmosphere or some other cooling medium. After much of the water has been removed from the original geothermal brine, the remaining fluid is greatly concentrated and disposed of from the plant. This residual brine is called blowdown. In the case of the power plant, the geothermal steam entering is turned into condensate, reject heat, and electricity. In the case of the desalting plant, the intake geothermal fluids are turned into fresh water, blowdown brine and reject heat.

The steam that is separated at the head of the geothermal well can also be used to power a desalting plant using feedwaters other than geothermal. Thus it would be possible to use the separated geothermal waters in one desalting plant and the geothermal steam in another plant to desalt local contaminated feedwaters.

Electric Power-Closed Cycle Design

It has already been pointed out that in geothermal water wells, the water will flash into a mixture of steam and liquid in passing up the well. In so doing, considerable energy is given up by the geothermal water in propelling itself and evaporating.

Another means exists of exploiting the thermal energy. This is to maintain a high pressure on the geothermal water in the well and to pump it out of the well as high temperature water.

In general, it will be necessary for these pumps to be in the well itself. (It should be recognized that it will require substantial power from the power plant to operate such pumps. Therefore, a careful trade-off is needed to determine if it is more efficient to use electric power in this fashion or whether one should simply accept the thermodynamic penalty of having the water propel itself up the pipe by the percolator method.)

The hot water then flows from the well at high pressure and is passed through a heat exchanger. Here a part of its heat is transferred to a working fluid which might be water or some other substance. It has been proposed to use isobutane or other organic substances commonly used as refrigerants for this purpose.

The geothermal water, now cooled, is returned to the earth through injection wells or otherwise disposed of. The heat exchanger acts as a boiler for the working fluid, which is evaporated and passed through a power turbine to generate electricity. The fluid is then condensed in a condenser and the working fluid, now once again liquid, is pumped to a high pressure and again run through the heat exchanger. This is exactly the fashion in which either a conventional fossil-fired steam boiler or a nuclear power plant operate. The condenser of course must be cooled with atmospheric air or outside water.

One motivation mentioned for using this cycle is to avoid the deposition of carbonates resulting when some geothermal waters are permitted to flash. The cycle does not avoid the depositions that are possible when saturated solutions of such materials as silica (which comes out of solution upon cooling) are cooled in the heat exchanger. The plant design specifically excludes consideration of the geothermal water supply. Deep well pumps would be required in most areas. No electrically driven pumps are available that can operate where water must be taken from depths of over about 1,000 ft. At shallow depths, it is unlikely that sufficiently hot water will be found in most known geothermal areas.

It may be found in some areas that waters will rise naturally from greater depth and may be pumped, while a pressure sufficient to preclude flashing is maintained in the well. The Magma Power Company is actively exploring the technology of this method. While the system offers many potential advantages, it is as yet in early development, and firm cost and performance estimates are impossible to make ("Magma Max Power Generating Plant-Feasibility Study and Preliminary Cost Estimate," Rogers Engineering Co., Inc., San Francisco, September 1970).

Cooling

In each of the power and water production methods discussed it is seen that a means of cooling the plant is required. This is a fundamental requirement of closed thermodynamic cycles and is accomplished in various ways. Generally, one of three types of cooling methods is used.

The first is to use the atmosphere as coolant, much as an automobile radiator is cooled by the air. Here in a power plant or water plant, the equivalent of the automobile radiator will have atmospheric air passed through it, thereby heating the air and cooling the fluid within the radiator. Such devices are usually referred to as dry cooling towers. In this case, no water is evaporated into the air and the performance of the cooling tower is a function of the temperature of the air as it is usually measured.

It is also possible to reject heat into the atmosphere using the wet cooling tower. In a wet cooling tower, coolant is circulated through lattice-like construction so that a flow of atmospheric air passes over the fluid droplets. The liquid is cooled by heat transferred to the air directly and by evaporation into the air. Such cooling towers involve a very substantial consumption of water, but on the other hand, are more efficient than dry cooling towers both in their cost and their cooling capability.

Both classes of cooling tower may be divided into two types. One is the type that uses large fans to force the flow of atmospheric air through the tower (forced draft). The other type (natural draft) relies on the heating of a column of air to cause the air to flow by natural convection, much as hot air flows up a chimney. At The Geysers, forced draft wet cooling towers are used.

Still another method of cooling is to use a source of outside water such as the ocean, river, lake or cooling pond and pass that cool water directly through the condenser and discharge it, heated to some extent, to the body of water from which it came. This is generally the least expensive and most efficient of all cooling methods. However, it is sometimes not possible to utilize it, either because of shortage of water for such purposes or because of the environmental effects of the discharge of such heated waters into the lake or river.

SYSTEM SYNTHESIS AND COST OF POWER FROM DRY STEAM

This system is relatively well understood, in particular owing to The Geysers' experience. No other dry steam field is known in California. At The Geysers, conventional oil field drilling methods are used. In the producing zones, however, compressed air must be substituted for drilling mud. After simple filters are used to remove particulate matter from the wellhead steam, the steam passes through insulated pipes to a power plant. The wells vary in their output, but recent deeper wells here averaged 8 Mw electrical capacity, and a single recent hole is reported to yield a 20 Mw equivalent.

Pacific Gas and Electric Company buys steam from the producers at the power plant. Conventional condensing turbines are used to drive the generators. The design of these turbines is adjusted to account for the comparatively low temperature and pressure of the geothermal steam. For example, Geyser No. 5 uses steam at 100 psi and 355°F. This can be contrasted to PG&E's Moss Landing fossil plant, which uses steam at 3,675 psi and 1000°F. The lower temperatures and pressures lead to lower efficiency, thus geothermal plants consume steam at two to three times the rate of conventional plants.

After passing through the turbine, the steam is condensed in a contact condenser. Here the cooling water is mixed with the steam, thereby condensing it. The liquid, coolant and condensate, is then pumped out of the condenser, which operates at vacuum conditions (4" Hg abs), to the forced draft cooling water.

Most of the water condensed from the steam is evaporated by the cooling tower. However, a fraction averaging about 20% (the exact amount depending on the weather) is not, and constitutes tower blowdown. This water contains traces of boron and ammonia, and is not suitable for disposition in the local creeks. Therefore it is delivered back to the steam producer, who disposes of it in injection wells.

The Geysers units being installed are purchased from Toshiba, a Japanese manufacturer, which is also supplying equipment for the Cerro Prieto installation. They are 55 Mw capacity each. The size is limited both by transportation difficulties into the area and by the economical length of steam lines between the wells and the plant. The various units at The Geysers are scattered through the producing area.

Much interesting and useful information concerning the financial and scheduling aspects of The Geysers power plants is contained in PG&E's application of November 1970 to the California Public Utilities Commission for a Certificate of Public Convenience and Necessity for Geysers 9 and 10. For example, it is shown that, for this field where circumstances are well understood, a period of only 28 months is required from approval and equipment purchase confirmation to commercial service. The cost breakout is illustrative of several important points.

Land and land rights	- -
Structures and improvements	$ 1,997,000
Boiler plant equipment*	456,000
Turbogenerator unit	6,191,000
Accessory electrical equipment	1,000,000
Miscellaneous power plant equipment	130,000
Communication equipment	6,000
Engineering, superintendence	716,000

(continued)

Construction plant, warehouse	$ 81,000
Overhead construction cost	1,770,000
	$12,347,000

*This refers to steam handling equipment: there is no boiler.

These costs do not reflect substations, transmission lines, or the cost of steam production and transportation. Thus a cost of over $100/kw capacity is involved in the power-plant alone.

In the application, PG&E goes on to estimate the total cost of delivered power to their distribution system. Fixed charges include return and depreciation on the investment, taxes, maintenance, and amount to $2,025,000/year for the two units. With an 80% capacity factor, which is not unreasonable considering the experience of The Geysers, and using a cost for steam of 2.1 mills/kwh (plus 0.5 mill for effluent disposal) the total cost of electricity produced is 5.33 mills. In "Economics of The Geysers Geothermal Field, California" by D.A. McMillan, Jr., *U.N. Symposium on Geothermal Resources,* Pisa, Italy, September 1970, it is noted that in 1968, the average cost of thermally generated power in California was around 7 mills.

Viewed in another fashion, the per Mw investment in a Geysers geothermal plant is substantially less than that for a modern fossil plant (whose costs are increasing, in part as a consequence of pollution control equipment) and its fuel (steam) and operating costs are about 85% of those for a modern fossil plant. Thus, conservatively, one can say that for the situation at The Geysers, geothermally produced electricity is somewhat cheaper than that available from fossil plants, and is produced without significant insult to the environment, except possibly in the immediate vicinity.

New nuclear plants are costing in excess of $300/kw, and cost of the nuclear fuel cycle is estimated to be above 2 mills/kwh. Thus it appears that geothermal power will be cheaper to produce than nuclear power also, for conditions as at The Geysers.

Supplemental Water for Reinjection

A discussion of system synthesis and costs involved in obtaining power and water from flashed steam as was expected in the Imperial Valley is given on pages 35-47 of PB 212,753. Owing to the withdrawal of geothermal water for power or fresh water production, land subsidence may possibly become a serious problem in some areas. Owing to the configuration of the water distribution and drainage system in the Imperial Valley, this could be particularly serious if it should occur there. A possible means of preventing subsidence is injection of water into or near the producing zones in such a fashion that production is not adversely affected. Lack of detailed knowledge of reservoir conditions precludes definitive statements as to the necessity or practicality of this action.

If, however, it is to be done, the question of the source of the injected water must be addressed. Because it is possible that subsidence may occur in the

Imperial Valley (natural subsidence goes on at the rate of about 1 cm/yr, owing to tectonic action) and because it is in the irrigated sections of the Valley that most harm could be done, the discussion will center on that area. Many, if not most, other geothermal areas would be unharmed by subsidence.

The most likely source, of course, are the waters remaining from the geothermal production process. Water from the production wells, or desalting plant or cooling tower blowdown will have to be disposed of in some fashion, and reinjection is the most likely. However, some of the original water will have been consumed, either as steam, distilled water or by evaporation. It is this balance that might have to be made up for by an outside source.

In some areas of the Imperial Valley, irrigation wastewater is present. Some of the problems involved in reinjecting such waters include the possible pretreatment required to remove organic material, and the transportation problem. The Valley has a complex net of water distribution and drainage canals. To move water across the pattern may be difficult and expensive. Also, as mentioned, this water maintains the Salton Sea, an asset of considerable importance for recreation. Without specifying particular locations, it is not possible to make cost estimates for wastewater use.

Likewise, some supplies of shallow, groundwater are available. The only water that could be made use of in the geothermal cycle would be subagricultural-quality brackish sources, but the extent and location of even these are not established. No well-founded estimate of cost can be made, but the acre-foot cost of reinjection water can be added directly to the acre-foot cost of distilled and cooling water that is withdrawn from the geothermal stream. Thus, if the outside water is delivered for $10/acre-foot, the price of the distilled water is increased by $10/acre-foot, as there is roughly a one-to-one correspondence between them. This cost would have to be added to the equivalent electrical production, if it were making up for coolant evaporation.

The only known, assured source of reinjection water is the ocean. Fortunately, the Gulf of California is separated from the Imperial Valley by less than one hundred miles, and the maximum elevation between them is only about 30 feet. This does not mean that obtaining the water will be cheap, or even possible. The Gulf lies within the Republic of Mexico. That country would have to grant permission for the water withdrawal and the right-of-way over its territory.

Granted that permission was given, the problem is still not trivial. Crossing the irrigated areas of the Mexicali Valley would involve serious problems of interference with on-going activity. The extensive salt flats at the head of the Gulf and the extreme tidal range will make for great problems in water intake design. No layout for such a water transport system has been made.

However, in *Salinity Control Study, Salton Sea Project,* by M. Goldsmith, Aerospace Corp. ATR-71 (S9990)-5, February 1971, estimates are given for the cost of a pipeline/canal system to transport water in the other direction, from the Salton Sea to the Gulf. A system for transporting 100,000 acre-feet/year was estimated to cost about $30 million; for 500,000 acre-feet/year, about $60 million. Even at the large flow rate, capitalization costs amount to over $12/acre-foot, and operating cost must be added to this figure. It is probable that pretreatment before flowing in the canal will be necessary to prevent fouling, and more extensive treatment before injection will be needed. Costs are presently unknown. It is unlikely, however, that the sea water could be made available for less than $20/acre-foot, and the number could be considerably higher.

It has been suggested by Rex and others that the Salton Sea salinity could be controlled by removing an amount of water from it each year. This amount is presently estimated to be 140,000 acre-feet. This will suffice once the salinity increase is brought under control; a larger rate is necessary initially. For the maintenance phase, however, this removal rate will keep the salinity at a reasonable value, and the level of the Sea will be about 5 ft below its present height.

This amount of water would make up for the condensate from 2,000 Mw of electrical production. While a useful idea, it does not cope with the original adjustment period for the Sea, nor would it provide for the very extensive geothermal developments sometimes suggested. If, however, reinjection were required in only parts of the Imperial Valley, this may prove to meet the need. The cost of pretreating the salty, organically rich water to prevent well plugging will perhaps be high. However, that cost, plus the transport and reinjection could be shared by the beneficiaries, the Salton Sea users and the power consumers. Pending detailed study, made with reference to particular situations, all these concepts of obtaining nongeothermal water for reinjection are highly speculative.

ENVIRONMENTAL EFFECTS

Land Use

The production of geothermal power or water will involve a number of environmental effects. The most obvious items are embodied in the intrusion of an industrial operation into nonindustrial land areas. For example, a geothermal well is drilled in the same fashion as an oil well. Problems included noise and the appearance of drill rigs. Once drilled and in production, the geothermal well can be made unobstrusive and hopefully will offer no severe environmental problems. However, the well products must be collected by pipe and transported to the using plant, often over distances of up to a mile.

These insulated pipes are costly to install below ground, and the tendency will be to run them above the surface. Such pipes can be obstrusive. The power and water plants are typically industrial, with a not unreasonable noise level, but noticeable, and with cooling towers. Thus commercial and residential usage would be generally incompatible with a geothermal field.

Wastewater

More general and destructive environmental effects are possible, but fortunately are capable of being controlled. When developing a field yielding a steam/water mixture, there is the matter of disposal of surplus waters. In some instances, these wastes will be high in mineral content, and cannot be discharged into surface waters. Unless very well mixed, even ocean discharge could lead to severe local effects, if the plant waste differed substantially from ocean water.

Let us review the magnitude of the problem. It is assumed that plant condensate, being distilled water, will be used all or in part, for consumptive purposes. Even when this is done, however, as at The Geysers, (the condensate water is used for cooling tower make-up) the surplus water generated contains trace chemicals which preclude its discharge into the local streams. The water often requires further treatment, or at The Geysers, is reinjected into the ground, in deep wells. There, about 20% of the condensate is reinjected. For 100 Mw of generating capacity, this amounts to over 1 million gallons/day. One large injection well can accomodate this flow.

The more difficult problem arises when the geothermal wells produce hot water, rather than dry steam. In this case the water may be highly mineralized. For example, at the Buttes Field, near Salton Sea, the water contains over 20% salts, an extremely high value. At Cerro Prieto, in Mexico, the waters contain about 2% salt (ocean water contains 3.3% salts). As has been calculated, for conditions like Cerro Prieto, for an equivalent electric plant of 1,000 Mw size, salt water is produced at a rate of approximately 150 million gallons/day, or over 150,000 acre-feet annually. For 2% brine, 12,000 tons/day of salts would result if the water was evaporated away. This poses a monumental solid disposal problem, and constitutes a real environmental danger.

Before disposal, the water might be concentrated in evaporation ponds, or used as feedwater to a desalting plant. In either case, limits exist to the concentration factor, thus the disposal of the brines is a problem requiring careful consideration. In most inland areas, transport of the brines to the ocean would be difficult and costly, therefore other methods must be sought.

The problem has been specifically examined for the Salton Sea district in *Geothermal Wastes and the Water Resources of the Salton Sea Area,* California Dept of Water Resources, Bull. No. 143-7, February 1970. The method that appears most promising is disposal into injection wells. An injection well is drilled to a depth where a porous formation will accept the water. To avoid contamination to groundwaters, these may involve depths of several feet. For the prevention of subsidence, where necessary, the water would probably be injected into the producing zone.

After overcoming original wellhead pressure, it is often found that the water is literally poured down the hole. This is often made easier by the greater density of the concentrated and cooled brine. This method is widely used for disposal of oil well brines and industrial wastes. One must take care to avoid aquifers that connect to areas where the waste will do harm, e.g., sources of agricultural or potable water. This is not thought to be a problem in geothermal areas. A potentially serious problem in injection well operation is the deposition of minerals from the water in the pores surrounding the well. Such deposition can cause rapid impediment to well flow.

It is unlikely that geothermal waste will be disposed of in California other than by well injection. In Mexico, at Cerro Prieto, the water simply flows in ditches to the Gulf of California, via the Rio Hardy. The present flows already affect the Rio Hardy, and it is anticipated that additional development will require the construction of a special waste canal to the Gulf. This course is not open in the Imperial Valley, which is land-locked and drains to the Salton Sea. It is already forbidden by the Regional Water Quality Control Board to discharge geothermal waste to the Sea or its tributaries. This is necessary to preserve the already threatened Sea from an uncontrolled increase in salinity. The same considerations will probably hold for the Mono Lake region.

Subsidence

A closely associated problem is that of land subsidence. If large quantities of fluids are removed from the underground reservoir, the land surface may sink, with sometimes disastrous consequences. This happened in the Wilmington oil field. It is, however, an unusual occurrence. If the aquifer consists of fractured rock, subsidence is unlikely. If it is a very porous medium, which can collapse when the water is withdrawn, subsidence can occur. The problem is sometimes coped with in oil field practice by injecting water into the reservoir.

Thus the disposal of geothermal wastes by injection may also be necessary to prevent subsidence. This effect is potentially very serious, and might have to be met by importing water for reinjection in a drained agricultural area such as Imperial Valley. An elaborate survey network is already being laid out in the Imperial Valley so that possible subsidence resulting from future geothermal development can be quickly detected. Subsidence has already been reported at the geothermal field at Wairakei in New Zealand and at Cerro Prieto, but that at Cerro Prieto may not be connected with geothermal production.

Seismic

Experience in Colorado, and some on-going experiments there, have indicated that seismic activity can be stimulated by the injection of water deep underground. The seismic effects of water withdrawal and reinjection in geothermal fields will be peculiar to the particular area and cannot be stated to be or not to be a problem at this time. There is even speculation that the induced microseismicity relieves strain on faults and tends to prevent major earthquakes.

Air Pollution

Noxious gases are often a by-product of geothermal wells. At The Geysers, for example, the odor of hydrogen sulfide (H_2S) is prevalent. It exists in the steam with other gases, most notably carbon dioxide. The noncondensable gases contribute from 0.2 to 1.8% of the steam flow at the Geysers. Of this 82.5% is CO_2, 6.6% methane, 1.4% hydrogen, 1.2% inerts, 4.5% H_2S and 3.8% ammonia.

What do these small percentages mean? For one thing the materials of a turbine, condenser or water plant must be chosen to avoid corrosion problems. Thus the PG&E condensers must be lined with stainless steel. The noncondensable gases must be separated from the steam or water flow to insure proper operation of condensers. This is simply done, and at The Geysers the eject gas is discharged to the atmosphere. Some gases dissolve in the condensate. This is the case of H_2S, where a portion dissolves in the condensate and later escapes to the atmosphere when the condensate water is evaporated in the cooling tower. If we assume that only ½% of the steam flow, on the average, is noncondensable gas, the above figures indicate that H_2S is present in the steam to the amount of 225 parts per million.

If a total of 1,000 Mw of power were produced there, this would require 430 million lb/day of steam. Thus 97,000 lb/day of H_2S will be released. This is roughly equivalent to the amount of sulfur released by a fossil-fueled power plant of the same size, burning low-sulfur oil. Technology is available to prevent the release of these gases, if it is found to be necessary, as is probable if extensive fields are developed. PG&E and Union Oil Company are presently investigating methods of emission control at The Geysers. The question is one of practicality and cost. To say that geothermal power intrinsically involves no air pollution is incorrect. With controls, however, it need not cause air pollution.

Heat Rejection

A possibly significant environmental effect to be expected in routine operation of a geothermal power plant is heat rejection. All power production cycles using thermal energy must reject heat, and the less efficient they are, the greater is the heat rejection. As previously pointed out, geothermal steam is available at low pressure and temperature, as compared to that from conventional boiler or nuclear plants. Thus the heat rejection will be high. This is clearly indicated by the comparative steam rates (which can be roughly equated to comparative heat rejection) given in "Engineering Aspects of a Geothermal Power Plant", by A.W. Bruce, *U.N. Symposium on Geothermal Resources,* Pisa, Italy, September 1970 for The Geysers and a modern fossil plant. The rate for Geysers 3 and 4 is given as 18.53 lb/kwh, while Moss Landing 6 and 7 is shown as 6.68.

Let us examine the heat rejection quantitatively. For 100 psi inlet conditions, and a water tower-cooled condenser, thermodynamic calculations show that 3,630 Mw of heat are rejected by a 1,000 Mw geothermal electric plant (a 1,000 Mw nuclear power plant rejects approximately 2,000 Mw of heat). This number means little unless there is a basis for comparison. Data exists for the total 24 hours average solar heat received in a desert area in summer.

This number, called the solar insulation, is 720 Mw/square mile. Thus the 1,000 Mw power plant produces heat equivalent to that received by 5 square miles of desert from the sun. If ten such plants or their equivalent were installed in the Imperial Valley the total heat added to the 1,000 square mile area of the Valley would be 5% of the total summer solar heat. The effect of this heat on the local weather is unknown at this time.

If air-cooled condensers are used, the reject energy will be larger and will go directly to heating the atmosphere. How this heated air would distribute itself and affect the local climate will require detailed consideration of local conditions. If water cooling towers are used, the temperature would be affected to a lesser extent, but substantial quantities of water would be evaporated, thus influencing the humidity. Considering the heat rejection rate, and for typical cooling tower performance, up to 50,000 acre-feet/year of water will be evaporated by a 1,000 Mw plant. Ten such plants will evaporate 500,000 acre-feet/year. Presently, approximately 1.5 million acre-feet pass into the atmosphere through evapotranspiration in the Imperial Valley each year by agriculture and another 1 million evaporates from the Salton Sea. This added burden of water vapor is a small fraction (20%) and in times past, more water has been used in Imperial Valley than is presently the case. Coincidently, this increment corresponds roughly to this cooling tower use.

Thus no overall detrimental effects should be expected. Local effects are very possible, and deserve detailed examination. On balance, for the Imperial Valley as a whole, the variations of heat absorption that presently occur owing to the state of the field (e.g., plowed, in full leaf) would tend to be greater than the effects of the power plant heat. Likewise, the real humidity variations that occur in the area seem mostly a function of air mass movement. Thus permanent valley-wide effects are unlikely. However, in the immediate vicinity of the plants, considerable environmental effects are possible, and appropriate investigations should be carried out.

Well Blowout

In any well drilling operation involving high pressure fluids, the possibility of a well blowout must be taken into account. Blowouts occur in a variety of ways. The classic oil well blowout is one type. This same class of accident occurred during the drilling of one of the early production wells at Cerro Prieto. Standard oil field methods were used to bring the well under control. Days were required to accomplish this, however, while the well geysered steam and salt water. Such a blowout might flow as much as 10 acre-feet/day. Clearly, such a release of salt water in an agricultural area would pose a major environmental problem. Means of affecting prompt blowout control must be provided, and are called for by state regulation.

Another class of blowout has occurred at The Geysers. Here the formation through which the well passes is unstable. Attempts to cap the flow cause steam to escape into the ground threaten to cause an eruption from the ground. Thus the flow continues and attempts at control are being made. This physical situation is akin to that found in the Santa Barbara Channel oil field, where control of the oil well tends to cause seepage from fracture in the seabed. Again, such a possibility must be avoided by careful well design. California State oil drilling regulations seem to offer good protection on this point.

RESOURCE POTENTIAL

Of the several geothermal areas identified in California, only two have received extensive exploratory attention; the first of these is the area of The Geysers in Northern California and the second is the Imperial Valley in Southern California. Let us first examine the potential for power production of The Geysers field, which is currently being developed by Union Oil and PG&E. PG&E has indicated its intentions for power plant construction during the balance of the decade of the 70's. They list geothermal units through Geysers 14, to be installed in the Fall of 1975, which lead to a total of approximately 600 Mw of geothermal electric production. During this same time frame, it is their intention to install 735 Mw of fossil power and 2,120 Mw of nuclear power. During the balance of the decade they indicate an intention to install 4,400 Mw of additional nuclear capacity. On this basis then, the 600 Mw of geothermal capacity appears noticeable, useful, but relatively small.

If on the other hand, The Geysers field is more extensive than presently postively established (if for example the estimate of 4,000 Mw capacity is found to be realizable), it is conceivable that it could replace half or more of the installation of nuclear capacity presently in planning stages by PG&E. It is worthwhile to note, however, that unless considerably more extensive exploration is carried out to identify areas presently not verified as economically and technically attractive geothermal fields (if indeed they exist), geothermal resource cannot substantially replace nuclear power in PG&E's plant.

The first of the now planned nuclear plants is to be on line in Spring of 1977. With current construction lead times it is highly unlikely this date can be met. If the electrical load continues to grow at a rate that would have necessitated this addition, then it is conceivable that further exploitation of the geothermal resource will enable the gap in generating capacity to be filled, thus overcoming the effects in delays of nuclear plant construction.

Several factors serve to inhibit more aggressive development of The Geysers geothermal field. First, the utility is already making a substantial investment in money in the area, and the utility is accustomed to investing money in plants whose useful lifetime can be accurately predicted in advance. In the case of the geothermal field, however, if the field should lose its capacity to produce, the investment would become useless; and it is not possible to accurately predict this period of useful production.

Second, if PG&E has only 600 Mw capacity at The Geysers, out of a system capacity of over 15,000 Mw, the possibility of loss of geothermal field productivity does not seriously affect their ability to supply their customers. Consider the reliability problem if they were employing 4,000 Mw of geothermal capacity. It could be argued that field productivity would not drop precipitously, but would gradually wane over several years. This is small comfort however, as the leadtime for new nuclear plants is now nearly ten years. Thus, for major exploitation of geothermal resources, developed, proven reserves will be required to insure system reliability. This emphasizes the need for aggressive exploration for additional producing areas, and not simply for exploitation of known areas.

Thus, it is to be expected that the utility would have a natural reluctance to invest too much of its money and to rely too heavily for generating capacity on this somewhat uncertain resource at this time. Projected load growth for the PG&E system indicates that even with completion of a two-unit nuclear plant (2,300 Mw) subsequent to Diablo Canyon, and continuous installation of geothermal capacity at the present rate, system reserves are not as great as desirable. Then in a realistic sense, one cannot state that intensive development of The Geysers field would enable PG&E to postpone the installation of nuclear units subsequent to Diablo Canyon 1 and 2 until even the end of this decade. Rather, geothermal is simply an addition to the generating mix for PG&E, and may permit a reduced pace of nuclear plant installations in the future.

Turning now to the Imperial Valley, the picture becomes even less clear. Here no proven economic production area has been positively identified. Moreover, no intensive program for clearly identifying and determining the extent of such an area is contemplated which would yield such results in the next year or two. The largest published estimates of the generating capacity of the Imperial Valley, those of Professor Rex, indicate a generating potential of 20,000 Mw.

Growth projections of electric demand in the Southern California area indicate that such a supply will satisfy expected needs for the next 10 to 20 years. However, other planned electric generating capacity expansions cannot be deferred at this time in the expectation of geothermal development because no proven geothermal capacity exists at this time in Southern California. No reasonable utility planner could at this time factor geothermal capacity into his overall planning strategy nor will he be able to do so until an intensive exploration program is conducted. It is still unclear when such an aggressive exploration program will be initiated, and by whom.

Even if the program is successful, utilities would probably proceed with development in a most deliberate fashion, for reasons of system reliability. Until the capacity and duration of yield of the fields are well understood and proven, the developments of the resource must be carefully paced.

It is interesting also to see how potential geothermal fresh water production might match increasing water requirements in the Imperial Valley area. It is known that the California Aqueduct system will supply the general needs of Southern California (apart from the Imperial Valley area) for the next 20 years or perhaps longer. However, no California Aqueduct water is intended to be used in the Imperial Valley and it is known that quality of the water being used by the Imperial Irrigation District from the Colorado River is steadily deteriorating. It is certain that this steady deterioration will have a deleterious effect on agriculture.

For example, at this time the water drawn from the Colorado River at the Imperial Dam contains dissolved solids to the amount of about 900 parts per million. By the year 2,000, projections of the Colorado River Board indicate that this figure will increase to in excess of 1,300 parts per million. It is known that the farmers in the Mexicali Valley are already suffering from inferior water quality at the level of 1,200 parts per million, that which is delivered to them at the Morelos Dam. The expected increase in water salinity at Imperial Dam is shown in Figure 6.1 as a function of time.

Assume now that the desalted geothermal waters would be used for blending with existing irrigation waters to maintain the total dissolved content at a reasonable level for agricultural purposes. If it is also assumed that a reasonable level is 900 parts per million, and that the total use of water by the Imperial Irrigation District remains at 2.7 million acre-feet annually, one can then plot the amount of distilled water needed, as a function of time, to maintain favorable water conditions. This plot is shown in Figure 6.2 on the following page. It is seen that by the year 2000, 900,000 acre-feet annually of distilled water would be required for blending.

FIGURE 6.1: PROJECTIONS OF SALINITY AT IMPERIAL DAM COLORADO RIVER BOARD 1971

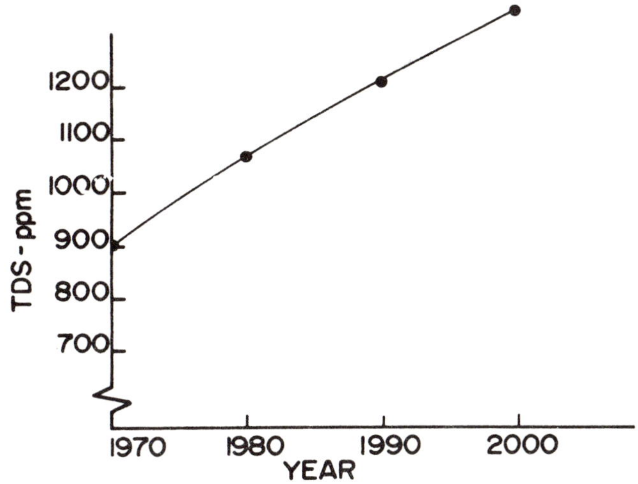

FIGURE 6.2: DISTILLED WATER REQUIRED FOR BLENDING

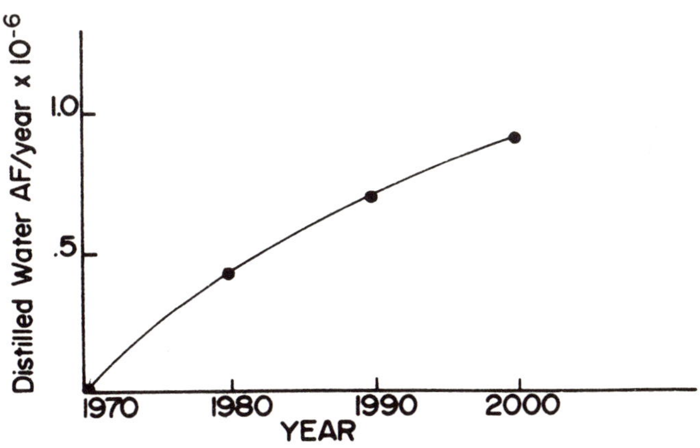

Source: PB 212 753

It has been noted that nearly 60,000 acre-feet of water would be produced annually by simply conserving the condensate of a 1,000 Mw power plant. Thus, if electrical generating capacity in the Imperial Valley should actually amount to 20,000 Mw by the year 2000 then an ample supply of distilled condensate would be available for water blending to maintain the quality of water in the Imperial

Valley, although the supply of wastewater for cooling would be stretched. As previously calculated, if a desalting plant is associated with the electric plant to process the waste brines from the well after the flashing process, perhaps 145,000 acre-feet per year of fresh water would be associated with 1,000 Mw of generating capacity, in addition to the 60,000 acre-feet already discussed. Therefore, it is seen that potentially an additional water supply of nearly 3 million acre-feet annually could result.

It should be remembered that the geothermal power plant itself produced the first increment, and, given the availability of cooling water, this water increment is available at fairly low cost. It should also be remembered that the second and larger increment must be produced by a distillation plant dedicated to the production of water, and this water may be fairly expensive as estimated in previous sections.

In summary then, present geophysical evidence indicates that the possibility for supplementing Southern California electrical growth over the next 10 or 20 years might be met by the geothermal resources of the Imperial Valley and that if such a development took place the associated water production could maintain the quality of water in the presently irrigated areas of the Imperial Valley until the end of the century and beyond. Also it is possible that a supply of additional water equivalent to or exceeding the capacity of the California Aqueduct as it flows into the Southern California area could be created for transport to water-short areas. There is an evident incompatibility between the potential of the Imperial Valley resource and the availability of public funds to conduct geothermal explorations in the area.

METHODS OF ENERGY RECOVERY FROM HOT BRINE DEPOSITS

TOTAL FLOW CONCEPT

The total geothermal energy resource base in the United States is very large. Geothermal energy exists in near-surface deposits as superheated steam, hot brine, and hot dry rock. A method is proposed for production of electrical power from the energy in hot brine, especially the brines of the Salton Sea geothermal area in California by A.L. Austin, G.H. Higgins and J.H. Howard of the University of California's Lawrence Livermore Laboratory in UCRL-51366, April 1973, *The Total Flow Concept for Recovery of Energy from Geothermal Hot Brine Deposits* from which the material in this section is derived.

This energy form appears to represent the largest potentially exploitable part of the geothermal energy resource base. Although superheated steam deposits, such as those at The Geysers, California, are more desirable forms of geothermal energy than are hot brine deposits, they are estimated to be only one-twentieth as common. On the other hand, although there is an immense amount of energy in dry hot rock, power output is limited by the inherent low thermal conductivity of rock, and no technology yet exists for its recovery and conversion.

The proposed method for producing power from hot brine uses the total fluid flow concept. It is considered to be economically competitive with all other electric power sources. Conceivably it can develop power from hot brine deposits in California equivalent to 100,000 Mw of electrical generating capacity. By comparison, the present electrical generating capacity of the United States is ~340,000 Mw.

Other approaches have been proposed for using hot brine. One consists of flashing the brine in a centrifugal separator, disposing of the liquid fraction, and operating a conventional reaction turbine with the retained vapor fraction. This system is used at Wairakei, New Zealand, and is the system under construction at Cerro Prieto, Mexico. The brines at these locations have salt contents about one-tenth that of the Salton Sea brines, and it is clear that high salt content has inhibited similar developments of the Salton Sea deposit.

A second approach to using hot brine is a binary fluid system in which hot brine is pumped under pressure to prevent flashing and passed through a heat exchanger to transfer the thermal energy to a noncorrosive working fluid such as isobutane. This fluid is then used in a Rankine cycle turbine-alternator system for producing electrical energy.

In view of the corrosive nature of the brine, it is believed that reliable operation of the downhole pump and the heat exchanger required in the concept is a difficult problem that currently limits application of the concept.

A third approach has been conceived that in theory requires a capital investment of ~\$200/kw and should produce energy at a cost of 3.2 mills/kwh. These estimates are lower than those for both the flashed steam and binary fluid systems. The proposed method consists of allowing the brine to expand to the surface as is now being done at Wairakei and at Cerro Prieto, and then expanding the total fluid stream through a nozzle to low pressure and high velocity.

The kinetic energy of the entire fluid stream would then be used in a corrosion-resistant impulse turbine, modeled after typical hydroelectric devices, to drive an electrical generating system. Theoretically, this method should produce ~1.6 times as much power from a given well as either the flashed steam system or the binary fluid system. Since fewer mechanical components are required at the surface, capital costs should be less, despite the need to protect against corrosive brines.

FIGURE 7.1: METHODS FOR RECOVERY OF ENERGY FROM HOT BRINE DEPOSITS

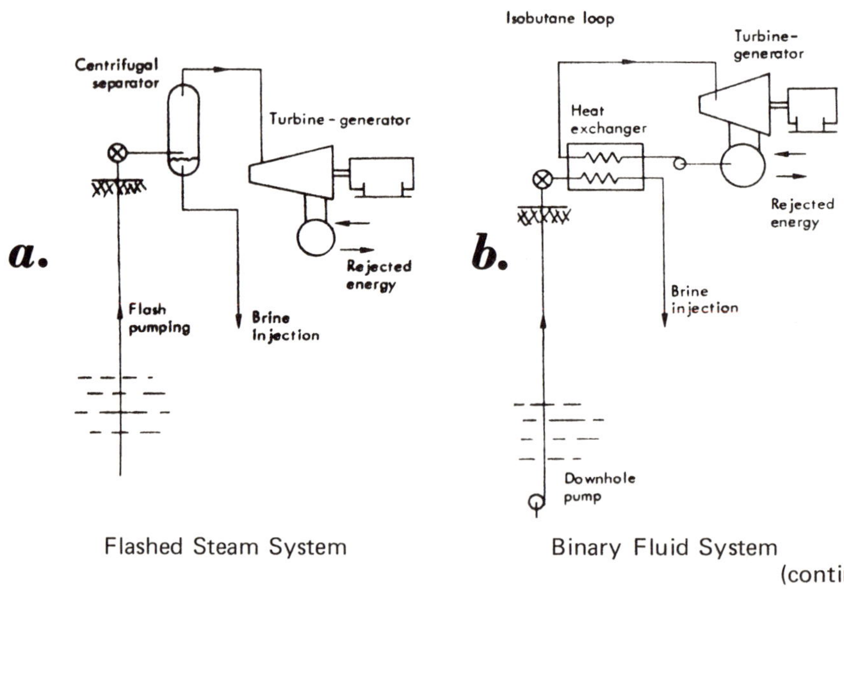

Flashed Steam System

Binary Fluid System

(continued)

FIGURE 7.1: (continued)

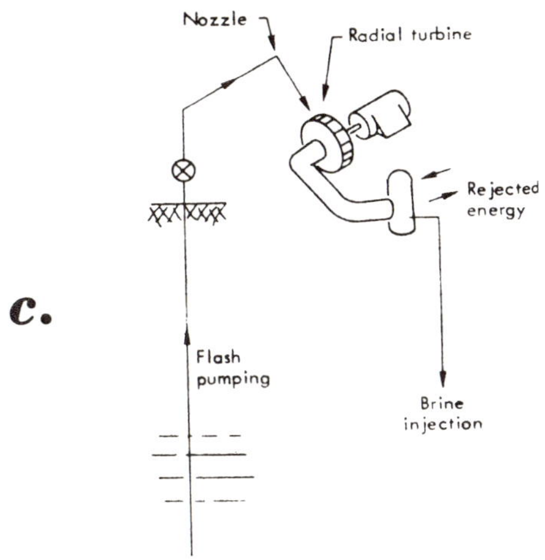

Total Flow System

Source: UCRL-51366

Figure 7.1 contrasts the three methods. Note that in the flashed steam system (Figure 7.1a) and the binary fluid system (Figure 7.1b) brine is discarded at temperatures higher than the turbine outlet and ambient temperatures for waste heat disposal. Thus, insofar as turbine efficiencies for all systems are comparable, the total flow system (Figure 7.1c) must always have a higher overall efficiency.

EXTENT AND CHARACTERISTICS OF THE HOT BRINE RESOURCE

Figure 7.2 shows the areal distribution of existing and potential geothermal fields in the western United States. Included in these areas are dry steam and hot rock geothermal resources as well as hot brine resources. The only facility presently recovering geothermal heat is at The Geysers, where dry steam is used to produce more than 100 Mw of electricity.

The area of interest lies within the western geothermal province in southern California and northwestern Mexico. Topographically the area is the landward extension of the Gulf of California. It includes the Coachella, Imperial, and Mexicali Valleys and the Salton Sea (see Figure 7.3). The entire low-lying area is referred to as the Salton Trough. Geothermal resources have been extensively explored in both the Imperial Valley (the Salton Sea geothermal area) and the Mexicali Valley (the Cerro Prieto geothermal area). Both these areas appear to be manifestations of the same geological system of heat flow in this part of the earth.

FIGURE 7.2: EXISTING AND POTENTIAL GEOTHERMAL AREAS OF THE WESTERN UNITED STATES

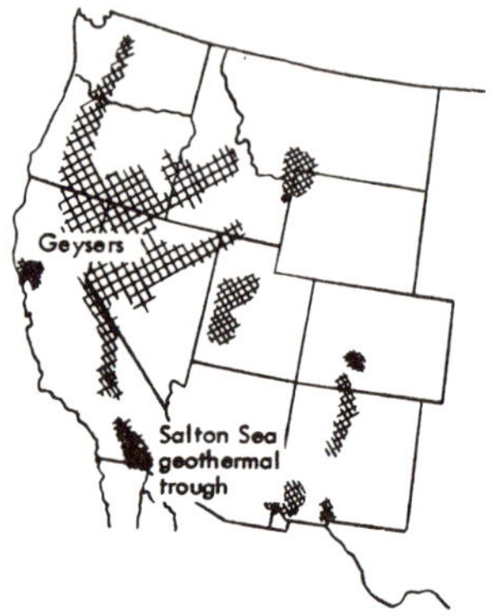

FIGURE 7.3: OUTLINE MAP OF THE SALTON TROUGH

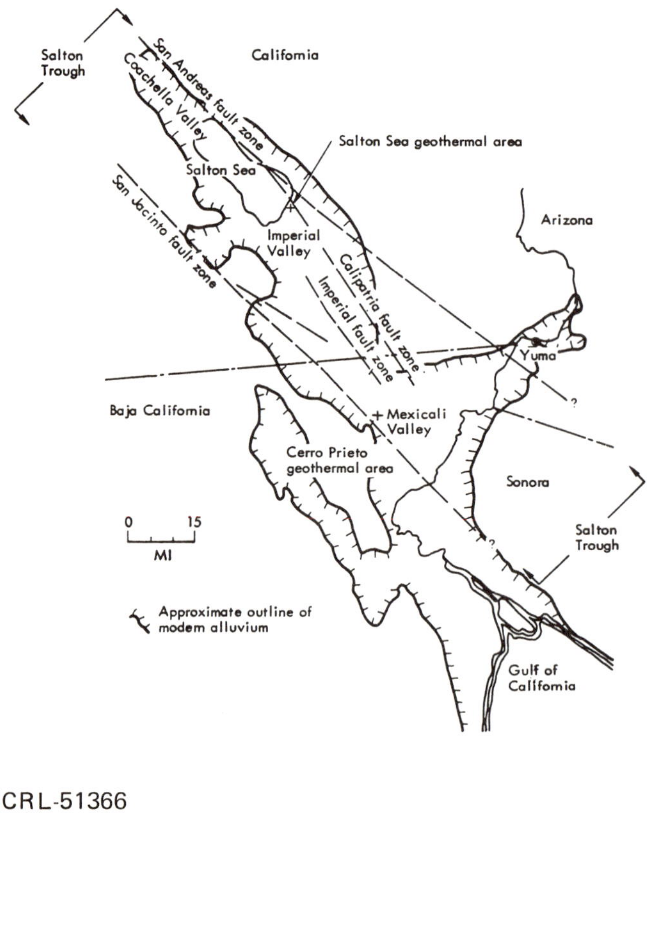

Source: UCRL-51366

Volume and Areal Distribution of Hot Brine in the Salton Trough

Certain characteristics of the geothermal resource may be defined even though its distribution in the Salton Trough is incompletely known. Direct observations show that brines with temperatures >300°C occur at the Salton Sea area and at Cerro Prieto. Knowledge of the distribution of brines of >300°C elsewhere in the area is based on temperature data in wells of various depths and from springs.

Figure 7.4 is an interpretation of the surface of the 300°C isotherm, based on maps and cross-sections provided by the U.S. Geological Survey. Some authorities consider the USGS figures to be conservative extrapolations of temperatures to be expected at depth. More liberal extrapolation of near-surface thermal gradients suggest that still higher temperatures could be expected at depth.

FIGURE 7.4: SALTON TROUGH—SURFACE OF 300°C ISOTHERM

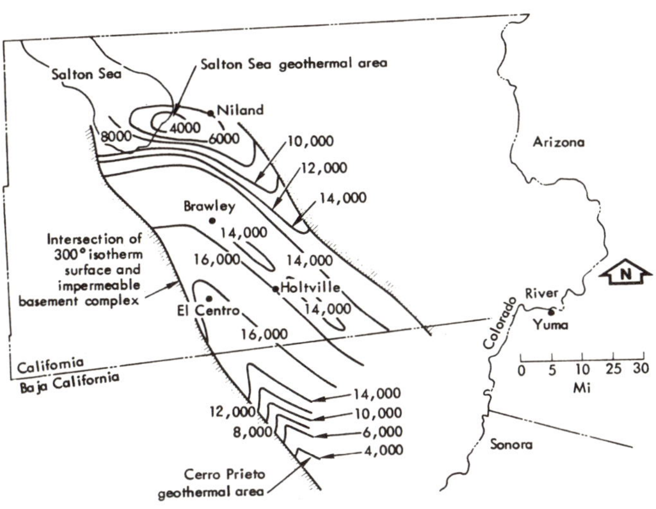

Source: UCRL-51366

Figure 7.4 shows that the 300°C isotherm lies at depths of <4,000 ft at the south end of the Salton Sea and at Cerro Prieto. Elsewhere it lies at greater depths. The surface has highs southeast of El Centro, east of Holtville, and near Brawley. These locations correspond to areas of high geothermal gradients reported by W.A. Elders, R.W. Rex, T. Meidar, P.T. Robinson and S. Biehler in *Science* 178 (4056), 20 (1972).

Figure 7.5 shows the surface of the basement complex. In principle, this is an important surface because it probably represents the lowest depths at which hot brines might be recovered. It is assumed that the basement complex has essentially zero permeability and that fluids will not flow in it.

FIGURE 7.5: DEPTH TO BASEMENT COMPLEX, IMPERIAL VALLEY-SALTON SEA AREA

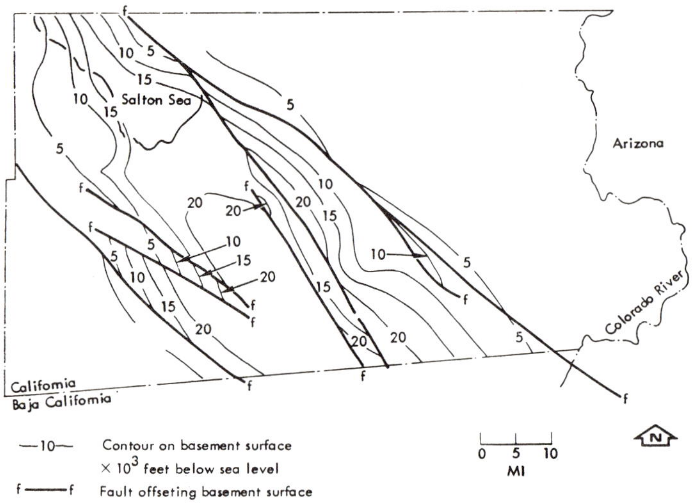

Source: UCRL-51366

Comparing the top and bottom of the rock volume containing brines with temperatures >300°C (Figures 7.4 and 7.5) shows that the volume of >300°C brine is enormous. Although a value of this volume has not been computed for the entire Imperial-Mexicali Valleys vicinity, an estimate has been made for that part that constitutes the drilled portion of the Salton Sea geothermal area. To a first approximation, this area includes a volume of rock with brines >300°C on the order of 36 square miles in areal dimension and on the order of 2 miles in height.

Assuming that the average porosity of this rock volume is 10%, the storage volume in the drilled portion of the reservoir amounts to 100 x 10^{10} ft^3. Although the Salton Sea geothermal area may justifiably be assigned a larger volume by defining its limits differently than the drilled portion, for these purposes it is sufficient to point out the very large resource base contained in the drilled portion. If one were to define the Salton Sea geothermal area as the rock volume with brine in excess of 300°C and with its upper surface more shallow than 10,000 feet below sea level, then its volume is on the order of 700 x 10^{10} ft^3.

To appreciate the significance of this volume it is instructive to compare it with production achieved in one year from a particular well in the area. Otte reports that the River Ranch well produced ~4.5 x 10^6 barrels of fluids (~25 x 10^6 ft^3) in a year [C. Otte, "Drilling, Production and Disposal Technology in Salton Sea Geothermal Area, Imperial County, California," in *Compendium of Papers, Imperial Valley-Salton Sea Area Geothermal Hearing, Sacramento, 1972* (State of California, Sacramento, 1972), paper R].

At this yearly rate, a single well like the River Ranch well would exhaust the 100 x 10^{10} ft^3 of brine in 40,000 years. Assuming that any new well penetrating the drilled portion of the Salton Sea geothermal area will flow at an average yearly rate of 12.5 x 10^6 ft^3/yr for 20 years, a system of 4,000 such wells would be required to exhaust the 100 x 10^{10} ft^3 of brine. It appears reasonable, therefore to conclude that the Salton Sea geothermal area reservoir could produce satisfactorily over the 20-year time span needed to justify the costs of construction of electric power plants.

If 4,000 wells were drilled, each with a ½-ft^2 cross-sectional area, they should theoretically be capable of supporting a 92,000-Mw capacity for 20 years. Inasmuch as the present total electrical capacity of the United States is about 340,000 Mw, it is clear that the Salton Sea geothermal area, as defined by drilling to date, is a very significant natural energy resource.

Chemical Composition of the Brines

Table 7.1 lists five analyses of chemical constituents produced in the brines (H.C. Helgeson, *Amer. J. Sci.* 226, 129 1968). Other data are also available. Data from Helgeson (column 2) are for brines with temperatures >300°C; presumably they are also for columns 1, 3 and 4. The analysis in column 5 is thought to be from cooler water. It is clear that the brines are very saline.

Furthermore, there are indications of a spatial variation in salinity (cf data from Arizona with data from the Salton Sea geothermal area). Elders et al suggest that, based on electrical resistivity studies, a regional salinity gradient exists such that salinity is lowest near the Colorado River near Yuma (column 5) and increases northwestward toward the Salton Sea (columns 1 through 3). Helgeson notes that salinity varies with depth and argues on theoretical grounds that salinities in the Salton Sea geothermal area increase sharply to depths of 3,000 to 4,000 feet and then maintain values in the 250,000 to 350,000 ppm range.

Otte reports other important features of the geochemistry of the brines system. The chemical composition of the brine produced over a period of one year in the River Ranch well was essentially unchanged. This observation indicated that the spatial variation of salinity in the reservoir tapped by the River Ranch well is small. This observation is consistent with Helgeson's conclusion that salinities are more or less constant in the Salton Sea geothermal area at depths >3,000 to 4,000 feet.

Otte also indicates that the permeability of the near-bore vicinity of a well can be adversely affected by pressure drawn down at the wellbore during production. Reduction in the formation fluid pressure can cause the fluids to flash into steam and lead to precipitation of minerals in the pore space. He states that spacing of wells and production rates of geothermal wells in the Salton Sea geothermal area should be planned to minimize or prevent such precipitation.

TABLE 7.1: ANALYSES OF WATER FROM GEOTHERMAL WELLS AT SALTON SEA, CERRO PRIETO AND YUMA

	Niland			Cerro Prieto	Arizona
Element	**Well 39**	**Well 36**	**Well 57**	**M 3**	**Musgrove 1**
Sodium	50,400	53,000	10,600	5,610	141[a]
Potassium	17,500	16,500	1,250	1,040	–
Calcium	28,000	27,800	1,130	320	148
Lithium	215	210	40	14	–
Magnesium	54	10	74	–[b]	43
Strontium	400	440	85	27	–
Barium	235	250	3	57	–
Rubidium	135	70	–	–[b]	–
Cesium	14	20	–	–[b]	–
Iron	2,290	2,000	0.7	–[b]	–
Manganese	1,400	1,370	6.4	–[b]	–
Lead	102	80	–	–[b]	–
Zinc	540	500	–	–[b]	–
Silver	–[b]	–[b]	–	0.05	–
Copper	8	–[b]	–	0.09	–
Silica	400	400	120	–[b]	18
Chloride	155,000	155,000	19,700	9,694	188
Boron	390	390	100	12[c]	–
Fluoride	15	–[b]	1	0.88	–
Sum of sulfur	–[b]	30	–	≈10	–
Dissolved solids	258,973	259,000	34,800	≈17,000	1,000

[a]Includes potassium.
[b]Not reported.
[c]Recalculated from H_3BO_3.

Source: UCRL-51366

Fluid Pressures of the Brines

Helgeson reports information on fluid pressures for the brines in the Salton Sea geothermal area. Information on six wells indicates that pressures in the fluid are normal hydrostatic for this area. Presumably this conclusion is applicable to other areas in the Coachella-Imperial-Mexicali Valleys vicinity. Figure 7.6 summarizes Helgeson's data.

Figure 7.7, which is also from Helgeson, summarizes the relationship between the temperatures and pressures of the brines at any depth in certain wells and the temperatures and pressures required to change the liquid brine to steam. If the pressure at a point in a well is reduced sufficiently by artificial means, the brine will flash. Subsequently, because of the reduced hydrostatic head in the well to that point in the reservoir, the reservoir should flow a two-phase mixture to the well head.

FIGURE 7.6: PRESSURE MEASUREMENT AT CASING PERFORATIONS IN SIX GEOTHERMAL WELLS

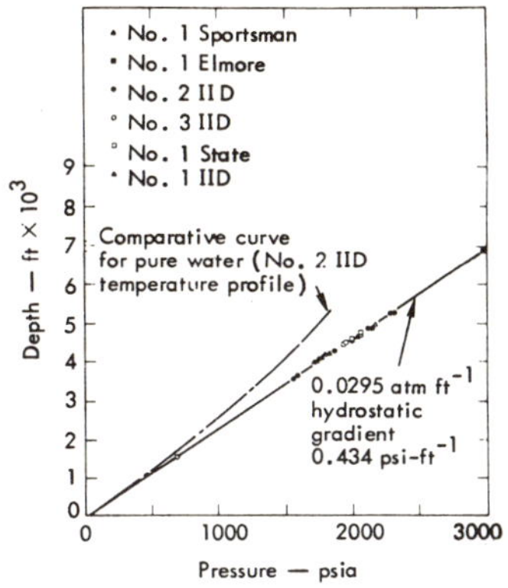

FIGURE 7.7: PRESSURE-TEMPERATURE DIAGRAM

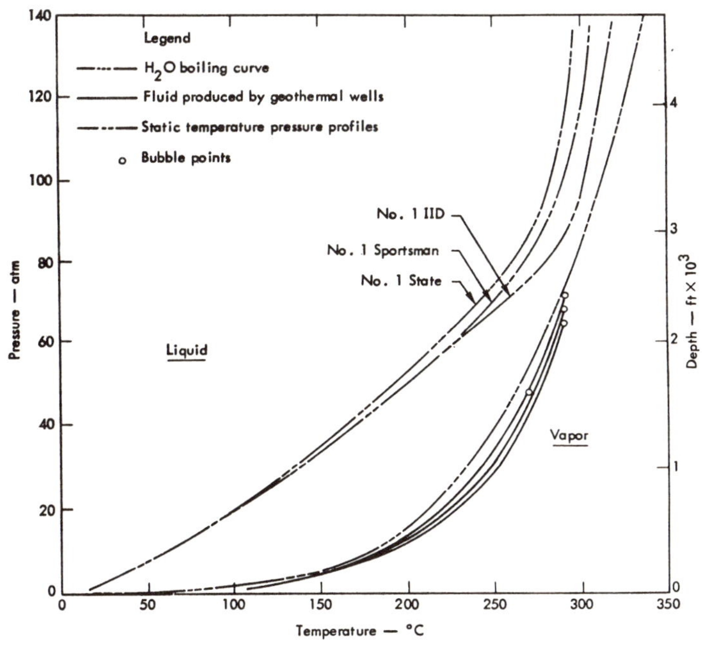

Source: UCRL-51366

Estimates of Production Capability of Wells

Production data for four wells other than the River Ranch well are shown in Figure 7.8. Otte's report for the River Ranch well and Helgeson's for the entire Salton Sea geothermal area are consistent. Productivities are high. Helgeson notes that no change in enthalpy, temperature, or brine composition occurred during sustained production tests (as long as 18 months) of the various wells.

Helgeson further indicates that natural fractures may serve as the major conduits for production of the brines. Although these fractures drain the interstitial pore space and are not the principal storage volume, they have high permeabilities.

FIGURE 7.8: GEOTHERMAL WELL FLOW RATES

a.

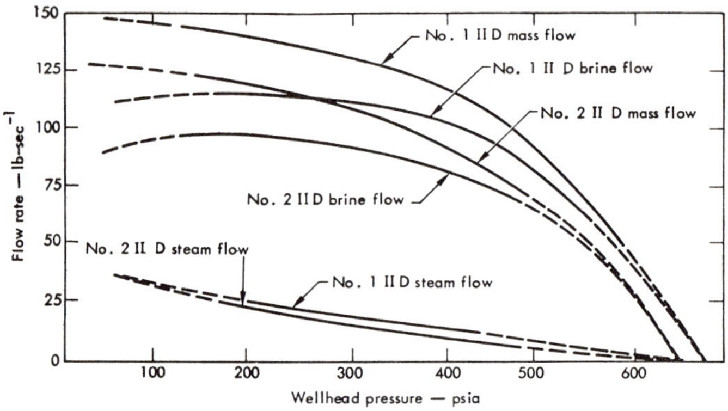

As a Function of Wellhead Pressure

b.

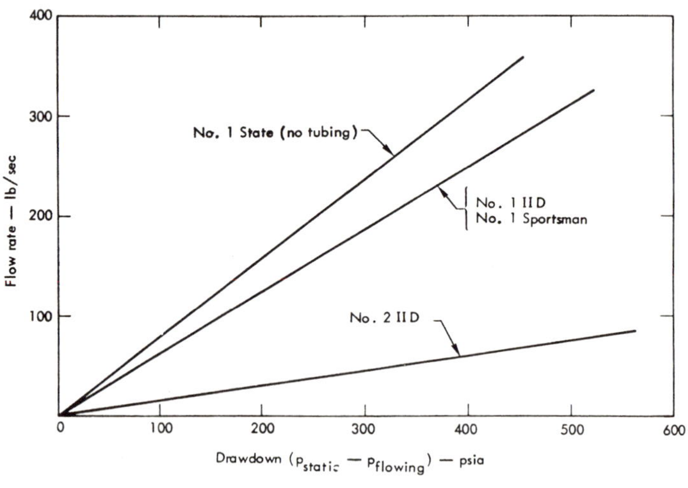

As a Function of Drawdown Pressure

Source: UCRL-51366

They are probably the cause of high production capacities. It may be very important that they be intersected by a well if the well is to produce at the rates noted so far. To date such intersection has not been difficult, so flow capacities to meet the described power plant requirements should pose no problem.

Another point about flow capacities of wells in the area is that Otte reports no problems in injecting waste brine back into the system. In fact, injection capacity exceeded production capacity.

Mechanism for Generating Geothermal Heat in the Salton Trough

The most important point regarding the mechanism for generating geothermal heat in the Salton Trough is that the motor driving the system is of global proportions. It is virtually certain that it will continue to maintain the fundamental geothermal system of the Salton Trough for an indefinitely long period of time despite any exploitation of this resource. The regeneration rate is presently unknown, however.

There are two parts to the mechanism for developing hot brines in the Salton Sea geothermal area; flow of water into and its accumulation in the Imperial Valley-Salton Sea area, and heating the water and concentrating the dissolved solids in the brine. Water input to the brine system of the Salton Sea geothermal area appears to come from two sources. Most recharge is from run-off from nearby mountains, and some may come from the Colorado River. Various workers (e.g., Helgeson) argue about the relative roles of these sources, especially insofar as they affect the chemical composition of the brines.

Recharge water in the Salton Sea geothermal area that descends to depths greater than ~3,000 ft is warmed by heat normally associated with the earth's mantle. This water apparently becomes part of a circulating convection cell (see Figure 7.9d) that taps this very deep heat source.

As noted earlier, the Salton Trough is the topographic extension of the Gulf of California. Both are, to a first approximation, astride the East Pacific Rise, an oceanic ridge of global importance. Ridges such as this are the principal zones separating the earth's crust and upper mantle and are the sites of upwelling hot mantle rocks. The East Pacific Rise intercepts the North American continent at the Salton Trough.

FIGURE 7.9: MODEL OF RIFTING AND MAGMA GENERATION DURING GROWTH OF THE SALTON TROUGH

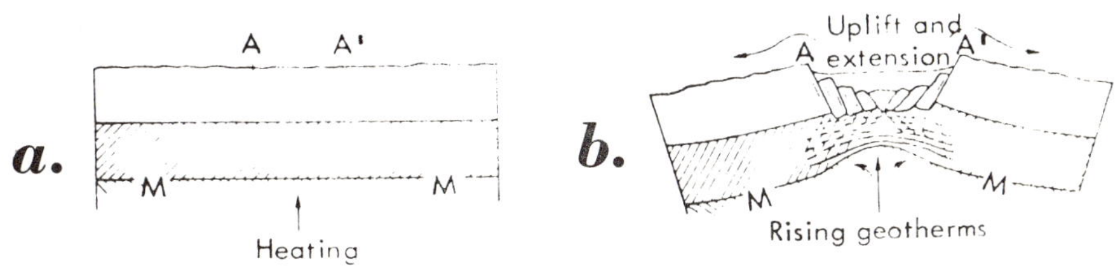

(continued)

FIGURE 7.9: (continued)

c.

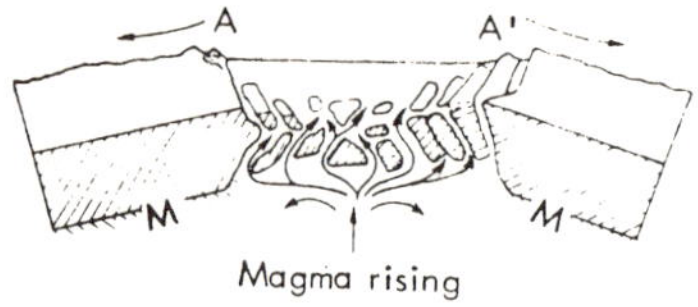

d.

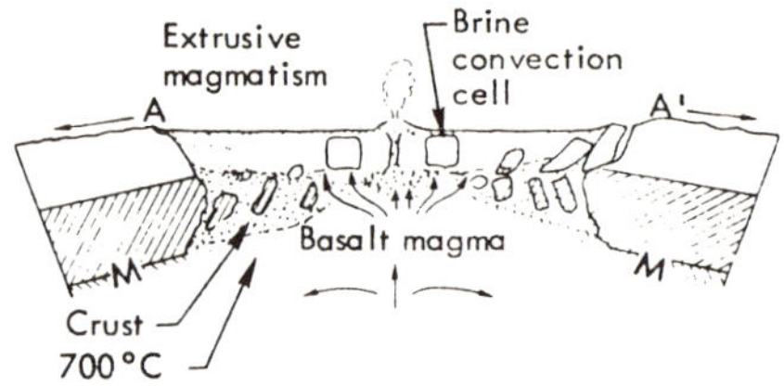

Source: UCRL-51366

Elders et al describe the progression of events following the initial interception, and Figure 7.9 shows their interpretation where the sections are drawn parallel to major strike-slip faults in the area. In Figure 7.9a, two layers of crust overlie a hot zone in the mantle. **M** is Mohorovicic discontinuity and **A** and **A'** are reference points for later movements.

In Figure 7.9b, there is upward and lateral expansion; a trough is initiated and partly filled in by sediments. In Figure 7.9c, the widening trough is invaded by basaltic magma; metamorphism of the sediments and gravitational sliding of the tilted walls occur. In Figure 7.9d, melting of the basement and extrusion of rhyolitic magma occur. Figures 7.9b and 7.9d clearly show that the ridge allows a great deal of heat to be transferred to the base of the sequence of rocks filling in the Salton Trough. Not only do the brines become hot (note the convection cell in Figure 7.9d), but volcanism also develops. Volcanism of this sort is seen at the south end of the Salton Sea, in the Obsidian Buttes area.

THERMODYNAMIC ANALYSIS

A thermodynamic analysis comparing the total flow system with the flashed steam and binary fluid systems is given in pages 14-21 of UCRL-51366. The following conclusions are drawn.

Since the overall turbine efficiencies are all ~70% it is evident that the total flow system is most efficient. Table 7.2 (p 263) shows the expected power outputs per square foot of well. System efficiency is defined in Table 7.2 as the ratio of the electric power generated to the recoverable energy from the wellhead product. Thus, 60% more power is expected from a given well from the total flow system than from either alternate.

This increased power is expected because the total flow system uses the entire mass flow rate and, hence, a higher conversion efficiency of thermal energy to kinetic energy. The flashed steam system uses only 29% of the mass as steam to drive the turbine while discarding the remaining as hot brine at 338°F. The two-fluid system rejects all of the brine at 300°F at optimum efficiency. Reinjection of the hot brine is possible with all systems, so part of the rejected energy is returned to the reservoir.

TURBINE SYSTEMS

There is a wide variety of turbines, but they generally fall into two classes. The impulse turbine converts thermal energy to kinetic energy by a pressure drop through a nozzle only. The fluid kinetic energy is then converted to rotational energy by directing the stream to impinge on some kind of vane or bucket attached to a wheel. There is almost no pressure drop through the rotating components. The reaction turbine, usually used to recover energy from high pressure gases, operates by allowing the gas to expand as it passes through the rotating vanes, and this expansion is accompanied by a pressure drop. To prevent bypass of the rotating components there must be very close tolerances between the vanes and housings or stators.

Therefore, there are two possibilities for recovering energy from the full fluid stream; expanding the two-phase mixture through a nozzle to obtain high velocity jets that are used to drive an impulse turbine, with the entire pressure drop occurring in the nozzle, or expanding the two-phase mixture through a reaction turbine where both a velocity change and a pressure drop occur.

The axial flow reaction turbine has a higher efficiency, normally operates at higher speeds, requires multiple stages, and is generally more complex, requiring very small clearances to maintain the pressure drop/stage. But a multistage turbine cannot be expected to function reliably when driven directly by mixture of vapor, brine, silica, and probably, sand and other extraneous materials.

This nature of the working fluid requires that the turbine be simple and easily maintained; the impulse turbine therefore appears to be the best choice. The most common impulse turbine is the axial flow of configuration in which the fluid is fed through a series of nozzles located around the wheel periphery.

The maximum efficiency of both single and multistage turbines is ~88%. The influences of blade friction, leakage past the blade, turbulence, and fanning will reduce this efficiency. In addition, large axial forces can be developed, and control of vibration is always a serious factor in axial flow turbine design.

The other alternatives are radial and tangential flow impulse devices; these appear to be most promising for this application. They are shown schematically in Figure 7.10. The radial flow device shown is similar to the Francis hydraulic turbine except that nozzles are used here in place of inlet guide vanes. Proper selection of the rotor inlet and outlet vane angles will direct the exit fluid radially inward with no tangential velocity component (whirl).

In principle, the blade efficiency for this condition can be 100%. This is unlikely, however, since turbulence losses will occur, and it may be necessary to select vane angles to give some whirl to the exit fluid. The advantages of the radial flow impulse turbine are high efficiency, reduced fanning losses, flexibility in choice of nozzle angle, reduced losses by jet dispersal, and minimal vibration problems. There are few vibration problems because the nozzle lies in the plane of the wheel, producing low axial forces.

FIGURE 7.10: TOTAL FLOW IMPULSE TURBINE CONCEPTS

a.

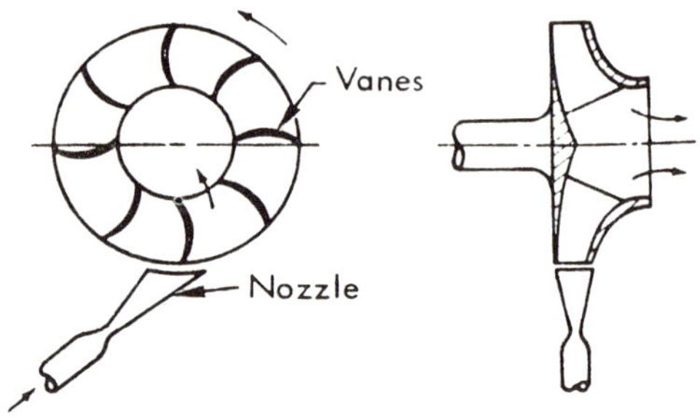

Radial Inflow Turbine

b.

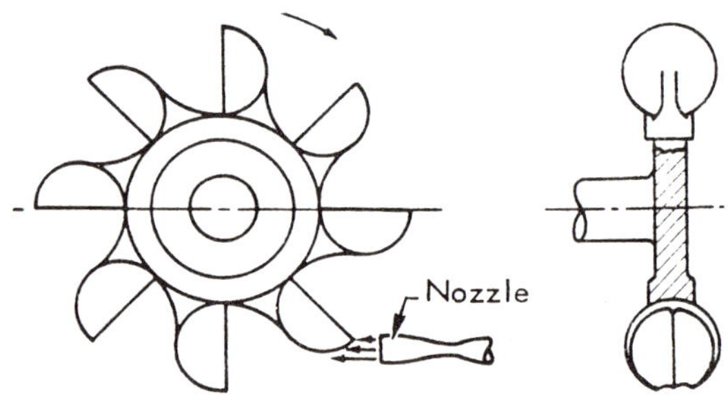

Tangential Flow Turbine

Source: UCRL-51366

The tangential flow turbine shown in Figure 7.10b is modeled after the Pelton wheel. Although it has only been used as a hydraulic turbine, there is no fundamental reason why it cannot be operated with a two-phase fluid. This device can have an efficiency as high as 95%, depending on bucket splitter angle and exit angle. Blade friction, of course, may be important, but the major losses (and disadvantage) will probably result from fanning and jet dispersal, since the nozzles must be placed some distance away from the buckets.

The major advantages, however, are inherent high efficiency, simplicity of fabrication (and resultant low cost), low potential for vibration, and the fact that the buckets may be readily replaced or relined. This latter point may be very important since the working fluid will be very corrosive.

Typically, hydraulically-operated devices of these types exhibit wheel efficiencies greater than 90%. No data exist for operation in the two-phase flow regime, and it is not possible to report efficiencies for this operating condition. Nevertheless, there is no fundamental reason why gross differences should exist, so it seems reasonable that a two-phase total flow turbine could be developed to give a 90% efficiency.

A TRIAL CONCEPT

To discuss the nature of the design problem, we shall examine one possible system. It should be recognized that this example is intended only for illustrative purposes, and is not necessarily the optimum. Figure 7.11 is an elevation view of an installation with a radial inflow turbine.

FIGURE 7.11: SCHEMATIC OF A TOTAL FLOW SYSTEM

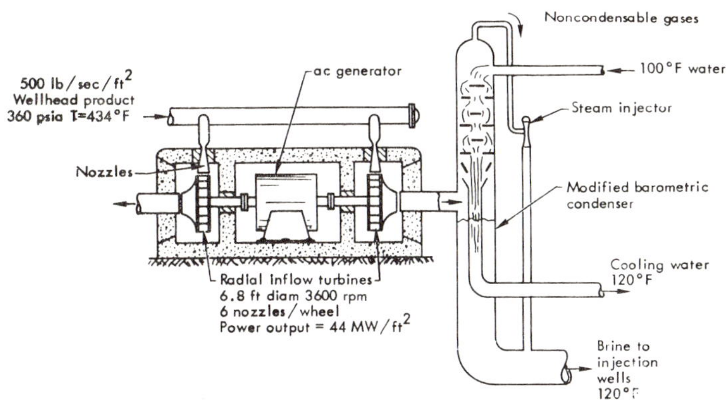

Source: UCRL-51366

Since the jet area is 17 times the inlet area, multiple nozzles and/or wheels will be necessary. The particular system chosen for this example is a single generator operated by two turbines with six nozzles/wheel. In this case the turbines have a diameter of 6.8 ft, rotate at 3,600 rpm, contain nine equally spaced vanes, and (with an assumed turbine efficiency of 90%) produce about 44 Mw/ft² of wellbore area. Hence, this power output requires two 9⅝" i.d. wells, each flowing at 250 lb/sec.

This is not a large system compared to conventional hydraulic installations; however, the peripheral wheel speed is much higher than usual and will result in high stress levels, somewhere between 50,000 and 70,000 psi for a steel wheel. The stresses would scale down linearly with density for lighter materials, but it may prove more effective to develop design alternatives that either use lower mass flow rates per turbine or smaller nozzle angles or require pressure staging as a means of lowering peripheral wheel velocity.

A two-stage device, for example, reduces the peripheral velocity enough to reduce the stress to 40% of its former value, but it makes a more complex and possibly less efficient system. Nevertheless, many alternatives exist, and arriving at an optimized system will be a fundamental goal of a research and development program.

The barometric condenser can do the cooling and could be modified as indicated to allow return of the cooling water without excessive contamination by the exhaust brine. The cooling water can be fed through either a cooling tower or a spray pond. Preliminary calculations indicate that for a cooling range of 20°F, each 44-Mw module (500-lb/sec flow) requires ~9,000 lb/sec of cooling water flowing between the condenser and pond. This includes 170 lb/sec added from the condensed vapor.

The evaporative cooling loss is calculated to be 172 lb/sec. Depending on drift losses, there will be either a small net gain or loss in cooling water flow. Since there are usually no strong prevailing winds in the area, a circular spray pond with nozzles spaced equally along the periphery of circles with radii every 25 ft would function effectively from vertical air movement by convection.

Detailed information on the composition of the condensed vapor phase is needed, however, to select the optimum cooling system. Disposal of the concentrated brine outflow from the barometric condenser will be done by injection through a number of wells equal to the number of production wells and extending to about the same depth.

CHEMICAL AND CORROSION FACTORS

The major problem with the brines in the Salton Sea-Mexicali Valley province results from their chemical composition (see Table 7.1). These brines are corrosive in the extreme, cause scale deposits, and under certain circumstances can saturate and deposit salts.

Brine Physical and Thermal Properties

The specific weight of the brine varies with pressure and temperature, as shown in Figure 7.12. This curve was calculated by assuming no flashing. The specific weight of the brine at the turbine outlet after 36.8% of the water is vaporized is 75.7 lb/ft^3. The average specific weight of the brine in the reservoir is ~62 lb/ft^3.

Therefore, the reservoir pressure is greatly exceeded by the hydrostatic pressure developed in the injection wells. For a 5,100-ft well, the bottom hole pressure difference will be ~556 lb/in^2. If silica does not plug the injection well and injection pressures are the inverse of drawdown pressures, the injection wells will accept up to 700 lb/sec/ft^2 of well before injection pumps are needed. Since production is estimated to be 500 lb/sec/ft^2 of well, it is probable that no significant energy will be needed for injection.

In the thermodynamics and turbine analyses, it was assumed that the properties of brine were approximately those of the same weight of pure water. Since 1 lb of mixture at the wellhead contains only ~0.743 lb of water and 0.257 lb of salt, the heat available from water is only 70% as great as indicated in the Steam Tables plus the heat available from cooling the salt.

Calculations show that the net heat available to do work as the brine passes through the nozzle is 88.0 Btu/lb, or ~9% less than from pure water. In this sense, the power calculations in the thermodynamic analysis may be too high by approximately this amount; however, in view of all the estimated factors involved, the expected power output shown in Table 7.2 is not in serious error.

FIGURE 7.12: BRINE PROPERTIES

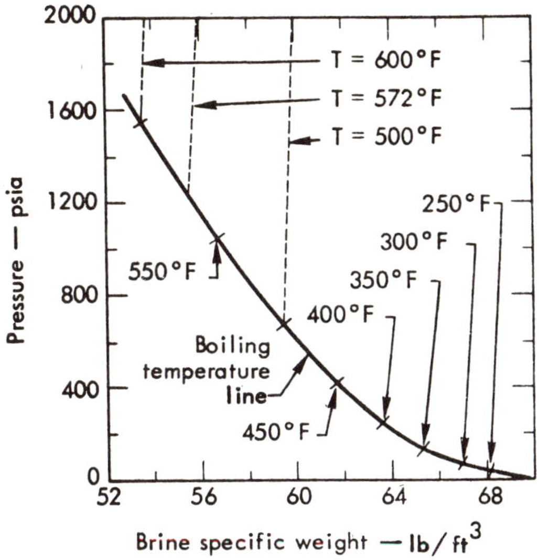

Source: UCRL-51366

TABLE 7.2: EXPECTED POWER OUTPUTS

System	Ideal Power (Mw/ft^2)	Expected Power (Mw/ft^2)	System Efficiency (percent)
Flashed steam	$39.5\ e_t$	27.7	11
Binary fluid	$40.4\ e_t$	28.3	11
Total flow	$63.1\ e_w \eta^2$	44.2	18

Source: UCRL-51366

Corrosion

Geothermal brines are notorious for their corrosive character. The Salton Sea brines are likely acid. Titanium is less reactive in neutral or alkaline solutions than in acid solutions and is particularly susceptible to crevice corrosion and hydrogen embrittlement in a highly acidic environment; therefore titanium may not be acceptable in the California brines. The 13% chrome ferritic stainless steels at Rockwell B-60 to C-28 hardness were found to be satisfactory in New

Zealand, and should be investigated for application in California. Monel metal, Inconel and Hastalloy are known to be unsatisfactory. The only materials surely corrosion-resistant to the Salton Sea brine are ceramics, tantalum, plastics, and perhaps zirconium.

Methods for tantalum plating have been developed, making tantalum-coated steel a possible production tubing, but costs will probably be high. The high temperature plastics, such as Teflon, used for cookingware coating may prove satisfactory. Glass lining or all glass may also be considered for use as well production tubing.

The nozzles and turbine components can probably be tantalum coated, but if extreme abrasion resistance is needed, tantalum carbide linings may be necessary. Even though much of the advanced materials technology exists, a development program will be required to apply this knowledge for commercial production of corrosion resistant components and to minimize scale formation.

Scale Formation

Scale comprised of magnetite (Fe_3O_4), Hausmanite (Mn_3O_4) and silica (SiO_2) has been observed to form in wellhead plumbing at every well tested. This scale is reported to form on any surface; even such inert materials as Teflon are soon deeply coated. It is believed that formation of such scales is a consequence of the material used for well casing. There is a series of reactions between iron ions and water of the form $2Fe^{+3} + Fe^{+2} + 4H_2O \rightleftharpoons Fe_3O_4 + 8H^+$.

Similar reactions occur between manganese ions and water. These reactions are reversible as long as the solution remains in contact with the scale, and when the acid strength reaches ~pH 3, no further scale should be formed. In the formation it is believed that no scale is being formed and the solution is stable.

If, however, the solution contacts a material that consumes acid (e.g., limestone) or a reactive metal (e.g., iron), the equilibrium is disturbed by removal of acid and more scale and acid are formed while iron is removed from solution. Thus, it is believed that scale formation can be minimized by meticulous avoidance of any material that reacts with acid at pH 3 in the well or surface equipment.

For example, ordinary cement will probably react with the solution and cause scale formation until sufficient scale is deposited to seal the cement from contact with the solutions. Cements used in well completion must be compatible with high acid concentrations (e.g., gypsum-based cements).

Brine Saturation

Using Helgeson's analysis (Table 7.1, column 2), the brine composition was recalculated as the sum of a group of chemical compounds. Table 7.3 shows these compound concentrations and their solubility at 100°C. In general, their solubility increases with temperature. As the right column shows, NaCl is closest to saturation, and as steam flashes from the brine, it will begin to precipitate when 61.1% of the water is vaporized. This concentration is never reached in the total flow system.

TABLE 7.3: COMPOUND CONCENTRATIONS AND SOLUBILITY

Compound	Concentration (g/cm^3)	Solubility at 100°C (g/cm^3)	Concentration/ solubility
NaCl	0.1345	0.346	0.389
KCl	0.0315	0.502	0.063
$CaCl_2$	0.0772	1.41	0.055
LiCl	0.00128	1.13	0.0011
$MgCl_2$	0.000039	0.374	0.00010
$SrCl_2$	0.000796	0.089	0.0089
$BaCl_2$	0.000375	0.523	0.00072
RbCl	0.000099	1.231	0.00008
CsCl	0.0000253	2.40	0.000010
$FeCl_3$	0.00581	4.75	0.00122
$MnCl_3$	0.00402	large	small
$PbCl_2$	0.000107	0.0296	0.0036
$ZnCl_2$	0.00104	5.45	0.00019
$CuCl_2$	0.0000169	0.957	0.000018
SiO_2	0.000400	—	—
H_2O	0.8862	∞	0

Source: UCRL-51366

The calculated steam quality at the points of highest concentration (nozzle outlet, turbine chamber, and barometric condenser) is 36.8%, which is the same as removing 0.326 lb of water per pound of original brine, or ~70% of saturation. At 25°C the solubility of NaCl is 0.316 g/cm^3, so if the brine were cooled separately from the vapor, salt would reach saturation when 57.4% of the water was removed. Concentrations never approach these values in the total flow system, so no problems are expected from salt formation.

The SiO_2 solubility is not shown in Table 7.3 because its solubility is a complex function of both temperature and pressure. In all likelihood, the SiO_2 concentration is in saturated equilibrium in the formations, and this should be very close to the saturated equilibrium value for quartz-water or amorphous silica-water corrected for the presence of other salts.

Figure 7.13 shows the lower two-phase region of the quartz-water system corrected for the presence of other salts on the assumption that there are no chemical reactions between the other salts and silica. If the actual solubility is more like amorphous silica in hot water, then the silica will be 2.5 times more abundant at the higher temperatures.

Below ~450°F the actual value will be close to that shown. Therefore, as the brine passes up the well and through the nozzle, silica will be continuously precipitated. Assuming 300°C initial reservoir temperatures, the brine will carry more than 600 ppm of SiO_2. At the nozzle outlet (120°F) essentially all this will be insoluble, so silica will be produced in the condenser outlet at the rate of 25,920 lb/day/ft^2 of well if the brine flow rate is 500 lb/sec/ft^2 of well. This amorphous silica will create a serious problem if it does not remain in colloidal suspension from production through injection. If the brine acidity is not disturbed, the silica should be expected to remain colloidal.

FIGURE 7.13: LOWER TWO-PHASE REGION OF THE SILICA-WATER SYSTEM

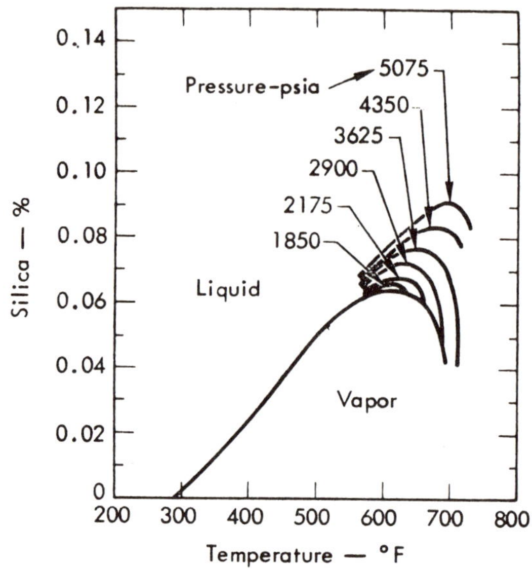

Source: UCRL-51366

Well Completion Design

Possible downhole salt precipitation may impose additional constraints. Several problems have been encountered during well completions. First, during casing at intermediate levels, the annular space between casings must not contain confined water. If water is left in the annulus during cementing, it heats up during steam production and collapses the inner casing or production tubing when it expands.

Second, flow velocities from the point of vapor formation to the wellhead must be high enough that appreciable slip of brine does not occur. At low flow velocities there is a tendency for the vapor phase, as bubbles, to rise faster than the liquid. This causes higher and higher salt concentrations in the well and the concentration at which salt could precipitate might conceivably be reached. The two-phase mixture can be produced without salt concentrating if the flow velocities are maintained above 100 ft/sec.

Figure 7.12 shows that the specific weight of the brine in the deep well averages

56 lb/ft^3. Thus, 500 lb/sec is equivalent to 8.9 ft^3/sec; since this mass flow is associated with 1 ft^2 of well, the brine velocity in the lower pipe is 8.9 ft/sec. To achieve 100 ft/sec the diameter should be reduced to 0.30 times its original value. A 9⅝" casing would be reduced to 2.89" at that depth, for example. As the brine-vapor mixture passes up the pipe and more vapor is formed, the diameter should be increased again to keep the flow velocity from becoming too high and increasing the pressure drop due to frictional losses.

Assuming 200 ft/sec as an upper limit to velocity, the production casing area can be determined directly from the continuity equation. Since v is ~0.25 ft^3/lb at the wellhead, the area is ~0.625 ft^2, or, for a 9⅝" well, the tubing diameter at the surface would be 7.6".

Figure 7.14 is a schematic drawing showing the entire well casing and production tubing in concept. As shown in Figure 7.14, 30" surface conductor would be set in a 36" hole to a depth of 200 ft. The hole would then be drilled at 24" diam to ~2,300 ft and an 18" diam casing set at that depth. Then the hole would be drilled at 17½" diam to 4,350 ft and a 13⅝" casing set and carefully cemented. The hole would be completed to 5,000 ft with a 12" bit. The production tubing string would consist of corrosion-resistant material with wipers set at ~4,000 ft on a section 9⅝" diam tubing that would extend from 5,000 to 3,650 ft and be perforated or slotted below 4,600 ft.

FIGURE 7.14: WELL CASING DIAGRAM

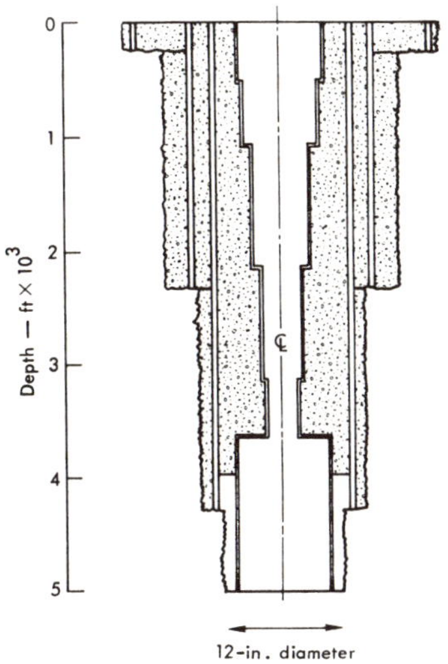

Source: UCRL-51366

From 3,650 ft to the surface there would be 500 ft of 3" tubing, then 1,000 ft of 4" tubing, 1,200 ft of 5⅝" tubing, 500 ft of 7⅝" tubing, and 450 ft of 8⅝" tubing. With such a production tubing, flow velocities will be between 100 and 200 ft/sec everywhere. It is not yet clear whether the whole annulus between the production tubing and casing should be cemented or left partly open to facilitate replacement.

ENVIRONMENTAL CONSIDERATIONS

The total flow geothermal power system will have little adverse environmental impact. Since fluids are to be produced and injected at the same depth there will be no subsidence or accumulation of salts or brine on the surface or in any other aquifer. Noncondensable gases are to be injected by aspirator into the disposal wells so there will be no atmospheric contamination.

It is necessary to have an inventory of less concentrated brine circulating between the barometric condenser and cooling system. It is expected that this might amount to ~5,000,000 ft^3 circulating at a rate of ~45,000 lb/sec for a 220-Mw power module. This brine could be contained in an impermeable earthen pond of about a 400-ft radius by 6 ft deep. Unless this tank were breached, there should be no adverse contact between the brine and the environment or ground waters.

The turbines will be noisy if unshielded, but since they possess so much stored energy it will be necessary to build massive protective cases anyway. It may be desirable to emplace the entire power station except the cooling water system and transformer bank just below ground level to provide earthen sound barriers and shields.

In all, a fully developed geothermal brine area would have transformer banks, cooling ponds, wells, and generators occupying only ~5% of the surface of the land. Such a development should be able to produce all the underlying geothermal energy.

COST ESTIMATES

Estimating costs of an operating geothermal total flow system is difficult, since neither the materials nor machine design problems are solved and the optimum cooling system is unknown. Nevertheless, it is desirable to make a crude economic evaluation to determine if geothermal power development should be undertaken.

In this analysis, an attempt was made to arrive at an estimate that is valid to within 30% by scaling costs of analagous plant equipment from the chemical industry, from other geothermal plants, and from the petroleum industry. Hopefully the estimate is as likely to be too high as too low.

Assumptions

It is assumed that plant module is served by 10 producing wells of 9⅝" nominal production diameter driving three to five alternators, each with two turbines. In case the radial flow machine is used, each turbine would be driven from one well and two turbines could be used to drive one alternator, so there would be five turbine alternator combinations in the module. With the tangential flow turbine, each wheel would be driven by one well, with four wheels for each alternator.

Peak capacity of the module is 220 Mw and with a load factor of 80%, the annual power production is 1.53 billion kwh. It is assumed that each such module will be constructed independent of all others, so there will be no shared shops, utilities, transformer substations, or management.

Finally, all estimates are based on the assumption that research and development has been completed and experience has been gained from a commercial scale pilot plant. The first plant will probably be the most expensive.

Capital Costs

The major elements of capital requirement are the production and reinjection wells, power house and plant that houses and services the wells and turbo-generators, the turbine-alternators, and a high voltage substation.

In estimating the production and reinjection well costs, they were assumed to be the same as offshore wells drilled to the same depth. In addition, the cost of an exotic production tubing was added as if it were fabricated of pure titanium, although it very well may not be titanium and may be more or less costly. A hole loss factor of 0.2 has been assumed; that is, that during drilling one of each five attempts to complete a hole will be unsuccessful. These lost holes were distributed against the drilling costs of the completed wells at a value equal to 75% of the base drilling cost without the completion tubing. The well life is assumed to be 20 years. The estimates are:

Drilling and casing 20 wells, 5,000 ft ea @$20/ft	$2,000,000
Lost hole allowance, 20,000 ft ea @$15/ft	300,000
Special tubing, 100,000 ft @$22/ft	2,200,000
Contingency @15%	675,000
Subtotal	$5,175,000

Capital investment for the remainder of the facility was estimated from New Zealand experience, appropriately inflated to 1973. Note that all these costs are higher than current fossil fired plants.

Power house and plant facility @$49.50/kw	$10,791,000
Turbine and generator @$56.54/kw	12,325,720
Cooling system @$17.62/kw	3,841,160
High voltage substation @$11.11/kw	2,421,980
Contingency @15%	4,406,979
Subtotal	$33,786,839

The drilling costs totaling $51.75/ft compare favorably with the costs in New Zealand, which are $51.10/ft when inflated to 1972-73 dollars. Table 7.4 shows the total capital costs. Note that the greatest uncertainty is the tubing cost. However, even if the cost were multiplied fivefold, the capital costs increase to only $214/kw, which is still less than current systems. All capital construction costs include architect and engineer fees, land, utilities, and services.

TABLE 7.4: CAPITAL COST SUMMARY

Drilling	$ 5,175,000	
Power house, plant, and high voltage substation	$15,194,927	
Turbines, generators, and cooling system	18,591,912	
Total capital	$38,961,839	(or $40,000,000)
Capital costs/kW	$180	

Source: UCRL-51366

Operating Costs

Operating costs were estimated by using operating chemical process plant figures augmented by an assumed well maintenance cost. Well production life was assumed to be 20 years, but it was assumed that the production tubing would be replaced every 5 years. To average this cost an annual operating cost was allocated equal to 20% of the initial capital cost of production tubing.

Maintenance and plant repair were estimated at 1%/yr of original capital cost of the power house, plant, and high voltage substation and 5%/yr for the units subject to corrosion: turbines, generators, and cooling system. To calculate operating labor and supervision, it was assumed that two operators would be required on each shift and that there would also be two craftsmen required on each weekday shift.

TABLE 7.5: ANNUAL OPERATING COST SUMMARY

Production well maintenance	$506,000	
Power house, plant, and substation maintenance	151,949	
Turbines, generators, and cooling system maintenance	929,596	
Labor and supervision	190,362	
State and local taxes	243,511	
Insurance	389,618	
Contingency @ 30%	723,310	
Total cost of sales	$3,134,346	(or $3,000,000)

Source: UCRL-51366

A 30% supervision cost was added. State and local taxes were calculated as 5% of the assessed valuation, which was estimated as 25% of the depreciated capital value of the whole facility. This is shown in the cost summary for the average year at 0.625% of the initial capital cost. Insurance is shown as 1% of the initial capital cost and a contingency of 30% is included because of the gross method used in calculating maintenance and operating expenses. Corporate management is included in the maintenance costs. Table 7.5 summarizes the operating costs.

Cash Flow and Profitability

There are a number of uncertainties that prevent a rigorous cash flow analysis. These include lease royalty rate, depletion allowance, regulated permissible rate of return, method of financing, and corporate tax structure for calculating federal income tax. In the example analysis below an 80% load factor was assumed; without prejudice to future legal or regulatory actions, lease royalty at 12.5% and 5% of gross sales; internal financing; and computed internal rate of return as a function of power sales price.

Table 7.6 shows the example using 6 mills/kwh, 12% royalty, no depletion allowance, and 48% federal income tax.

TABLE 7.6: CASH FLOW

Power sales @ $0.006/kWh	$9,180,000
Royalty @ 12.5%	1,147,500
Cost of sales	3,134,347
Gross profit	$4,898,153
Depreciation	$1,948,092
Net profit before federal income tax	2,950,061
Federal income tax	1,416,029
Net profit after taxes	1,534,032
Total cash flow	3,482,124
Years to return investment	11.9
Internal rate of return	6.4%

Source: UCRL-51366

Figure 7.15 shows the rate of return on the investment for various power sales prices. The upper curve is calculated with 5% royalty and no federal income tax. The lower curve is calculated with 12½% royalty, no depletion allowance, and 48% federal income tax. The actual profitability should fall between these two curves. Depreciation has been calculated as straight line for 20 years and the investment and plant life are assumed to be 20 years, with no recoverable value at the end of the 20th year.

As Figure 7.15 shows, there is no return on investor capital at a power sales price of 3.2 mills/kwh, and, if bank rate interest is assumed to be 7.5%, this investment yields in excess of bank rates for all power prices greater than 6.5 mills per killowatt hours.

FIGURE 7.15: EFFECT OF POWER SALES PRICE ON INTERNAL RATE OF RETURN

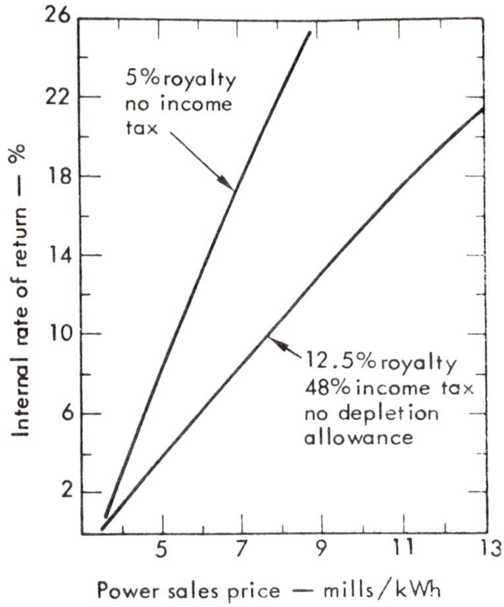

Source: UCRL-51366

The profit rate is not terribly sensitive to the royalty rate. For example, assuming a 5% royalty instead of 12.5% increases the rate of return at 8 mills/kwh sales price from 11 to 12.5%. At the same power sales price, eliminating federal income tax increases the internal rate of return from 11 to 19.1%. If a 22% depletion allowance were permitted at 8 mills/kwh sales price, the internal rate of return would increase from 11 to 14.5%.

In view of these calculations and considering that straight line depreciation and generous contingencies were used, development of the total flow geothermal powe system should return profits to the investor commensurate with the risk at a rate competitive with fossil fuel or nuclear power generating systems.

The state and local taxes should approach $100 million/year if the whole resource is developed, and federal taxes would be 10 times as large. This represents a significant revenue to the whole public.

SPERRY RAND METHOD

The method of *H.B. Matthews; U.S. Patent 3,824,793; July 23, 1974; assigned to Sperry Rand Corporation* provides means for efficient power generation employing energy derived from geothermal sources through the generation of dry superheated steam and the consequent operation of subsurface equipment for pumping extremely hot well water at high pressures to the earth's surface.

Figures 7.16 and 7.17 illustrate the structure of the portion **1** of the geothermal energy system which is immersed in a deep well in strata far below the surface of the earth or other planet and which is located particularly at a depth such that a copious supply of extremely hot water under high pressure is naturally available, the structure being supported at the second station within a generally conventional well casing pipe or conduit **2**.

FIGURE 7.16: DEEP WELL PUMPING APPARATUS

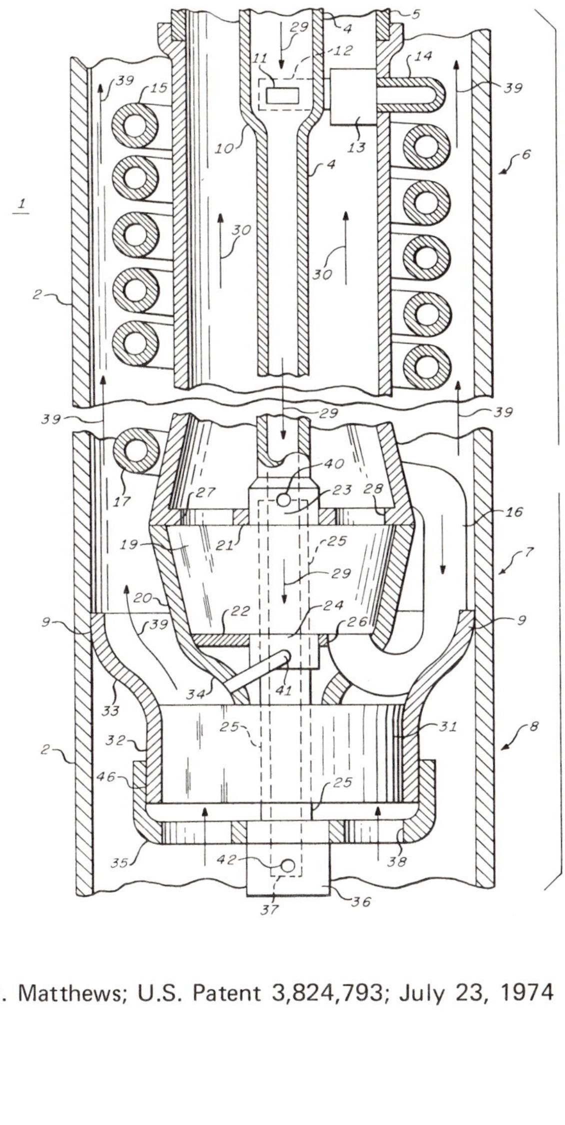

Source: H.B. Matthews; U.S. Patent 3,824,793; July 23, 1974

FIGURE 7.17: SURFACE COOPERATING APPARATUS

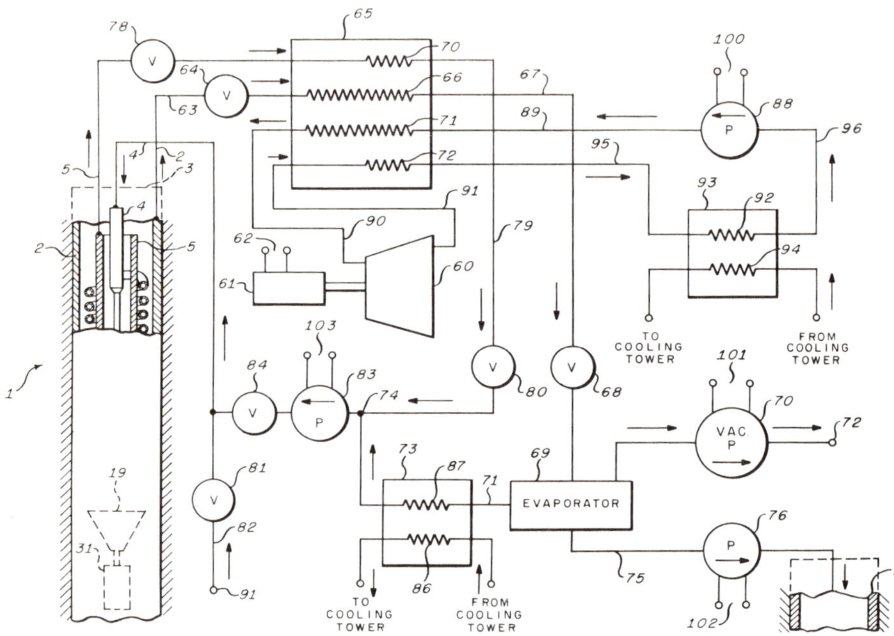

Source: H.B. Matthews; U.S. Patent 3,824,793; July 23, 1974

Extending downward from the well head **3** (seen in Figure 7.17 and located at or near the earth's surface), the casing pipe **2** surrounds in preferably concentric relation an innermost stainless steel or other quality alloy steel pipe or conduit **4** for supplying relatively cool and relatively pure water near the bottom of the well for purposes which will be explained.

A second larger internal diameter pipe or conduit **5** of similar quality and surrounding pipe **4** is also provided within casing **2**, extending from well head **3** to the energy system apparatus at the bottom of the well for permitting the hot turbine exhaust steam to flow to the surface of the earth, as will be described.

The geothermal energy extraction apparatus **1** located at the bottom of the well comprises, as will be explained, several cooperating elements, including a steam generation section **6**, a steam turbine or other rotary motive section **7**, and a rotary hot water pumping section **8** driven from motive section **7**. The several elements of the three sections are suspended in cooperative relation adjacent the bottom of the well within casing **2** and relative to the inner pipes **4** and **5**, partial support being supplied at seal **9** from the inner wall of casing pipe **2** and further support by concentric pipes **4** and **5**. Additional support or alignment means may be supplied, as will be evident to those skilled in the art, by additional support elements which may be of conventional nature. It will be seen in Figure 7.16 that relatively clean and cold water is pumped down inner pipe **4** from the first or surface station to the region of constriction **10**. On the upper side of constriction **10**, one or more openings are formed in pipe **4**, such as opening **11**, coupling cool water through branching pipe **12** to a conventional pressure regulator and reducer **13**.

Openings **11**, pressure regulator and reducer **13**, and the actual pressure of the clean water present in pipe **4** are arranged to permit clean water to flow through pipe **14** into the steam generation section **6** without undesired backflow of steam. The clean water flows from the conventional pressure reducer and regulator **13** located between pipes **4** and **5**, through pipe **14** and into a long alloy steel heat exchanger or boiler tube **15** formed as a plurality of turns which may be located in a cylindrical stack between pipe **5** and casing pipe **2**.

Extremely hot water flows upward in the region between pipes **2** and **5**, as will be seen, for converting the clean water injected into coiled tube **15** into highly energetic dry superheated steam. The clean water, before flowing through pressure regulator and reducer **13**, is at a very high pressure due to its hydrostatic head. Pressure reducer **13** drops this pressure sufficiently so that the clean water can be vaporized and super-heated by the well water. The resultant steam is directed through a downwardly oriented pipe **16** extending from the last turn **17** of coiled pipe **15** of steam generator section **6**, pipe **16** lying adjacent the inner wall of casing pipe **2**. In the vicinity of the last turn **17** of steam generator section **6** and of the downwardly extending pipe **16**, the pipe **5** is expanded in diameter so as to form an expanded section generally commensurate in cross section with the shape of steam turbine **19**.

Steam turbine **19** is supported within the confines of an oppositely tapered alloy steel cylindrical wall **20**, which wall serves as an extension of the expanded wall section and the opposed spaced parallel circular steel walls **21** and **22**. Turbine **19** is supplied with water-lubricated bearing housings **23** and **24** located at the centers of the respective circular walls **21, 22**. Bearing housings **23, 24** respectively contain bearings for direct support of the main shaft **25** of turbine **19**, as will be further explained.

Walls **20, 21** and **22** are solidly bonded together as by welding or other fastening means for serving as a supporting enclosure of turbine **19**; however, the high pressure steam pipe **16** is permitted entry into the input section of turbine **19** through aperture **26** in bottom wall **22**. Likewise, an array of large openings, such as apertures **27** and **28**, are provided in the upper circular wall **21** so as to permit exhaust of used steam from turbine **19** into the space between concentric pipes **4** and **5**.

Thus, relatively clean water flowing downwardly in the sense of arrow **29** within pipe **4** may flow after conversion as partially spent dry steam upward to the well top in the sense of arrows **30** in the region bounded by pipes **4** and **5**.

Pump **31** is supported within a cylindrical housing **32** concentrically located within casing pipe **2**. The upper portion of housing **32** is smoothly expanded in diameter to form a tapered wall section **33** having a shaped upper part suitable for sealing at **9** to the inner wall of casing pipe **2**. The tapered wall **20** surrounding turbine **19** is extended to form bell shaped section **34**, tapered wall **33** and bell-shaped wall **34** being generally similar in curvature so as to form an annular hydrodynamic fluid flow region between them, as will be explained.

Pump **31** is retained in part within housing **32** by a steel end wall **35**, welded or otherwise fastened to housing **32** at annular surface **46**. Wall **35**, at its center, accommodates an end thrust bearing housing **36** for supporting the end **37** of shaft **25**; it will be seen that bearing housings **24** and **36** contain bearings which may be clean water lubricated and which may cooperate in supporting shaft **25** and the impeller of hot water pump **31**.

End wall **35** has an annular array of apertures, such as opening **38**, which admit hot water to pump **31**; the latter fluid is accelerated by the impeller of pump **31** and exits therefrom into the annular region defined by the respective concentric tapered and bell shaped walls **33, 34**. Since these sections are respectively and smoothly joined to casing pipe **2** and to pipe **5**, the hot water is pumped upward to the top of the well in the sense of arrows **39**.

It is observed that the size of openings **11**, the characteristics of pressure regulator **13**, and the pressure applied to the cool water at the top of the well are adjusted appropriately for the supply of bearing-lubricating water through constriction **10** downwardly through pipe **4** to bearing housings **23, 24** and **36**. In the several bearings, the lubricating water is forced between bearing surfaces, maintaining their separation in the conventional manner while serving as a lubricant. Having served these purposes, the used water is permitted to exit from the respective bearing housings, as via the respective orifices **40, 41** and **42** of the housings **23, 24** and **36**.

The bearing lubricating water leaving orifice **40** in bearing housing **23** has sufficient pressure, for example, to flow into the exhausting steam from turbine **19**; being of small volume, it has insignificant effect. From orifice **41** in bearing housing **24**, the lubricating water simply flows into the rising hot water stream indicated by arrow **39**. The lower bearing housing **36** and the bearing it contains are lubricated by water that afterwards exits orifice **42** into the hot water being pumped by pump **31**.

In each case, it is seen that the pressure of the clean lubricating water is such as to prevent undesired reverse flow. For example, at orifice **42**, the clean water pressure is sufficient to prevent contaminated well water from entering housing **36** and destroying the bearing therewithin. Since the operation of turbine **19** and pump **31** may in practice either tend to lift or to lower the shaft **25** common to them, the bearings within housings **23** and **36** necessarily have both alignment and thrust absorbing functions. As is well-known by those skilled in the art, such bearings may be designed to operate to maintain the fluid gap in a substantially balanced hydrostatic thrust condition.

As shown in Figure 7.17, the objective of the deep well apparatus of Figure 7.16 is to generate large quantities of electric power at the first or surface station using vapor turbines and electrical generators preferably located at ground level, such as vapor turbine **60** and the electrical alternator **61** of Figure 7.17, at power output terminals **62**. For this purpose, the hot water pumped to the earth's surface is fed by pipe **5** and its extension (pipe **63**) through the normally open valve **64** to element **66** of the conventional boiler-heat exchanger device **65**. Device **65** is of conventional closed tank-like nature and is designed to exchange heat between the several heat exchanger elements **66, 70, 71, 72** contained therein. The elements may take the form of lineal or coiled pipes exchanging heat energy by direct conduction through their metal walls or through a suitable interposed fluid in the well-known manner.

Heat from the hot water of pipe **63** is a major source of heat for supply to device **65**. The hot water, having been relatively dropped in temperature within boiler-heat exchanger **65**, is then fed via pipe **67** through the normally open valve **68** to the conventional evaporator **69**. Valve **68** may be a throttle valve adjusted for the purpose of dropping the pressure of the fluid flowing through it so that the fluid will readily flash at low temperatures when supplied to evaporator **69**.

Evaporator **69** is of conventional nature and is supplied in the usual manner with a conventional vacuum pump **70** which considerably reduces pressure within evaporator **69**, causing the water therein to boil, and releasing steam via exhaust **71**. Vacuum pump **70** serves to remove the most volatile gasses at terminal **72**, some of which gases are undesirable, being corrosive, from circulation past evaporator **69** into condenser **73**. Valuable gases, such as helium, and other noble gases, may be extracted at terminal **72** and utilized, if desired.

Evaporator **69** performs two functions, as suggested in the foregoing paragraph, it generates clean steam which is condensed by the conventional condenser **73** and is supplied as water at junction **74** for augmenting the clean water supply. A further major portion of the water originally flowing upward in pipe **5** is returned by pipe **75** through pump **76** to the earth well formed by pipe **77**. Thus, a major portion of dissolved mineral salts pumped to the surface in solution in the hot water in pipe **5** is returned by pump **77** into the ground. The well formed by pipe **77** may be reasonably remote from the well of the thermal system **1** and may serve more than one such system. It may pass the liquid from pump **76** into an earth stratum different from that associated with the portion **1** of the system, if desired.

A second source of energy supplied to boiler-heat exchanger device **65** is the steam exhausted from the deep well turbine via pipe **5**. This steam is permitted to flow through normally open valve **78** to the heat exchanger element **70** of boiler-heat exchanger device **65**. Element **70** is arranged so that the steam therein is exposed to the thermal interchange at the coolest end of device **65** (adjacent the cool clean water input to heat exchanger element **71**).

Accordingly, the exhaust steam from pipe **5** is largely condensed within heat exchange element **70**. The water thus condensed is supplied through pipe **79** and the normally open valve **80** to the aforementioned junction **74**. The water from pipe **79** and that from condenser **73** arrives at junction **74** in relatively pure state and may therefore be supplied to the cold water input pipe **4** of the apparatus of Figure 7.16. With valve **81** in branch line **82** closed, the water at junction **74** is fed by a conventional feed pump **83** through the normally open valve **84** and pipe **85** into pipe **4**. It will be appreciated that a variable capacity storage tank may be inserted at the general location of junction **74**, so that any fluctuations in supply of clean water for injection by pipe **4** may be smoothed. Also, such water may be supplied by opening the valve **81** from any available source coupled at terminal **91**. It will further be understood that condenser **73** may be water cooled, as by supply of cool water from a cooling tower (not shown) to heat exchanger element **86** in heat exchanger **73**. Alternatively, element **87** may be cooled in many locations simply by forced air.

The major elements for supply of heat into boiler-heat exchanger device **65** have been described. The heat stored therein is removed and used in a substantially conventional manner to operate the surface-located vapor turbine **60**. For this purpose, liquid is supplied by a conventional feed pump **88** via pipe **89** to the heat exchanger element **71** of boiler-heat exchanger **65**. Flow of the liquid is counter to the direction of flow of heat into device **65** in elements **66, 70** and **72**. The liquid evaporates and consequently generates extrmely high temperature vapor that is coupled via pipe **90** to the input stage of turbine **60**.

After performing useful work therein, the turbine exhaust vapor is fed by pipe **91** into boiler-heat exchanger **65**, where part of its remaining thermal energy is abstracted near the input of element **71**. The vapor then flows to a conventional condenser device **93** having heat exchanger elements **92** and **94** and then flows again as a liquid via pipe **96** to the feed pump **88**.

Condenser **93** may be cooled by flow of water from a cooling tower (not shown) through heat exchanger element **94**. The exchanger **93** may alternatively be air cooled in the conventional manner. A fluid such as water may be used for the generation of high temperature vapor within boiler-heat exchanger **65** and its associated surface-located loop or certain organic fluids affording best use in Rankine cycle operation may alternatively be employed.

FEASIBILITY STUDIES FOR THREE AREAS OF THE UNITED STATES

TEXAS

Summary

Below depths of 6,000 to 10,000 feet, sediments in Tertiary basins are commonly characterized by abnormally high pressures and temperatures and low salinities. Such zones are called geopressured zones and are known to occur worldwide. Geopressured areas occur in continuous belts, are commonly bounded by regional faults, and extend hundreds of miles. The belt in the northern Gulf of Mexico basin is about 750 miles long, extending from the Rio Grande of Texas to Mississippi Sound. It underlies the Coastal Plain inland from 60 to 100 miles, and underlies the Continental shelf up to 150 miles offshore.

The overpressured waters located at depth in the Gulf Coast and in similar Tertiary basins throughout the world represent a possible source of electrical power. It is the purpose of a study by E. Herrin of Southern Methodist University to determine the feasibility of locating a pilot project in the Texas Gulf Coast area for the purpose of tapping the overpressured aquifers and transforming the thermal and mechanical energy into electrical power. The study, AD-766 855 from which the material in this section is derived, is entitled *Development of Geothermal Reservoirs from Over-Pressured Areas Beneath the Gulf Coastal Plain of Texas* was prepared for the Air Force Office of Scientific Research.

Three areas in south Texas were given particular attention for their feasibility of being the site of the pilot project. These are the Sebastian area in northwest Cameron County, the Port Mansfield area in eastern Willacy County, and the Corpus Christi area. Logs taken from deep oil wells in these three areas were analyzed to determine formation pressure, temperature, and salinity as a function of depth. This analysis indicated that large aquifers were present beneath both the Sebastian and Port Mansfield sites. At about 15,000 feet, aquifers were detected which have temperatures in excess of 300°F, formation pressures greater than 10,000 psi, and salinities less than 20,000 ppm. Growth faults compartmentalize each of the areas into structural traps of areal extent in excess of 300 sq mi.

Careful environmental studies of both the Sebastian and the Port Mansfield sites indicate that the construction and 5 year operation of a pilot plant would have no significantly bad environmental impacts in either area.

It was concluded that it is feasible to construct and operate a pilot plant for electrical power production from a single well in south Texas. Over 4 Mw of power could be produced from a single well; 2 billion ft^3 of methane would be produced as a by-product. Based on geological and environmental considerations, it was considered feasible to locate the project in either the Sebastian or Port Mansfield areas.

Geopressured Reservoirs

Tertiary (age less than 80 million years) basins filled with clastic sediments (sand and clay or shale) are generally undercompacted below depths of 6,000 to 10,000 feet. The interstitial fluid pressure reflects a part of the overburden load, and the deposits are said to be geopressured. Aquifer systems within the geopressured section are compartmentalized by regional faults into blocks of horizontal extent ranging from tens to thousands of square miles. Interbedded clay or shale commonly has a porosity 6 to 8% greater than it would have if fully compacted at its depth of occurrence.

The geostatic ratio (the ratio of fluid pressure to pressure of the overburden load) is commonly 0.7 to 0.9 in the geopressured zone, which extends downward to the zone of metamorphism. Geopressured sections have been penetrated by thousands of wells.

Geopressured deposits occur in continuous belts, are commonly bounded by regional faults, and extend hundreds of miles. The belt in the northern Gulf of Mexico basin is about 750 miles long, extending from the Rio Grande of Texas to Mississippi Sound; it underlies the Coastal Plain inland 60 to 100 miles, and underlies the Continental Shelf wherever drilled up to 150 miles offshore.

Geopressured deposits are hotter than normally pressured deposits because upward loss of the included water has been essentially stopped for millions of years. Water is a poor conductor of heat (thermal conductivity about 20% of that of the associated mineral grains) and undercompacted clay is an excellent thermal insulator. In addition, the specific heat of water is about 5 times greater than that of the associated mineral grains. Thus, geopressured deposits greatly reduce the geothermal flux above them, and store geothermal heat.

The geothermal gradient is sharply increased near the top of the geopressured zone (the hydraulic boundary), and the geopressured deposits form a sort of pressure cooker. In this setting, thermal diagenesis of expandable clays liberates the bound and intracrystalline water, and the free pore water thus formed may equal 30% of the volume of the unaltered clay. This new free pore water is fresh. As it drains into adjacent sand-bed aquifers, it flushes the more saline water upwards towards the top of the geopressured zone. Aquifers a few thousand feet below the top of the zone commonly contain water having less than 10,000 mg/l of dissolved solids. In some places, the water is potable (less than 1,000 mg/l).

Sand-bed aquifer systems in the geopressured zone have permeabilities ranging upwards of 25 millidarcies; the viscosity the the water is commonly 0.2 to 0.3 cp; and the water is a chloride-bicarbonate type, slightly alkaline (pH 7.5 to 8.5). Because the solubility of hydrocarbon gases in water increases rapidly with decreasing dissolved solids and because the high temperatures and pressures have resulted in a natural cracking of petroleum hydrocarbons, the geopressured reservoir waters commonly contain 10 to 16 standard cubic feet of natural gas per barrel of fluid (approximately ½ ft^3/gal of water). Dissolved hydrocarbon gas would be a valuable by-product of fluid production.

Production history of geopressured reservoirs (mainly for natural gas production) indicates that there is replacement of produced fluid by water from undercompacted shales that bound the reservoir. There is no consistent relation between volume of produced fluid and resultant reservoir pressure. Repressuring of the reservoir occurs, to some degree, whenever production is stopped. It is apparent that production of water in geothermal developments would effectively drain the undercompacted shales bounding the aquifer tapped, but that reservoir pressures would be appreciably depleted only over a long production history, provided the size of the reservoir were at least several hundred square miles.

Temperatures of produced water would range from 150° to 180°C; well head pressures would range from 4,000 to 6,000 psi; and production rates would be several million gallons of water per day for each well.

Western Gulf Coast Region: The Western Gulf Coast Region has been selected as the region for location of the pilot geothermal well. Reasons for selecting this area are several. (a) The area is representative of the thick Tertiary clastic sections which are developed in numerous basins around the world. (b) Onshore and, more recently, offshore, petroleum exploration has been incredibly intensive in the Tertiary of the northern and western Gulf Coastal Plain. Subsurface data are correspondingly very abundant and accessible. (c) Geopressured zones are relatively common, both stratigraphically and regionally, thus simplifying the task of locating a site for initial experimentation. It is believed that information derived from studies of the Western Gulf Coast Region can be used to construct a model which sbould apply to overpressured, geothermal reservoirs in other areas.

Three specific areas in the Western Gulf Coast Region were selected for determination of their feasibility as the site for a pilot geothermal well. These areas are shown in Figure 8.1, where they are designated as (1) the Sebastian site; (2) the Port Mansfield site; and (3) the Corpus Christi site.

The subsurface temperatures and salinities of the northern Cameron County area in which the Sebastian site is located have been previously studied by the United States Geological Survey (Jones, P.H., 1969, *Hydrology of Neogene Deposits in the Northern Gulf of Mexico Basin,* Louisiana Water Resources

Research Institute Bulletin GT-2, Louisiana State University, Baton Rouge, 106 pages). Jones has identified the area as one of high temperature and low salinity. The area surrounding the Port Mansfield site has been identified by the United States Geological Survey as one of the spots in the Western Gulf Coast Region where the 300°F isogeotherm most closely approaches the surface (about 13,000 feet). The Corpus Christi site was included for study because of the potential logistical value of being able to use Cabiniss Naval Air Station for the pilot program.

FIGURE 8.1: MAP OF TEXAS GULF COAST

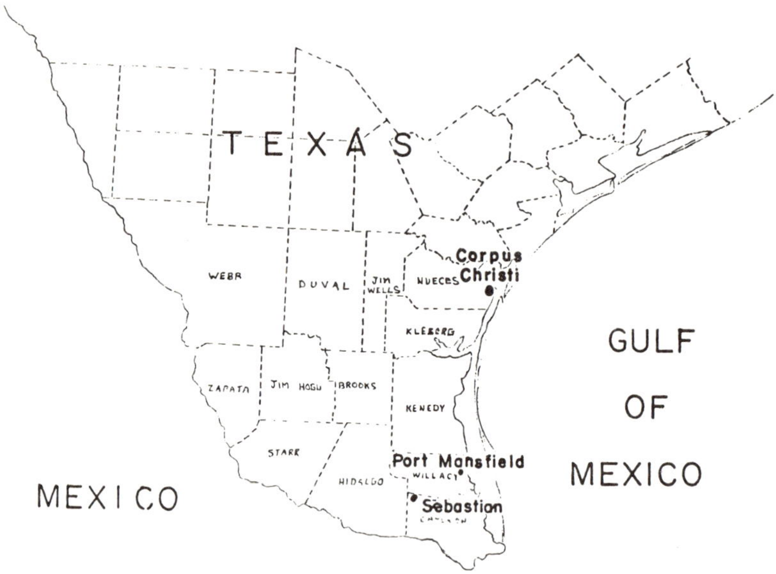

Source: AD-766 855

Subsurface Evaluation of the Test Areas

Survey maps showing the location and total depth of every oil well in the vicinities of the Sebastian, Port Mansfield, and Corpus Christi sites were obtained from Tobin Aerial Surveys. Electric logs from all wells with depth greater than 10,000 feet were visually analyzed, and aquifers were picked using the spontaneous potential and resistivity curves. Information such as depth, thickness, bottom hole temperature, spontaneous potential, mud weight and type, and identification for each aquifer was punched on computer cards.

A salinity computation procedure developed by the USGS (R. Wallace and P. Jones, personal communication) was programmed and, along with routines to calculate the corrected aquifer temperature and the formation pressure, was used to process the punched data. From these results, wells containing high-temperature, high-pressure, low-salinity aquifers were easily recognized. Porosity and permeability values for these prospects were obtained from core sample information (Johnson, R.B. and Mathy, H.E., 1957, The South Texas Frio Trend, *Gulf Coast Association of Geological Societies Transactions,* volume 7, pages 207-218). Structural control on the aquifer-bearing Frio formation was furnished by the USGS (Jones, personal communication). This information was used to delineate the particular growth fault blocks containing the better aquifers and to estimate the aquifer volume.

On the basis of the above analysis, the Corpus Christi area was eliminated from consideration. There are no aquifers in this area below a depth of about 10,000 feet. The Frio sands, occurring at 7,000 to 10,000 feet have salinities of 50,000 to 100,000 ppm, and their temperatures are not as high as at comparable depths in areas to the south.

The Sebastian site appears to be an excellent prospect. The pertinent subsurface data is listed in Table 8.1. These data were obtained primarily from a 15,000 foot well located less than a mile from the geographically-ecologically optimum site described in the following section. Another well located 10 miles away, but in the same fault block, shows similar low-salinity, high-pressure conditions, and indicates that the over-pressured aquifer system likely extends throughout the block.

TABLE 8.1: SEBASTIAN SITE

Approximate depth to geopressured aquifers	14,300 ft.
Thickness of aquifer series	700 ft.
Corrected temperature (°F)	320-325
Pressure (PSI)	11,600
Salinity (PPM)	2000-6000
Areal extent of faulted block containing aquifers	10 miles by at least 30 miles
Existence of evidence that low-salinity geopressured conditions exist elsewhere in block	Yes
Porosity of aquifers	20%
Permeability of aquifers	100-135 millidarcys

Source: AD-766 855

The subsurface data pertaining to the area between Port Mansfield and Raymondville is listed in Table 8.2. There is a group of sands at a depth of 12,650 feet with a temperature of about 267°F and a formation pressure of 10,000 psi.

Salinity is about 20,000 ppm. At a depth of 15,660 ft, an 800 foot series of sands occur with pressure and temperature conditions similar to Sebastian but the fluid salinity appears to be 3 to 4 times greater (i.e., about 20,000 ppm).

TABLE 8.2: PORT MANSFIELD SITE

Approximate depth to geopressured aquifers	12,650-15,660 ft.
Thickness of aquifer series	800 ft.
Corrected temperature (°F)	267-329
Pressure (PSI)	10,000-14,381
Salinity (PPM)	20,000
Areal extent of faulted block containing aquifers	About same as San Sebastian
Existence of evidence that low-salinity geopressured conditions exist elsewhere in block	Yes
Porosity of aquifers	20%
Permeability of aquifers	100-135 millidarcys

Source: AD-766 855

The data in Table 8.2 were taken primarily from a 16,122 foot well. This is the well closest to the site (26° 29' 00"N, 97° 30' 45"E) which penetrates the geopressured zone; however, it is about 15 miles to the west. Wells with total depth of 10,000 to 13,000 feet which are much closer to the site do not appear to penetrate the geopressured zone.

Environmental Information for Sebastian Site

Location: The Sebastian site is on a tract of land located in the Lower Rio Grande Valley of Texas in the extreme northwestern corner of Cameron County, at approximately 26° 19' N and 97° 50' E. The site, which is presently being used for dryland cotton production, is immediately to the north of the northern levee of the North Floodway of the Rio Grande River. It is approximately a half to one mile west of the 200 acre Langoria Unit of the Las Palomas Wildlife Management Area.

The nearest large market center is Harlingen (population ~41,000) 14 miles to the southeast via excellent, paved roads. Several small communities, of which Sebastian (population ~1,000) 3 miles to the northeast and Santa Rosa (population ~1,500) 4 miles to the south are the nearest, are situated within a radius of 5 miles and are accessible by bituminously surfaced state highways. The site is about 24 miles west of the northern end of the Laguna Atacosa National Wildlife Refuge.

Groundwater Hydrology: Two sources of groundwater suitable for irrigation, public supply, or industrial use have been recognized in the vicinity of the site: the Lower Rio Grande groundwater reservoir and the Mercedes-Sebastian shallow groundwater reservoir.

The Texas part of the Lower Rio Grande groundwater reservoir is in southeastern Starr, southern Hidalgo, western Cameron and a small part of southwestern Willacy counties. The lateral limits of the reservoir in Texas encompass an area of about 1,150 square miles, of which about 950 square miles is productive. The groundwater reservoir consists of beds of water bearing material in the Goliad, Lissie and Beaumont Formations and also post-Beaumont alluvium. The permeable beds are hydraulically connected so that they behave as a unit; however, locally they are separated by beds of less permeable material.

The general limits of the Lower Rio Grande groundwater reservoir in a northeasterly and easterly direction are marked by the limits of water suitability for irrigation and industrial use. In southeastern Hidalgo County and western Cameron County, the shallow deposits are usually treated as a separate reservoir on the basis of the chemical quality of the groundwater, discussed later as the Mercedes-Sebastian shallow groundwater reservoir.

The maximum thickness of the Lower Rio Grande groundwater reservoir is about 700 feet; however, the thickness is irregular and generally less than 500 feet. The dissolved-solids content of the water tends to increase with depth so that for most uses an effective lower limit to the reservoir can be defined on the basis of the chemical quality of the water. In general, water of the best quality in the Lower Rio Grande groundwater reservoir is near the Rio Grande and the water tends to be of increasingly poorer quality going north from the river.

Water in the upper part of the Lower Rio Grande groundwater reservoir generally is under water-table conditions. However, as the water moves downward and laterally it may pass under beds of relatively less permeable material so that locally it is under artesian conditions. This is true, for example, in the Harlingen area.

Apparently most of the recharge into the Lower Rio Grande groundwater reservoir is by the downward percolation of water from the land surface. The amount of recharge fluctuates with differences in precipitation, being largest during periods of above normal rainfall. Prior to the development by man most discharge from the reservoir was by evapotranspiration. The rate of discharge by evapotranspiration was reduced as land was cleared for cultivation.

During periods of high precipitation such as accompany hurricane storms in the area, the reservoir may be filled to near capacity so that water logging of the soil occurs. The amount of water available in storage is not large compared to the total potential capacity of the wells. During protracted periods of below normal rainfall, when the rate of pumping is at a maximum and the rate of recharge is at a minimum, the water available in storage could be depleted in a relatively short time.

A number of wells in the area peripheral to the Sebastian site were drilled into the Rio Grande groundwater reservoir to depths of 300 to 400 feet. Seven of the wells within a radius of one mile from the site used the water for irrigation on land holdings of 50 to 300 acres. These wells are known to have operated throughout the 1950's. During this period records show that the water level in the wells declined during the initial years but increased rapidly between 1957 to 1959 and in many places was within 5 feet of the surface.

The total dissolved solids content of the water was between 3 and 4% and was alkali with a sodium absorption ratio of between 25 and 35. The U.S. Department of Agriculture would classify this water as very high salinity water which is not suitable for irrigation under ordinary conditions and as a very high alkali hazard. The boron content ranged from 0.002 to 0.008%. A boron content greater than 0.00375% in irrigation water is thought to be unsuitable for tolerant crops.

Water from the Mercedes-Sebastian shallow groundwater reservoir contains considerably less dissolved solids than the underlying Rio Grande reservoir. It consists of permeable deposits, less than 100 feet below the land surface, of the Mercedes-Raymondville distributary floodplain and flanking Beaumont Formation. The reservoir extends through southeastern Hidalgo, western Cameron and southwerstern Willacy counties but its lateral extent is poorly defined and is best delimited on the basis of the quality of the water from the wells tapping it.

The water from shallow wells near the Sebastian site is noticeably less saline than that from the deeper wells and the data suggests that the site is located near the source of some of the freshest water in the Mercedes-Sebastian shallow groundwater reservoir. The U.S. Department of Agriculture would classify most of the shallow groundwater reservoir as very high salinity water which is not suitable for irrigation under ordinary conditions; the area immediate to the site would be classified as high salinity water which cannot be used on soils with restricted drainage.

The sodium absorption ratio (sar) is low in the vicinity of the site and is indicative of a low alkali hazard. The dissolved nitrate concentration is the highest for the reservoir and may indicate bacterial contamination.

The yield of individual wells tapping the Mercedes-Sebastian shallow groundwater reservoir is small and is used for public supply, domestic, irrigation and stock use. In the area of the site none of the Mercedes-Sebastian shallow groundwater is used for irrigation.

Land Use: The site of the proposed power plant is presently under dry-land cotton acreage. This is a land use for which the soil is well suited and a good yield per acre can be obtained without using irrigation water. At a distance of approximately $^5/_{10}$ to $^6/_{10}$ of a mile east of the proposed site the land use changes to that of wild life management of the Langoria Unit.

The site of the proposed cooling pond is a depression that abuts the northern levee of the floodway. This depression has not been cultivated because of poor drainage conditions and is presently vegetated by native grass and shrub.

The Northern floodway has a width of approximately ½ mile and artificial levees that are about 10 feet above the surrounding land. A pilot channel centrally located between the two levees has been dredged to carry local runoff from the surrounding fields. This pilot channel has a capacity of 1,200 to 1,500 cfs. The floodway is designed to carry excess water from the Rio Grande to prevent flooding of the river channel in the delta. Entrance to the floodways is provided by two levee openings immediately above and below the Anzaldus Dam. The floodway is designed to pass a peak flow of 75,000 cfs. However, a constriction in the floodway where the Willacy Canal Siphon crosses it, some ¾ mile down the floodway from the site of the proposed cooling pond, would locally reduce this figure.

Prior to building floodways, the U.S. Government acquired levee and floodway easements at the expense of the counties through which the floodways pass. These instruments permit borrowing of material for construction as well as use of land for carrying flood and drainage water. In effect, floodways are in private ownership to be used as prescribed by the U.S. Government.

Except in the immediate vicinity of the Willacy Canal Siphon, floodway owners are cultivating acreage within the floodway, but in compliance with the easement regulations. In the area of the site much of this acreage is under cotton production. Several factors contribute to the intense use of floodway land. Although flood risk will always exist, the probability of flooding decreases as reservoir capacity increases upstream.

Furthermore, floodways are among the best drained agricultural land in the Texas delta, as they provide the only drainage channel at a distance from the Rio Grande. New structures may be erected only by permission of the International Water and Boundary Commission, and must be portable and less than 225 square feet in surface area.

Other land uses in the neighboring area include two cemetaries, El Azadan, ¾ mile to the east of the proposed cooling pond, and Santa Rita, 1 mile to the north of the site of the proposed power plant. Both may date back to the era of Mexican sheepherding before the division of the Santa Rosa Ranch. South of the floodway there is a network of irrigation and drainage ditches and the land between the floodway and the town of Santa Rosa is irrigated farmland, with a sparsely scattered population. To the north of the site of the proposed power plant local wells supply water to small localized irrigation projects.

At a greater distance but within a radius of 10 miles there are several small population centers dependent upon local agriculture for their existence. Seven miles to the southwest is the Lacy Mercedes oilfield which is of very limited extent.

Environmental Impact: There is a possibility of significant impacts in the following areas:

Well Blowout — A blowout during drilling of the pilot well would have serious effects on almost every aspect of the environment in the vicinity of the well. However, a great backlog of experience in drilling into high pressure zones has been developed in recent years and, in fact, high pressure wells have been successfully completed in the immediate vicinity of the pilot site. Using state-of-the-art drilling techniques, the probability of a blowout must be considered minimal.

Effect on Aquifers — Using approved oil field techniques of drilling and casing, no contamination of fresh water aquifers should be expected during drilling. Nor should the reinjection of water into aquifers at the 5,000 to 6,000 foot level in any way degrade these waters. In fact, the waters at this depth are normally so salty that reinjection will tend to freshen them.

A lined surface holding tank will be constructed near the well site to serve as a retainer for water that might be accidentally spilled during operations. Spilled water will no doubt be very hot and possibly salty. If salty water were allowed to stand in an unlined holding tank for an extended period of time, it could have a contaminating effect on the water table which is very near the surface. However, water would be present in the holding tank only in case of an accidental spill and even then would be retained only long enough for it to cool before it was routed to the North Floodway and then to the Gulf.

Noise — The maximum noise will occur during drilling when a large diesel engine will be in operation. After drilling is completed, and when proper operation is underway, the only noise will be due to a large turbine. No venting of gases will be necessary; thus, a prime source of noise and a possible source of atmospheric pollution is omitted.

Effect on Sea Water — The water reaching the North Floodway and the Laguna Madre will have been cooled in the holding tank. Thus, there will be no adverse temperature effect on sea plant and animal life. The salinity of the Laguna Madre varies between 35,000 and 100,000 ppm, so any well water introduced into it will actually be fresher.

Subsurface Strain — Withdrawal of water by one well over a period of 5 years will cause some subsidence. However, the Gulf Coast is an aseismic area, and neither the subsidence nor the reinjection of water should trigger earthquakes.

Land Use — The land area composing the Sebastian site is presently under cultivation. Removing the area from cultivation and constructing the pilot project will cause no conflict with existing transportation networks, utilities, residences, or recreational facilities.

Environmental Information for Port Mansfield Site (Tenerias Area)

Location: The Tenerias area is situated in eastern Willacy County approximately 6 to 7 miles inland from the Laguna Madre and is described by the coordinates 26° 29'N and 97° 30' 45"E. The site which is located on the El Sauz subdivision of the King Ranch is 4 miles east of the small ranch community of El Sauz. 10 miles to the south-southeast is the northern boundary of the Laguna Atacosa Wildlife Refuge and 3 miles to the southwest lies the Willamar Oil Field.

The site is some half mile south of a bituminous surfaced road (FAS 497) which connects the area to two small communities; San Perlita (population 348) which is 8 miles to the west and Port Mansfield (population 200) which is 7 miles to the northeast on the coast of the Laguna Madre. Raymondville (population 8,000), the largest town in Willacy County, is 16 miles to the west on route FAS 497.

Groundwater Hydrology – The groundwater table in the Tenerius area is very close to the land surface. In the lower lying area of the Lomalto clay loam soil, the water table is usually at a depth of less than 2 feet during periods of heavy precipitation and intermittent lakes form in the topographic depressions. The water is very highly saline with a high boron content and a high sodium alkali hazard. The groundwater table is at a slightly greater depth in the area of the Victoria fine sandy loam. The water is too saline for human consumption or for growing crops.

Water for rangeland stock is tapped from the Goliad sands, a Pliocene aged formation consisting mainly of clay and sands, at a depth of 1,300 to 1,600 feet. The total dissolved solids content of this water ranges from 3 to 5,000 ppm in the vicinity of San Perlita, several miles to the west, to nearly 10,000 ppm at Port Mansfield. Boron content varies between 6.5 and 11 ppm which even exceeds the limits of boron tolerant crops. The water is classified as having both very high sodium alkali and salinity hazards. However, the water from the Goliad sands is thought to be preferable to the groundwater of the Lissie and Beaumont formations.

Land Use – High salinity and frequent water-logging of the soil has rendered the land completely unsuitable for crop growing. The area has traditionally been used for cattle rangeland.

In recent years this particular part of the El Sauz Ranch has been leased by the Federal Government and is used as a U.S. Naval Research Station for satellite detection.

Several miles to the southwest is the Willamar oil field which was at peak productivity during the 1940's and early 1950's at which time it ranked sixth among Texas oil fields. Today productivity has declined. East of the Willamar oil field is the Sauz Ranch-Nopal oil field. There is no organized community associated with any of the fields and population is sparse.

South of the Sauz Ranch-Nopal oil field is the Laguna Atacosa Wildlife Refuge. 6 or 7 miles to the east of the Tenerias area is the Laguna Madre. This inland sea is an important recreational and commercial fishing area. Much of the Texas shrimp industry is located along this coast and the shallow waters are ideal nurseries for the brown shrimp. Port Mansfield, an isolated community on the eastern shore of the Laguna, is the base for a rapidly growing recreational fishing industry. It is a center for both Laguna and Gulf fishing since it has access to the Gulf via the Port Mansfield Pass. The population of Port Mansfield has grown rapidly in the last decade which is a notable exception to the more general trend of rural and small urban population decline throughout Willacy County since the early 1950's.

The Physiography and Hydrology of the Laguna Madre: The Laguna Madre is a long, narrow coastal lagoon which extends from Corpus Christi Bay southward to the Rio Grande delta. It is sandwiched between the shore of the mainland and a narrow sand barrier known as Padre Island. The lagoon is shallow, averaging less than 3 feet in depth and in its natural state nowhere deeper than 9 feet. The bottom of the lagoon has a very gentle gradient but is irregular with shallow flats and relatively deeper basins. The width of the lagoon is from 3 to 5 miles but varies considerably with small seasonal and meteorologically induced changes in water level.

Much of the coastline is inundated intermittently and at times it is difficult to know where the shoreline really is. Midway down the length of the Laguna Madre, south of Baffin Bay, there are extensive mud and sandflats which are submerged only at the highest water levels. This bar effectively divides the lagoon into two separate units as far as marine life and hydrographic conditions are concerned. The northern or upper section of the Laguna Madre will be considered in this section. At the northern end of the Lagoon where it joins Corpus Christi Bay there is a long and shallow transverse bar which effectively cuts off water exchange with the Gulf at all times except at the periods of highest water levels.

Normally Corpus Christi Bay is linked with the Laguna Madre only by the Intracoastal Canal and the narrow entrance to the naval boat basin. Several small naturally formed sand islands occur in the Upper Laguna Madre of which the most noteworthy are North and South Bird Island and Pita Island. Perhaps the most outstanding single feature of the Lagoon is the Intracoastal Canal, a manmade ditch 125 feet wide and 12 feet deep. Deposition of sand and mud from the canal has formed a 16 mile long dike in the extreme northern end of the Laguna and allows few passageways for the exchange of water between eastern and western sides of the Laguna. South of the dike numerous islands have been formed by staggering of spoil dumps along each side of the channel.

In general the bottom sediments of the Laguna near Padre Island are chiefly sand and those along the mainland sand and clay. Waves are low as a result of limited fetch and shallow depths and the waters are generally turbid. Little of the suspended matter can settle in the shallow areas owing to constant wave agitation. Continuous strong winds, averaging 9 to 14 mph, especially regular from the southeast quadrant for about seven months of the year cause strong waves throughout most of the day abating only slightly at night. As a result of the predominant southeast winds, waves break mainly on the west and northwest shores and beaches develop only in these areas.

The currents follow the winds and are related to wind controlled tides, i.e., when north winds blow, currents flow southward and conversely with southerly winds. Countercurrents have been observed in the deeper water of the Intracoastal Waterway.

There is negligible periodic tide. Rise and fall of water at any particular locality is dependent upon wind condition. Extreme wind can cause a tidal range of 3 to 4 feet. Those meteorological tides are nonperiodic and are important controls over water exchanges with the Gulf. Spring and fall celestial tides can raise the water level as much as 18 inches.

The Laguna is a hypersaline body of water. The salinity of the water varies seasonally and annually and from north to south depending on meteorological conditions and runoff and channel discharge from the mainland. Salinity has been known to vary from ~110,000 ppm to ~10,000 ppm but probably averages between 35,000 and 45,000 ppm.

The Laguna is shallow and water temperatures closely parallel the temperature of the air. This means there is considerable daily and seasonal variation. In summer, temperatures between 75°F and 90°F are common in water that is >2 feet deep and can be >95°F in the very shallow coastal flats. Daily ranges can be great as 10°F in the deeper water and exceeds this in the very shallow reaches. In winter cold winds known as northers can bring near freezing temperatures.

Environmental Impact: Areas of potential environmental impact are as were observed for the Sebastian site. The discussions previously given for impacts at Sebastian in the areas of well blowout, effect on aquifers and sea water, noise, subsurface strain, and land use apply also to Port Mansfield.

The only difference between the sites in an environmental sense is that Port Mansfield is located adjacent to a Gulf inlet and contents of the holding tank would be emptied into the inlet rather than into a floodway as is planned at Sebastian. Either situation should cause no environmental disturbances.

Reservoir Production Calculations

Detailed maps of the geologic structure and sediment distribution in the Texas Gulf Coast indicate that the geopressured region is divided by growth faults and facies changes into water-filled sand units which must be regarded as separate reservoirs. The following calculations are concerned with a model reservoir.

Consider a geopressured zone 1 mile thick, beginning at a depth of 12,000 feet, and consisting of half sand and half shale. The horizontal dimensions are 20 by 30 miles. Essential parameters are as follows:

Bore hole radius	0.3 ft
Water viscosity	0.2 cp
Sand porosity	0.25
Sand permeability (k)	
Good sands	0.3 Darcy
Average sand (Gulf Coast)	0.05 Darcy

It is determined, from the above parameters, that a well drilled into this reservoir should produce 50,000 bbl (2.1×10^6 gal) of hot water per day. Calculations have been made to determine the optimum number of wells and the expected cumulative production in gallons for two assumed values of average permeability (Table 8.3).

TABLE 8.3: WATER PRODUCTION (IN GALLONS)

Number of years	k = 0.3 Darcy 190 wells	k = 0.05 Darcy 35 wells
1	1.5×10^{11}	2.7×10^{10}
2	2.8×10^{11}	5.1×10^{10}
5	6.9×10^{11}	1.2×10^{11}
8	1.1×10^{12}	1.9×10^{11}
10	1.3×10^{12}	2.4×10^{11}
20	2.6×10^{12}	4.7×10^{11}

Source: AD-766 855

These calculated production figures do not include the effects from expansion of dissolved gases or from dewatering of overpressured shales, both of which would tend to increase total production. The figures, therefore, are very conservative. Power production from hot water can be calculated as follows:

Initial Temperature	325°F
Final Temperature	212°F

The temperature drop of 113°F would provide 942 Btu/gal. Converting watts (electrical) and assuming a 10% conversion efficiency, 910 gallon per day would be required to produce 1 kw of continuous power. A 50,000 bbl well would provide 2.5 Mw. Neglecting the energy available from water pressure at the surface, the power potential for the reservoir is found to be, depending upon the permeability:

190 wells	k = 0.3 Darcy	475 Mw
35 wells	k = 0.05 Darcy	88 Mw

Additional power output of about 30% could be obtained by utilizing the energy from the average pressure of the hot water at the well head. Now consider the problem of subsidence resulting from a 20 year production history. As a worst case the calculations are based on the maximum rate of production; that is, for k = 0.3 Darcy.

The total volume of water in the reservoir is (at reservoir pressure) 7.5×10^{13} gallons. With a 0.8 geostatic ratio, the bottom hole pressure is 10,000 psi. Note that the 20 year cumulative production (2.6×10^{12} gal) is about 3.5% of the total water. The coefficient of elastic expansion for the water is 5×10^{-6}/psi and for the pore space collapse ratio 3×10^{-6}/psi giving a total reservoir volumetric decrease of 8×10^{-6}/psi. The pressure drop, assuming no dissolved gases or influx of water from the shale required to produce the 20 year cumulative amount of water is

$$\frac{3.5 \times 10^{-2}}{8 \times 10^{-6}} = 4{,}375 \text{ psi}$$

Initially the well-head pressure would be about 4,500 psi and would drop to about 100 psi after 20 years of production, again assuming no gas drive or influx of water from the shale. A pressure drop in the reservoir of 4,375 psi implies a pore collapse of ($3 \times 10^{-6} \times 4{,}375$) or 1.3% volume in 2,640 feet of sand. Assuming approximately ⅓ of pore collapse to be in vertical direction we get about 11 feet of subsidence at depth in 20 years of production. In practice, wastewater would be reinjected into normal pressured reservoirs at depths of 5,000 to 6,000 feet thus decreasing the surface effects of subsidence at a depth of 12,000 feet.

Calculations of power production from a single well are more pertinent to this feasibility study than are production figures for the entire reservoir. The parameters used in the following calculations are from the Sebastian Site in south Texas.

Depth of sand	14,300 to 15,000 feet
Thickness	700 feet
Temperature	320°F
Reservoir pressure (average)	12,000 psi
Total salinity	2,000 to 6,000 ppm
Permeability	0.10 to 0.14 Darcy
Porosity (approximate)	0.25
Area of reservoir	300 square miles or more
Hydrostatic pressure for 14,600 foot head	6,330 psi
Well-head pressure	over 5,000 psi

Again assume the daily production from a well to be 50,000 bbl or 2.1×10^6 gallons per day. Electrical power can be derived from both the heat and mechanical energy stored in the water. Considering 5 years production from a single well, no appreciable decrease in well-head pressure would be expected. If the water temperature is dropped from 320° to 200°F, then 1,000 Btu/gal would be available. Assuming 10% efficiency in the heat exchanger and electrical power generator 34 gal/kwh or 816 gal/day would be required to produce one kw. A single well should then produce over 2.5 Mw from the heat energy alone.

From mechanical energy, with a 5,000 psi pressure drop, about one kwh can be produced from 50 gal of water. The mechanical power production would be $2.1 \times 10^6/(50)(24) = 1{,}750$ kw giving a total electrical power production of over 4 Mw from one well.

The water produced would have a total salinity of about 5,000 ppm or less, and the salinity should decrease with production depending upon the amount of water given off by the shales as the reservoir pressure begins to drop. It is assumed, based on data from other wells in the south Texas-Gulf coast region, that the reservoir water would contain about 0.25 cu ft of methane per gallon and about an equal amount of carbon dioxide. The methane would be recovered and the carbon dioxide could be collected or vented. Total methane production should be about 2×10^8 cu ft/year. The slightly saline water produced by the well can be reinjected into very saline, normal pressured reservoir sands at depths of 5,000 to 6,000 ft.

Pilot Project

This study shows that it is feasible to construct and operate a pilot plant for electrical power production using water from a single well in south Texas. Over 4 Mw of power could be produced from a single well; two billion cu ft per year of methane would be produced as a by-product. Based on geological and environmental considerations, it is feasible to locate the project at either the Sebastian or Port Mansfield sites discussed in this report. Based on more complete subsurface information, the Sebastian site is preferred. The principle objectives of a 5 year pilot project would be as follows:

(1) Demonstrate the feasibility of power production from thermal and mechanical energy stored in a geopressured subsurface reservoir.

(2) Determine the production-pressure history of the well. Evaluate the contributions to production from gas drive and dewatering of the shale.

(3) Study the change in water chemistry with production as an indication of the change in shale composition in the reservoir.

(4) Develop optimum methods for converting the mechanical energy stored in the over-pressured water to electrical energy.

(5) Investigate the use of the facility as a stand-by power facility. Determine the effect of shutting-in the production for long periods of time.

(6) Determine, by use of sensitive instrumentation, the surface effects resulting from withdrawal and reinjection of the large amounts of water required for power production.

There would appear to be no alternative to the construction and operation of a pilot project if the above objectives are to be achieved. Although the basic principal of extraction of electrical power from hot, high-pressure water appears to be straightforward, there are no doubt many engineering problems which can only be recognized and solved by actually building the plant. The operational history of the project for the first few years will be critical in evaluating the roles of gas drive and dewatering of shale in maintaining well-head pressures. There does not seem to be any reliable a priori technique to evaluate these factors. Although finite element theory may offer some predrilling information, the effect of reinjection on preventing subsidence can be finally resolved only by operating the pilot project.

The pilot project in the south Texas Gulf Coast would serve as a world model. The pertinent structural and stratigraphic characteristics of the Gulf Coast are not unique. On the contrary, as has been pointed out earlier, the distribution of deep abnormally pressured sedimentary basins is world wide. Information obtained from the Gulf Coast pilot project would be directly applicable to most other abnormally pressured basins.

The Gulf Coast is unique in that extensive exploration for gas and petroleum have resulted in a comprehensive knowledge of the geopressured zone. The top of the geopressured zone has been mapped over most of the region, and thousands of well logs have provided information on salinities and temperatures of the formation waters. In addition, the Gulf Coast is logistically very handy for experimentation, both in nearness and climate. Thus, this region seems ideal to serve as the site of an experimental program which would have global implications.

CALIFORNIA

The material in this section is derived from AD-117 016, *Feasibility Study for Development of Hot-Water Geothermal Systems,* March 1973, by J.B. Combs of the University of California. This investigation prepared for the Air Force Office of Scientific Research, was directed toward a feasibility study for the development of hot-water geothermal systems for potential Department of Defense use as an energy source.

In the original research plan, it was proposed to use the Coso Thermal Area, located within the instrumented test ranges of the Naval Weapons Center, China Lake, California, as a type area for the study of hot-water geothermal systems; however as the research progressed, it was decided to investigate two other military installations because of the abundance of published data for the Coso Thermal Area. The two other areas were chosen for both their potential geothermal indicators and because of geographic expediency.

The first alternative area chosen to be studied was the Point Mugu and Port Hueneme, California areas. This area was chosen because of the occasional high temperature oil wells that are drilled in both the Los Angeles Basin and the Santa Barbara Channel. A logical connection of these two areas along the grain of the major geological features of the region suggested the Point Mugu area as a potential hot-water geothermal area. As can be seen from the data presented and discussed later this does not appear to be the case.

The Marine Corps Base at Twentynine Palms, California, was the second alternative area that was chosen. The primary geologic features on which this decision was made are the two very young and almost totally unweathered volcanic craters, Pisgah Crater and Amboy Crater, with their associated extensive lava flows. Very hot, greater than 50°C, water wells that have been drilled in the Twentynine Palms area also provided additional supportive information.

Point Mugu Site

The areas under investigation, the Naval Air Missile Test Center Point Mugu and the Naval Reservation near Port Hueneme, lie within the area known as the Oxnard Plain on the coast of California. The Oxnard Plain is bordered on the west by the Pacific Ocean and extends northward from Point Mugu to the city of Ventura. On the northeast and east the plain is bordered by the Camarillo Hills and Santa Monica Mountains respectively.

The area is, therefore, a part of the Pacific coastline of southern California. It is a moderately inhabited region with a mean annual rainfall of about 32 cm and a mean annual temperature of 16°C. The climatic conditions range from as low as 0° to approximately 39°C.

The water bearing deposits in the vicinity of the test center vary in thickness from about 242 meters (1N/21W-32A1) to about 450 meters (1N/21W-31L1) along the coast (Page, R.W., 1963, *Geology and Ground Water Appraisal of the Naval Air Missile Test Center Point Mugu, California:* U.S.G.S. Water-Supply Paper 1619-S, 40 pp). The well locations, water temperatures and geologic well logs are presented in the study. A log of well 1N/21W-27F1, 2.5 km north of the test site shows bedrock at a depth of 248 meters.

Data from the California Department of Water Resources shows that no well logs indicate abnormal water flow. According to Page, except for local seawater encroachment there is no evidence of movement of groundwater to the Oxnard Plain from sources other than precipitation that feeds local drainage streams.

Seawater encroachment in the vicinity of Port Hueneme has resulted in the destruction of wells 1N/22W-29A1, 29A2 and 29C1 that yielded salty water. Well 1N/22W-29A2 had a chloride content of only 41 ppm in 1947 so it is inferred that encroachment has occurred since that time. A high chloride content for well 1N/22W-28B1, 850 ppm, was measured in May, 1958. Samples on well 1S/21W-4E1 since 1949 indicate an increase in chloride from about 300 ppm to more than 5,000 ppm in 1953 and about 12,000 ppm in 1958. (It is believed by Page that contamination could have caused the high chloride content detected in 1948 because of faulty equipment.)

The chemical quality of the groundwater in the test center area differs with depth and locally seawater has intruded the groundwater body. A study of E-logs by Page revealed that the water bearing deposits could be divided into three distinct zones: (1) a shallow zone extending from the water surface to a depth of approximately 45 meters in which the water is salty, (2) an intermediate zone 45 to 300 meters below land surface in which the water is brackish; (3) a deep zone 300 to 450 meters below land surface in which the water is salty.

Water temperature analysis does not appear to have been the concern of previous studies on the hydrology of the area as data is sparse. The temperatures available on various wells however, indicate a variability between 30° and 24°C. The majority of available temperatures were about 19°C.

The geology and the existing data for the Point Mugu are not particularly encouraging as far as the potential for a hot-water geothermal reservoir in the area. However, this lack of thermal manifestations may be caused by the encroachment of cold seawater in the subsurface. Nevertheless, additional research in this area in the immediate future is not recommended.

Twentynine Palms Site

The Twentynine Palms Marine Corps Training Center occupies a portion of the southeastern Mojave Desert region, California. The nearly 2,500 sq km encompassed by the reserve lies almost entirely between 115° 45' and 116° 30' west longitude and 34° 15' and 34° 45' north latitude. The area includes within its boundaries the Bullion Mountains, parts of the Sheep Hole and Lava Bed Mountains, and portions of the Dale, Twentynine Palms, Bessemer, Lavic, Bristol, Ames and Surprise Valleys.

The southeastern Mojave Desert is an arid, very sparsely inhabited region with a mean annual rainfall of 8.9 cm and a mean annual temperature of 20°C (68°F). The climatic conditions, however, are extreme, ranging from over 46°C (115°F) in the summer months to as low as -7°C (19°F) in the winter months. The major portion of the annual precipitation occurs during the period of November through April; the infrequent occurrence of thunderstorms during the rest of the year accounts for the remainder of the precipitation and the total surface runoff.

Existing data on the groundwater conditions and water wells at the Twentynine Palms Marine Corps Training Center come mainly from two sources. Reports on the water conditions at the base during several consecutive years were compiled in the following studies: Dyer, H.B., 1960, *Ground-Water Conditions During 1960 at the Marine Corps Base, Twentynine Palms, California:* U.S. Geol. Survey open file report, U.S.G.S. Ground Water Branch, Long Beach, California, 32 p; Riley, F.S., and Bader, J.S., 1961, *Data on Water Wells on Marine Corps Base, Twentynine Palms, California:* U.S. Geol. Survey open file report, U.S.G.S. Ground Water Branch, Long Beach, California, 72 p and Weir, J.E., Jr., and Dyer, H.B., 1961, *Ground-Water Conditions During 1961 at the Marine Corps Base, Twentynine Palms, California:* U.S. Geol. Survey open file report, U.S.G.S. Ground Water Branch, Long Beach, California.

These studies, prepared at the request of the Department of the Navy, recorded data including temperature, chemical analyses and depth, and are included in the report. Data on the water wells of the Twentynine Palms, Yucca Valley and Dale Valley areas are found in Bader, J.S., and Moyle, W.R., Jr., 1960, *Data on Water Wells and Spring in the Yucca Valley-Twentynine Palms Area, San Bernardino and Riverside Counties, California:* Calif. Dept. Water Resources, Bull. 91-2, 163 p and Moyle, W.R., Jr., 1961, *Data on Water Wells in the Dale Valley Area, San Bernardino and Riverside Counties, California:* California Dept. Water Resources, Bull. 91-5, 55 p.

Although these reports are restricted to areas outside the reserve boundaries, several wells owned by the Navy were monitered. Their information is included in the report.

The major water-bearing units of the reserve are the alluvial deposits which underlie fans extending into the valleys or basins to varying depths. These units consist primarily of lenticular, intergradational beds of clay, silt, sand and gravel of Cenozoic age. Near the mountains, however, the deposits consist typically of coarse-grained, angular rock detritus.

According to Bader and Moyle, movement of groundwater through the area is impeded locally by groundwater barriers (presumably major faults). These barriers separate main valley areas into smaller groundwater basins. The displacement of the water level across the barriers is locally as great as 75 meters (e.g., across Mesquite Dry Lake Fault).

The analyses of base and near base wells show groundwater having a wide range of chemical quality. Fluoride content ranges from 0.4 to 100 ppm; chloride content from 10 to 725 ppm; and boron content from 0 to 1.3 ppm. The extremely high fluoride content of some of the wells in Mesquite and Deadman basins have rendered the water unpotable.

Although no wells surveyed in the reports record unusually high water temperatures, it should be noted that temperatures have been reported for only 22% of the wells in the area. Abnormally hot wells, registering 50°C and higher, have been located on privately owned land within several miles of the reserve boundaries (Donald W. Klick, personal communication, 1972). These wells, so far unexplored, represent potential areas of investigation for possible geothermal resources.

Because of the high temperature wells located in the vicinity of the Marine Corps Training Center at Twentynine Palms, and because of the very recent Amboy and Pisgah Craters and their associated extensive lava fields, it was recommended that further research be initiated in the area to attempt to detect any potential hot-water geothermal resources that exist in the region.

China Lake Site

The Coso Hot Springs area of the Naval Weapons Center at China Lake, California, was the original area that was chosen for the present study. It is a fairly rugged area with a combination of gentle valley and steep scarp terrain approximately 1,000 sq km lying within latitudes 35° 40' N to 36° 10' N and longitudes 177° 35' W to 117° 55' W in Inyo and Kern Counties, California.

The climate is arid desert. The Naval Weapons Center is located in the northern Mojave Desert. The mean annual rainfall is about 11 cm and the mean annual temperature is 18°C (65°F). The temperature extremes range from a low of (-11°C) 12°F in the winter to as high as (48°C) 118°F in the summer months.

There is an abundance of geological, geochemical, and geophysical data available for the Coso Hot Springs area in a number of Naval Weapons Center documents (see Austin, C.F., Austin, W.H., Jr., and Leonard, G.W., 1971, *Geothermal Science and Technology—A National Program,* Naval Weapons Center, China Lake, California, NWC TS 45-029-72; Austin, C.F., 1966, *Selection Criteria for Geothermal Prospects:* Nevada Bureau of Mines, Report 13, Part C, Reno, Nevada, pages 93 to 125; Austin, C.F., and Pringle, J.K., 1970, *Geologic Investigations of the Coso Thermal Area:* Naval Weapons Center, China Lake, California, NWC TP 4878, and Koenig, J.B., Garvarecki, S.J., and Austin, C.F., 1972, *Remote Sensing Survey of the Coso Geothermal Area, Inyo County, California,* Naval Weapons Center, China Lake, California, NWC TP 5233).

The existing water well locations, geologic logs, and chemical analyses are included in the report. The following future studies are recommended. First, a detailed surface geologic map should be prepared. The former geochemical and geophysical survey which has been made should be completed and published in technical journals.

The drilling of from 4 to 6 heat flow boreholes which should be no less than 100 meters total depth is suggested. These should be located using the seismic ground noise and the electrical resistivity data. At least one of the heat flow determinations should be made in a potentially nonproductive area in order to establish the magnitude of the background heat flow for the area.

Further research in the Coso area beyond those studies suggested above should be devoted to exploratory drilling to determine the extent of the reservoir as indicated by the surface measurements.

IDAHO

The following information is derived from ANCR-1155, March 1974, *Idaho Geothermal R & D Project Report for Period December 16, 1973 to March 15, 1974,* by J.F. Kunze and L.G. Miller.

Background Information

The area of Southern Idaho is one of the most promising regions within the U.S. for near surface, economically recoverable geothermal energy. It is identified by geologists as the younger end of a major Volcanic Rift Province. The Raft River Valley is a faulted, north-south trending sedimentary basin intersecting the major volcanic rift known as the Snake River Plain.

It is in the Raft River Basin that a number of wells, drilled for irrigation purposes, unexpectedly yielded warm to hot water. Two such wells bottoming at 400 and 540 ft yield boiling water under artesian flow. These wells are near the Malta, Idaho headquarters of the Raft River Rural Electrical Cooperative, Inc., an REA financed power company serving 10,000 sq mi of Southcentral Idaho, Northwestern Utah, and Northeastern Nevada.

The occurrence of the boiling water in the Malta, Idaho area prompted the Raft River Rural Electric Cooperative to conduct preliminary geological investigations pertinent to the possibility of establishing a geothermal power plant in the area for the production of electricity. Simultaneously, the Coop management began securing geothermal leases on the private land owned by its members.

Aerojet Nuclear Company, prime contractor to the U.S. Atomic Energy Commission at the National Reactor Testing Station (40 miles north of the Raft River Electrical Coop service area), entered into a preliminary engineering and feasibility study, including assistance in gathering additional geophysical information from the Raft River Valley.

Unlike the anomalous Geysers hot dry steam field in Northern California, the Raft River Valley is a hot water field (approximately 150°C) very similar to most of the identified geothermal potential resources in the Northwest. These temperatures are not currently attractive for commercial production of electric power. Neither the economics nor even the most economical power extraction techniques are yet known for these low temperature heat sources. Though the salinity of these waters is generally low (compared to the Cerro Prieto field, for instance), still the use of this water has not been mastered for long-term heat exchanger application.

Should further geological and geophysical work confirm the suitability of the Raft River resource for an R & D project and should funding be made available, the AEC and its contractor will design and install the geothermal plant. Raft River Rural Electric will arrange for leasing of the private-land geothermal rights on which it now has an option. The utility will also provide for the electrical connections and use of the power. The AEC will actively monitor and manage the geothermal field, which though highly convective exists in a region where groundwater supplies have been shown to be quite sensitive to withdrawal rates. Reinjection of both geothermal and condenser cooling-water (from the cold-water aquifer) is considered a likely necessity.

The Raft River area appears ideal for demonstrating the harnessing of the power producing potential in warm, not hot, water. With only 150°C water anticipated, conventional steam turbine systems appear only marginal for normal applications. But, the Raft River Valley's near surface aquifer is quite cold, approximately 10°C (50°F). With such a cold temperature heat sink, 150°C water becomes quite attractive, from a Carnot efficiency viewpoint. A low boiling point organic fluid is a prime candidate for the working fluid, but the economics of it compared to steam has not been established, as yet.

The demonstration power plant will contain a steam and/or a binary (organic fluid) cycle depending upon final temperatures and mineral content of the hot water. Initial operation of the plant is anticipated to be mid-1977.

The overall goals of the project are basically as follows:

(1) To demonstrate the extraction of energy from a hydrothermal field utilizing full reinjection of the waste fluid. The reinjected fluid is to contribute to the fluid recharge of the hydrothermal field, with minimum environmental degradation and minimum degradation of the thermal heat source.

(2) To demonstrate the design and operational characteristics of a geothermal electric power plant operating from a reservoir temperature well below the limits (170°C or 338°F) now considered to be economically attractive for commercial development.

A summary of the status of the project follows:

(1) U.S. Geological Survey data taken in the fall of 1973 was released to open file February 1, 1974 and has been examined by coworkers at Boise State University. Subsequent geological and geophysical investigations by the USGS were to resume in the spring of 1974.

(2) The shallow well drilling program commenced to study the groundwater patterns possibly feeding the geothermal reservoir. As of March 15, 1974 18 holes to a depth of 100 feet were completed.

(3) Discussions have been held with two drilling companies, and efforts to secure a sufficient quantity of long-lead-time drilling pipe are now underway.

(4) Power cycle studies and conceptual design were begun on both low

temperature steam flash and low temperature water-to-organic-fluid heat exchangers.

(5) Contacts have been made with both Utah Power and Light Company and Idaho Power Company to ascertain how this project would best serve utility needs of the future. Also, several venders of equipment were contacted to determine their interest in the project.

Status of the Project

The U.S. Geological Survey has open filed the geology and geophysical data for the Raft River Valley in the fall of 1973. The drilling of shallow (100 ft) hydrology wells (5½" diameter) through the Raft River Narrows area is progressing satisfactorily with 18 wells completed. Two well positions may have to be rotary drilled due to the large rock encountered at 10 to 15 ft depths. Highest temperature encountered to date was 48°C (118°F) at the bottom of a 30 ft deep and a 40 ft deep wells east of the narrows. Though these shallow holes are intended to study the near surface water flow pattern, temperatures of this magnitude this close to surface also provide clues concerning the circulation of heated water from depth to the surface.

Petrological examinations were performed on drilling chips supplied to Boise State University (work sponsored by Idaho Nuclear Energy Commission). Subsequently, drilling records, mud logs, lithology logs, and analysis of the well samples from the two deep oil and gas wells were supplied to REECO well drilling personnel for use in estimating drilling costs in the Raft River Valley.

Drilling time was very short (on the order of three weeks) for the Malta (Griffith-Wight) well to 6,787 ft. The two oil and gas wells were drilled in the fall of 1973 by a commercial gas exploration company. Both wells have since been abandoned. One began to spout hot water three weeks after being abandoned. That flow has since been controlled. Time estimates for the first test well should also be short but considerably more logging will be done substantially increasing the actual drilling period.

Power Cycle Studies and Power Plant Design: Efforts have concentrated on system parametric studies with the aim of establishing criteria from which one can arrive at an optimized (minimum) cost system. The work has been largely with the single stage flash steam cycle, but brief studies on the binary (organic fluid) cycle have also been made on heat exchanger design.

As for the latter, the results of the preliminary work show that the supercritical fluid cycle (e.g., Freon-12) will involve long counterflow heat exchangers (over 100 ft) unless the pinch points are significantly relaxed from the otherwise desired 10° to 15°F value. Relaxing the pinch point to larger values in turn results in less temperature drop for the geothermal fluid (less heat removed) and lower operating temperatures and efficiencies for the organic fluid turbine.

PROPOSED RESEARCH

The ever-increasing demand for energy and the recent pressures brought to bear through concern for the environment have led the Nation to consider energy alternatives to the conventional sources.

One of the new alternatives is geothermal energy. There is a great scarcity of reliable data on the geothermal resources of the Nation and this has led planners to ignore geothermal energy in comparison to other, more adequately explored, energy resources.

Nevertheless, other countries, including Italy, Iceland and New Zealand, have utilized geothermal power for some years, and Pacific Gas and Electric has built and is operating geothermal power generation equipment in Northern California.

In response to a proposal submitted by the University of Alaska with Mr. Walter J. Hickel as principal investigator, the National Science Foundation, through its RANN Energy Research and Technology Program, granted funds in support of a geothermal resources research conference to be held in September, 1972.

The objective of the conference was to develop an assessment of the state-of-the-art of geothermal science and technology and to recommend a research program to provide the requisite knowledge for establishing the proper role of geothermal resources in providing additional energy to alleviate the Nation's impending shortage. A further objective was to investigate the potential of geothermal water to supplement present national fresh water supplies.

The conference was a working conference, as contrasted to one for the presentation of scientific papers, and attendance was by invitation only. The setting for the conference was the Battelle Seattle Research Center in Seattle, Washington, an installation designed for just this type of working conference.

For convenience, the conference was organized into six substantive panels with co-chairmen of each panel responsible for assembling panel members who could speak authoritatively on the substance of the panel.

The panel topics selected were: resource exploration, resource assessment, reservoir development and production, utilization technology and economics, environmental effects and institutional considerations. Both industrial and university viewpoints were represented among the co-chairmen. Federal scientists were invited to contribute to each of the panels.

PB 222,326, *Geothermal Energy, A National Proposal for Geothermal Resources Research,* from which the material in this chapter is derived presents the results of the Geothermal Resources Research Conference which was held September 18-30, 1972. The chairman was W.J. Hickel, editor, J.C. Denton and Secretary D.D. Dunlop.

RESOURCE EXPLORATION

Objective and Summary

The objectives of the resource exploration panel were to identify the research programs necessary to develop, improve, and refine geothermal exploration techniques. These techniques are used to locate areas underlain by high temperature heat sources; to estimate the reservoir volume, temperature, porosity, and permeability at depth; and to determine the nature of any fluids producible from the geothermal resource.

Even though there are several different classes of geothermal resources with each having different physical properties, the panel directed its attention to the individual techniques. The combination of techniques which will be most useful in specific explorations is different for each category of resource. Deep drilling technology is not considered as a part of exploration.

The geothermal exploration techniques reviewed by the panel are: geological, geochemical, electrical, seismic, gravity, magnetic, and thermal methods. In addition, some possibly fruitful areas of basic research were considered, and some research projects concerned with present drilling technology improvement presented.

Some important geological research problems are concerned with determinations of: age, size, and magmatic type of igneous occurrences related to convective hydrothermal and permeable rock systems; and the relationship of convective hydrothermal systems to broad regions of elevated heat flow.

Geochemical research is needed in: chemical, physical, and thermodynamic properties of aqueous solutions at temperatures ranging from 100°C to 400°C; determination of chemical compositions of in-situ geothermal fluids as a function of in-situ temperature; and isotopic relations among water and various dissolved constituents, particularly gases.

Electrical resistivity research should be directed toward: understanding the variation of porosity, water salinity, and temperature in actual geothermal reservoirs;

improving electric field techniques and procedures for extracting true resistivity values from field data; and developing complementary exploration techniques which will improve the interpretation of resistivity data. Other electrical exploration research should include further work on DC resistivity, self-potential, electromagnetic, telluric, and magnetotelluric techniques.

Active seismic methods should be studied to characterize energy absorption and attenuation, and frequency shifts in known high temperature systems. Seismic noise studies should be undertaken to evaluate: temporal and spatial noise variations; characteristics of recognized noise sources, noise spectra, noise coherency, location of source, and cause of noise; and direction of noise propagation and apparent velocity.

Additional research is needed to determine the source of the gravity and magnetic anomalies associated with known geothermal areas and to decide whether these anomalies can be used as indicators of the internal temperature of the system.

Thermal research is needed to determine the relationships between temperature gradients, subsurface distribution of isotherms, and the geometry of geothermal systems. Laboratory temperature experimentation on model geothermal systems is needed.

Hydrologic studies are required to understand more fully the effect of ground water movement on local geothermal gradients. Basic research should be pursued on physical properties of geothermal water-steam mixtures, thermal geophysics and hydrology, modern geoscience theories and geothermal anomalies, data processing and general interpretation theory, and computer modeling.

Present drilling technology needs improvement in techniques for: isolating tested intervals in unconsolidated sands and in fractured reservoirs; inexpensive core recovery with reservoir fluid in place; logging at temperatures above 180°C; and transmission of information from the bore hole face to the surface. Development of cheap, low density, low viscosity, nonthermally sensitive, high thermal conductivity, and high surface tension drilling fluids would lower drilling costs and leave the bore-hole face in a more nearly undisturbed state.

Research Areas

At the present time, the most reliable exploration technique for new geothermal reservoirs is to find areas containing hot springs, a situation similar to that in the petroleum industry in the early 1900's when petroleum exploration consisted of finding areas of surface oil seeps. A great deal of research is needed to develop, improve, and refine geothermal exploration techniques. Several pertinent research areas are discussed below.

Geological Techniques: Geological techniques are used throughout all phases of geothermal exploration, from reconnaissance to site evaluation to detailed interpretation of drilling data and drill logs. Geological considerations provide the framework in which the various geochemical and geophysical techniques are interpreted, and, as such, are an integral part of all phases of geothermal exploration.

Geological research required to provide the framework for interpretation of data from other exploration techniques includes: determination of the age, size, and magmatic type of igneous occurrences related to convective hydrothermal systems; the nature and cause of structural features controlling the location of convective hydrothermal systems and impermeable rock systems; and the relationship of convective hydrothermal systems to broad regions of elevated conductive heat flow.

Geochemical Techniques: Geochemical exploration for geothermal resources involves sampling waters and gases from hot and cold springs, steam vents, shallow drill holes, and surface streams. The resultant data are used to: determine the state of the fluid at depth; estimate the minimum temperature expected at depth; estimate the homogeneity of water supply; infer the chemical character of water expected at depth; and determine the source of recharge water. Geochemical data collected during exploration is also valuable in the production and utilization phases of geothermal development and therefore should be carefully preserved for future use.

Although several hydrogeochemical indicators have had considerable use and success in geothermal exploration, others can be used only qualitatively at present or are not interpretable in the present theoretical and experimental framework. Research is needed in: chemical, physical, and thermodynamic properties of aqueous solutions at temperatures of 100°C to 400°C; chemical compositions of in-situ geothermal fluids as a function of original (predrilling) rock temperatures; and isotopic relations among water and various dissolved constituents (particularly gases).

Electrical Methods: Up to the present time electrical resistivity surveys have been one of the primary geophysical means used for discovering geothermal fields. Geothermal systems are characterized by thermally induced changes in electrical resistivity. Evaluation of the actual temperature and volume of a system is complicated in many cases by resistivity variations associated with changes in porosity, water salinity, rock texture and partial boiling.

It is most important that research be directed to understanding how porosity, water salinity and pressure vary in actual geothermal reservoirs so that interpretation of electrical surveys can be made more certain. In addition, research to improve electrical field techniques and procedures for extracting true resistivity values from field data is needed. Research on the use of other exploration techniques to provide input for improving interpretation of resistivity data may be particularly fruitful.

Electrical exploration methods which should be evaluated or further developed include DC resistivity, self-potential, electromagnetic methods, telluric and magnetotelluric techniques; airborne methods may be of special interest in large scale reconnaissance.

Seismic Methods: Reflection and refraction studies can be used to determine the structural setting of potential geothermal areas. In addition to these classical arrival time studies additional work should be done in the area of energy absorption and attenuation, and the effects on frequency in known high temperature systems. Geothermal areas do exhibit high attenuation and a shift to lower frequency for transmitted seismic waves. Perhaps these data can be used as a monitor of the internal reservoir temperature.

Detailed studies in Iceland, El Salvador and California have shown the existence of earthquake activity in close association with geothermal fields. The earthquakes occur at shallow depths (6 km) and are small in magnitude. Accurate determination of the earthquake pattern delineates faults that may channel hot fluids to drillable depths.

Although most authorities agree that relieving geothermal pressure will dimish earthquake potential rather than stimulate it, the causes of these tremors should be understood, and the effects of fluid and high temperature on fault strength and slip characteristics evaluated. Additional potential and developing geothermal areas should be studied in this regard.

Areas of high background ground motion (seismic noise) have been correlated with geothermal areas in New Zealand and in the Imperial Valley of Southern California. Seismic noise surveys are being conducted in potential geothermal areas in many countries. Studies should be undertaken to evaluate: temporal and spatial noise variations; characteristics of recognized noise sources, noise spectra, and noise coherency; direction of noise propagation and apparent velocity; location of noise sources; and cause of noise.

Gravity and Magnetic Techniques: The gravity and magnetic signatures of geothermal systems vary tremendously from one geologic province to another. There is no simple answer to the question of what type of anomaly should be expected from a geothermal reservoir. However, these methods have been very useful in the exploration of known geothermal reservoirs for determining gross structural features and in the location of potential geothermal systems. They should be routinely evaluated as a reconnaissance exploration method in outlining suspected geothermal areas.

Additional studies should be made to determine the source of the gravity and magnetic anomalies associated with known geothermal areas and if these anomalies can be used as indicators of the internal temperature of the system.

Thermal Techniques: Thermal exploration techniques provide a direct method of assessing the size and potential of a geothermal system. The thermal techniques can be ordered in terms of increasing cost and usefulness as follows: surface temperature measurements, geothermal gradient surveys, and heat flow determinations.

Additional regional heat flow determinations are needed in order to refine estimates of available geothermal resources. Much research is needed to determine relationships between temperature gradients, subsurface distribution of isotherms, and the geometry of geothermal systems. Research in the area of laboratory temperature experimentation on model geothermal systems is also required. Hydrologic studies must be made in order to more fully understand the effect of ground water movement on local geothermal gradients.

Drilling: Drilling techniques are developed in the United States to a point of high reliability for depths from a few hundred feet to 25,000 feet and for temperatures up to 400°F. Temperature gradient holes are drilled by adapting seismic shot hole drilling methods. Wells required are shallow and of small diameter. Current methods are effective and inexpensive. No research is indicated. Heat flow measurement holes are drilled with techniques used in mineral exploration. These methods are reliable though more costly than the temperature gradient holes. The number of heat flow holes which must be drilled is not judged of such magnitude that research to lower the cost should be funded, rather, it is recommended that heat flow holes be drilled using present technology.

Formation testing in uncased holes can provide useful input to the exploration for geothermal fluid reservoirs. Problems of isolating tested intervals in unconsolidated sands and in fractured reservoirs have not been solved. Research effort in this area would benefit geothermal, oil, gas, and water development projects. Inexpensive core recovery with reservoir fluid in place would aid in model definition of a geothermal cell.

Instrumentation for logging devices now available is not satisfactory at temperatures above 400°F. Research on methods of transmission of information from the bore hole face to the surface is imperative. The development of low density, low viscosity nonthermally sensitive, high thermal conductivity and high surface tension drilling fluids would improve drilling times thus lowering costs and at the same time leaving the bore hole face in as nearly an undrilled state as possible.

Documentation of data recovered is necessary for future programs. The preservation and storage of drill cuttings, cores and logging data will require funds that are not a research item but are necessary for the effective prosecution of research on geothermal systems.

Basic Research: There is urgent need for studies of the physical properties of water-steam fluid mixtures typical of those found in geothermal systems. This information is necessary to improve the interpretation of geophysical exploration results and to assist in determining optimum production practice.

Research is needed in thermal geophysics and hydrology to obtain a better understanding of the nature of the ultimate sources of energy in developable geothermal systems, and of the main modes of energy transport (conduction, convection, radiation) in the crust and the upper mantle. Moreover, the mode of energy transport within individual geothermal systems (conduction, one- and two-phase convection) is of considerable interest.

The setting of the exploitable geothermal systems within the framework of the modern geoscience theories (sea-floor spreading and plate tectonic models) should be investigated in order to establish global patterns of distribution of geothermal anomalies.

Research efforts should be directed to the solution of the many mathematical problems encountered in the exploration and utilization of geothermal systems. Of particular interest are the problems arising in thermal geophysics, hydrology, data processing, and general interpretation theory. This will involve research in applied mathematical analysis, numerical analysis and computer modeling.

New techniques of geothermal exploration such as in the fields of seismology, potential field and elastic strain should be researched and developed. Development of theory for optimization of exploration methodology and techniques should be undertaken.

RESOURCE ASSESSMENT

Objective and Summary

The objectives of the resource assessment panel were to identify the research programs which are prerequisite to assessing adequately the quantitative characteristics of geothermal resources of all types and, in addition, to assess the magnitude of the resource which underlies the entire United States. These programs do not include research on exploration techniques, and stop short of research on development and production techniques.

The process of estimating the quantitative characteristics of the resource includes the geographical distribution of known reserves, probable reserves, and potential reserves under specified conditions for each different type of geothermal system. Assessment research involves the development of the requisite theory and mathematical models which constitute the assessment methodology. These models must be based on the geological, geophysical, geochemical, and thermal characteristics which can be expected to be available from exploration work preceding the assessment.

In order to assess all different types of geothermal resources it will be necessary to drill into the formations. For this reason, the improvement of present drilling technology as well as research on new drilling technologies for great depths is considered a part of resource assessment research.

A primary research objective is to define the extent of the resource. During the progress of the resource assessment program a national inventory of geothermal resources should be established, and all goethermal resources should be assessed with respect to multiple use: the production of energy, desalinated water, minerals, and process heat.

In order to assess the hydrothermal convection systems, the dynamics of representative convection systems and geological, geochemical, hydrologic, and geophysical characteristics of geothermal systems must be studied. The objective of the first topic is to understand the size, life, origin, and dynamics of natural hot water geothermal systems, whereas the second is to provide the basic data on which the assessment model must be founded.

Three forms of energy are available in the produced fluid from geopressured aquifers. These are the intrinsic thermal energy, the mechanical energy available from the high pressure, and the methane contained in the formation fluid.

A regional resource study should identify three suitable sites for field research leading to a determination of the size and total energy content of typical geopressured reservoirs. This study should be conducted to compile data on temperature, structure, salinity, sedimentary characteristics, pressure potential distribution, and possible waste water disposal methods. In order to validate the geopressured resource model it will be necessary to drill and test the formations selected. This approach should lead to adequate assessments as well as assessment methodology.

Adequate assessment of the impermeable rock resources will draw upon explorations to establish the occurrences, age, volume, temperature, and permeability of the igneous intrusions under consideration. This will require study and modeling of rock mechanics and fracture systems, and greater understanding of the hydrology and the energy transfer characteristics of fracture systems.

The geochemistry of fractured hot impermeable rock systems is not well understood. Research will have to be conducted on geochemical kinetics to provide key characteristics of the heat transfer medium which brings the energy to the surface.

In order to determine the three-dimensional characteristics of representative geothermal areas, it will be necessary to undertake a deep drilling program into geothermal formations. A minimum of four wells must be drilled to depths in excess of 20,000 feet.

The difficulty and great expense involved in drilling to these depths in hot formations makes it imperative that new innovative approaches to drilling receive high priority in the research program. Erosion drills, electric melting drills, turbine drills, and other techniques should be explored.

Research Areas

The program envisioned includes actual assessment studies as well as research on assessment methodology. In all cases, the assessments much consider energy, water, and minerals. During the resource assessment research program a national inventory of geothermal resources should be established. A deep drilling research program is vital to the adequate evaluation of many geothermal resources. The several research areas directed toward resource assessment are discussed briefly below.

Hydrothermal Convection Systems: Present geothermal resource utilization depends entirely on hydrothermal convection systems, in which most energy is transferred upward in deeply circulating fluids. These systems thus provide high temperatures relatively near the surface and also natural fluids for transferring thermal energy to a power plant, desalination plant, or mineral production plant.

Two major types of convection systems are recognized. The relatively rare vapor-dominated (dry steam) systems account for most present world-wide geothermal power production. The much more abundant hot water systems must account for most of the near-future increases. The problems that must be solved to assess the hydrothermal convection systems of the Nation are described briefly on the following page.

In order to understand the dynamics of hydrothermal convection systems it is necessary to determine the size, natural life and circulation rates, origin, and flow patterns of several representative natural hot water geothermal systems. This information establishes the constraints and the main outlines of a comprehensive hydrothermal convection system model which must be validated on fields previously not included in the determinations. Once the modeling methodology is developed and validated, it can be applied to the assessments of such resources and the actual assessments can be supplied to the national inventory.

The modeling effort and the actual assessments referred to above are based on the geological, geochemical, geophysical, hydrologic, and thermal characteristics of the hydrothermal convection systems. Because of the diversity and geological complexity of hydrothermal convection systems, the assessments require the collection of a great deal of data which will be very useful in the subsequent development of these resources. Thus all of the data should be retained in the national inventory.

Geopressured Resources: Abnormally high fluid pressures in the northerm Gulf of Mexico basin are associated with high porosity deposits in which the confined water has stored geothermal heat for millions of years. These deposits extend in a 750-mile belt beneath the coastal region and underlie the coastal plain and adjacent contintal shelf. The aquifers are discontinuous, occurring in partitioned volumes bounded by faults generally paralleling the coastline. Similar aquifers have been found in Wyoming and in California.

A reservoir suitable for power production would contain a large volume of permeable sandstone. Since formation water is saturated with methane, three forms of energy are available in the produced water. These are the intrinsic heat, the obtained gas, and the mechanical energy available from the high pressure of the discharged water. The research and assessment program for these resources is discussed below.

The first step in assessing the geopressured resource is to conduct regional resource studies based upon data which presently exist in petroleum company files. Because of the proprietary nature of the petroleum company files, even though the data on the geopressured resource itself is normally not considered proprietary, it will be necessary to derive these data through a cooperative endeavor in which each company extracts the information from its own files and provides it to the resource assessment group. A neutral part, in respect to the petroleum companies, will be required to carry out the resource study.

The objective is to compile the available data concerning temperature, geological structure, salinity, sedimentary facies, pressure potential distribution and methane content for the geopressured resource. Regions of interest are the Gulf Coast, Wyoming, and California.

As with hydrothermal convection systems, it will be necessary to develop models to characterize the dynamics of the geopressured reservoir, and then apply these models to assess the resources of this type. The geological setting of geopressured resources appears somewhat simpler than that of the typical hydrothermal convection system and therefore the work necessary to assess them is less extensive.

The models will need to relate to the size, recharge rates from neighboring formations, origin, natural life, and flow mechanics of representative geopressured resources. The actual assessments must be added to the national inventory.

The modeling effort and the actual assessments are based on the geological, geochemical, geophysical, hydrologic and thermal characteristics of the geopressured resources. It is expected that a significant amount of data must be collected from field tests even though companies' files are thought to contain much of the required data.

Hot Impermeable Rock: The thermal energy in the crust, mantle, and core of the earth constitutes an immense energy resource base. The minute fraction of this energy which is actually being used by man is only that delivered to the earth's surface, or at least to shallow depths, by the convective processes represented by natural steam and by hydrothermal convection systems. The rest is not yet established as an accessible resource because the means to deliver it to the surface as usable energy have not yet been developed and demonstrated.

Various approaches have been proposed to develop this new major source of energy, all involving the creation of an artificial circulation system by drilling two holes into the formation and fracturing the rock between these holes. Several fracturing techniques have been discussed, such as hydraulic, chemical explosive, acid treatment, nuclear explosive, and thermal fracturing.

The research discussed below is directed at developing the methodology to assess and actually assessing these resources. No assumption is made with regard to the viability of one approach relative to another.

The hot impermeable rock can occur in two substantially different configurations. First, in a reasonably uniform sense, it is expected that hot impermeable rock underlies the entire United States, even in regions of subnormal heat flow. This expectation rests upon the idea that meteoric water will only descend to some rather consistent depth in the earth, and that the rocks at great depth have either crystallized from a magma or are so highly recrystallized that they have little or no permeability.

Secondly, during geologic time, magma has forced its way up into the earth's mantle and outer crust as a result of geological processes that are not well understood. Essentially the magma was trapped and either has crystallized or is crystallizing in place as a magmatic intrusive. The development of assessment methodology and the actual assessment of the intrusive resources is the object of this portion of the research program.

It is anticipated that the resource assessment for the very deep resource can be accomplished without further research or field work once the deep drilling technology is developed. Such a development would truly make geothermal energy available to all sectors of the United States. In areas of average temperature gradient (20°C/km) a temperature of 200°C is expected at a depth of 10 kilometers (about 33,000 feet).

In order to determine the methodology which is necessary to assess the hot impermeable rock resource, several research tasks must be completed: studies of rock mechanics, modeling of hydrodynamics of a fracture system, studies of site and bore-hole hydrology, studies of nature of allowable injection fluids, selection of test sites, and drilling and fracturing of formation (both injection and extraction wells). Assessment must include the intermediate grade (older) eastern deposits as well as the western ones.

Magma Systems: A continuing history of magmatic activities in the Hawaiian islands is surface evidence of the continuing intrusion of magma into the earth's mantle and upper crust. These very large sources of energy at quite high temperatures present some unique problems and opportunities concerning new approaches to derive useful energy from magma.

Including Hawaii and Alaska, there are fifty or more live volcanoes in the United States. Unobserved intrusives at depth, however, probably outnumber the live volcanoes. In order to assess the magmatic resources it will be necessary to conduct geological, geophysical, geochemical, hydrologic, and thermal tests on typical deposits. Drilling into the formation surrounding the magma chamber will be necessary. The information gained in this manner will be used to establish a model for assessing magmatic geothermal energy resources.

National Inventory of Geothermal Resources: Several times previously, it has been stated that all of the assessments actually performed in the nation's geothermal resources research program should be retained in a central depository for subsequent use in the reservoir development and production phases of exploitation of the geothermal resources. In addition, all of the data used as input to the assessment models should be retained. Not only should the data be retained, but key core samples should be stored for subsequent analysis from different viewpoints. Although this type of activity is not normally considered research, the national inventory is perhaps one of the most vital keys to making the more inaccessible parts of the geothermal resource base available as a true resource.

Deep Drilling: Deep drilling into hot formations is essential to the determination of the three-dimensional properties of geothermal resources. The deep drilling phase of the geothermal resource assessment program will involve a minimum of four wells to depths in excess of 20,000 feet and into extraordinarily hot formations.

The data acquired will provide an understanding of the three-dimensional characteristics of representative geothermal areas to depths previously unattainable. The frontier of geothermal science and technology will be pushed into entirely unexplored regions which will probably provide indispensable guidelines for further exploration and expansion of the general energy resource base of the country. It may open up a much larger reserve of geothermal energy for future development.

The difficulty and great expense involved in drilling to these depths in hot formations makes it imperative that new innovative approaches to drilling receive high priority in the research program. Erosion drills, electric melting drills, turbine drills, and other techniques should be explored.

RESERVOIR DEVELOPMENT AND PRODUCTION

Objective and Summary

The objectives of the reservoir development and production panel were to describe the research programs which are necessary prerequisites to the optimal development of a geothermal reservoir and production of the resource in its several manifestations: energy, desalinated water, minerals, and process heat.

The data gained from resource exploration and the models and data used for resource assessment form the starting point for reservoir development and production. However, the models will need to be extensively modified to encompass the behavior of the reservoir under different distributions of withdrawal and injection sites and schedules of production.

The models must address the problem of the probable life of the reservoir under these distributions and schedules. In addition, the data gained from exploration and assessment will need to be supplemented by further reservoir data gained from exploratory and test drilling. For this reason some research on drilling techniques is in the province of reservoir development and production.

There is a great deal of research required to form a sound basis for optimal reservoir development and production methods. This research is required in: geological formation evaluation methods, reservoir engineering, production engineering and production management, well stimulation technology and techniques, and drilling technology and techniques.

Formation evaluation tools include electrical, acoustic, and nuclear logging devices, bore-hole samplers, sidewall and conventional coring devices, bore-hole flow metering, and measurement of chemical properties of components and chemical behavior of fluid-solid systems. Improved evaluation tools, especially for high temperatures, are badly needed.

In order to support the practice of reservoir engineering for geothermal applications, fundamental research is needed into the physical, chemical, and thermal behavior of multiphase aqueous systems (steam-liquid solution-solid) in porous rock. Once these physical principles are established, meaningful physical and mathematical models of geothermal fluid systems can be constructed. These studies would include thorough transport phenomena studies.

There is a severe need for field reservoir engineering studies to establish requirements for constructing mathematical models for optimization studies of the effects of well spacing, liquid recharge from aquifers, reinjection of condensate and cooled liquids, cooling due to in-situ boiling of liquids, changes in pore space and fluid conductivity due to precipitation of salts and solids, and other important parameters. Field geothermal research sites are necessary to get this information and to validate reservoir engineering and hydrologic theory.

The problems of concern in production engineering and management are those of well workovers, well stimulation, corrosion and scaling, geothermal fluid separation and metering, critical velocity of flow in pipes and valves, deep well pumps, and other similar problems. The impact of these problems varies with the use of the resource, whether for production of electrical energy, desalinated water, minerals, or process heat.

The well stimulation techniques of interest included hydrofracturing, high explosive, acid treatments, thermal fracturing, and nuclear explosive applications to improve the permeability of the formation. The large void volumes produced by nuclear explosives could provide space for mineral deposition to collect from flashing hot water. Nuclear explosives might dry up a wet flashing system by providing space for the flashing to occur underground.

Research in drilling technology and techniques embraces improvements in current technology for application at higher temperatures. Problems are encountered with rubber seals, valves, cements, drilling muds, heat shields, sound mufflers, etc. Better high temperature instrumentation is needed to measure temperature and pressure and to sample geothermal fluids. Further efforts should be expended to improve the rotary drill.

Research Areas

There is much research required to establish firmly an adequate basis for geothermal reservoir development and production. This research is prerequisite to a final evaluation of the potential contribution of geothermal resources to the Nation. The research needs outlined below are broad in scope. In some cases, it is not possible to be completely specific about the details of the research.

Formation Evaluation: There is a strong need for research aimed at development of tools and techniques for geothermal formation evaluation. Conventional formation evaluation tools include wire-line electrical logging devices, bore-hole samplers, temperature and pressure measuring devices, sidewall and conventional coring devices, bore-hole flow metering, radioactivity and acoustic property measurement, and measurement of chemical properties of components and chemical behavior of fluid-solid systems.

Conventional tools are available for temperatures less than 350°F. Extended temperature range devices which are reportedly good to 500°F are sometimes available, but are not reliable.

In addition, there is a need for funding of formation evaluation measurements. Drilling objectives are often at odds with formation evaluation objectives. Because drilling costs are often in excess of 60% of total field development costs, it is important to drill a well rapidly. Formation evaluation data acquisition takes drill rig time that must be funded. Further, the need for research information may not be apparent to a private steam producer. In some cases, it may be necessary to drill special slim holes purely for formation evaluation purposes.

Commercial firms which are presently involved in developing formation evaluation tools and techniques should be encouraged to expand their efforts to development programs which are presently not feasible due to limited markets. Additional institutions with appropriate expertise should be funded to expand formation evaluation research.

Although existing physical laws govern geothermal systems, it should be recognized that measurement of many important physical, chemical, and thermal properties of components of geothermal systems have never been conducted at elevated temperatures. It should also be recognized that unknown physical mechanisms might be at work in geothermal systems. Thus, basic research in this area should be conducted.

Reservoir Engineering: Reservoir engineering involves evaluation of the size and deliverability of the geothermal resource, and planning for the optimum development of the resource. As such, this field is inextricably coupled with utilization technology and economics, and with environmental concerns. A proper reservoir engineering development plan cannot be obtained without a utilization plan, which in turn determines environmental effects which must be considered in the reservoir engineering plan. Nevertheless, the following will deal with only the reservoir engineering problems which must be solved to facilitate development of geothermal fluid production.

Basic research must be done in the field of reservoir engineering in order to get the most use from the resource, and to insure that apparently marginal resources will not be overlooked. Fundamental studies of the physical, thermal, and chemical behavior of multiphase aqueous systems (steam-liquid solution-solid) in porous media are needed to establish the physical principles for constructing meaningful physical and mathematical models of geothermal fluid systems. These studies would include thorough transport phenomena studies. There is a definite need for field reservoir engineering studies to establish the field requirements for constructing mathematical models.

These models should include sufficient detail to permit optimization studies of the effects of well spacing, liquid recharge from aquifers, reinjection of condensate and cooled liquids, cooling due to in-place boiling of liquids, changes in pore space and fluid conductivity due to precipitation of salts and solids, and other important parameters.

Proper field studies require careful, detailed plans for data acquisition, and realistic banking of useful reservoir data (in the national inventory mentioned previously) to permit development of reservoir engineering procedures without unnecessary duplication. To this end, there is a need for development of field pilot geothermal reservoirs operated to acquire such information and to test reservoir engineering and hydrologic theory. This approach would permit performance forecasting from actual performance matches with mathematical models.

Production Engineering and Production Management: Sound production engineering and management must be based upon resource use. Electrical power generation is considered most important currently, with space heating, mineral production, and water desalination recognized as other important uses. Current problems not involving production equipment, but nevertheless requiring study include: land use planning (surface and groundwater, mineral rights); legal regulation of geothermal fluid production; numerous environmental issues (sound, chemicals, seismic activity, subsidence, etc.); and reinjection of cooled liquids. Many other, more conventional problems exist concerning well workovers, well stimulation, corrosion and scaling, geothermal fluid separation and metering, and critical velocity of flow in pipes and valves.

One important problem impeding geothermal production development is the time lag between completion of a well and successful siting of the power plant and associated transmission lines and construction. Because of the likelihood of damage to wells with reduced flow as well as the waste of resources, interim operation of power plants should be considered to provide a market for this sort of developed steam production.

Well Stimulation Technology and Techniques: The effectiveness of well stimulation techniques has been demonstrated in the oil and gas industry, and many of these same techniques could prove effective in geothermal fields. Marginal wells might be brought to a level of economic production.

Hydrofracturing, high explosive, acid treatments, thermal fracturing and nuclear explosives should all be considered. They all have potential for improving the permeability of the formation and should prove equally effective in both production and reinjection wells. The large void volumes produced by nuclear explosives could yield additional advantages. They might provide space for mineral deposition to collect from flashing hot water and/or dry up a wet flashing system by providing space for the flashing to take place underground.

Large geothermal areas are known which are not economically productive and where scientific studies might yield a wealth of information. Some of these might benefit from field stimulation experiments.

The potential energy available from hot dry rock is apparently very large, although the existence of these resources near the surface is yet to be proved. One area near Marysville, Montana is expected to be over 500°C at a depth of 2 km. Correlation of heat flow measurements and gravity maps define an area which is estimated to contain billions of dollars worth of energy above 5 km in depth if the rock could be cooled by only 300°C. If this proves true it is important to tackle the problem of economic production of this energy.

Two methods for extracting energy from hot dry rock involve injecting water into the rock through fractures artifically produced. One proposal is to develop a reservoir by hydrofracturing the rock probably enhancing the fractured area by thermal stress cracking. The other uses nuclear devices to produce fractures. Field experiments are required on both.

Thermal energy in lava from volcanoes and in magma also is an intriguing potential source of energy which demands long-range study of energy production techniques.

Drilling Technology and Techniques: Drilling research contemplated in this section is directed at improvement in current technology. Such a small market exists for special geothermal equipment that the manufacturer simply cannot recover his research and development costs. Conventional equipment will not stand up to the temperatures encountered. Critical areas include rubber seals, valves, cements, muds, heat shields, sound mufflers, etc. Federal funding of research and development contracts on these problems with private industry could contribute to the safety and efficiency of the drilling operations. Research in Federal laboratories could also undertake this work.

Drilling technology and cost are important to the discovery and utilization of natural resources. Further efforts should be expended in improving the rotary drill, erosion drills, electric melting drills, and other techniques. These are recommended under resource assessment studies. A breakthrough in drilling techniques would be a giant step in resource assessment and reservoir development.

A similar situation exists in regard to logging equipment. Better high temperature instrumentation is required to measure temperature and pressure and to sample geothermal wells. Much standard well logging equipment is not satisfactory for geothermal applications. Until good measurements are readily available, reservoir analysis will continue to be very difficult.

UTILIZATION TECHNOLOGY AND ECONOMICS

Objective and Summary

The objectives of the utilization technology and economics panel were to identify the research and development programs, including demonstration projects where appropriate, which are needed to develop, improve, and demonstrate the technology to exploit the different categories of geothermal resources over the different temperature regimes.

The technology discussed in this section is principally the surface technology for utilizing the resource. In those cases where deep well pumps may be needed to lift the resource to the surface, such equipment is a logical topic for research on utilization technology.

A very important category for research is the technology for the best utilization of the different types of geothermal resources. For the vapor-dominated systems and the higher temperature (above (180°C) hot liquid-dominated convective hydrothermal resource, the technology for economic production of power appears well developed.

In the case of vapor-dominated systems, such as The Geysers field, steam is available at the well head in a slighly superheated state. After filtering, it is fed to low pressure turbines which drive the generator.

In the liquid-dominated systems, the liquid in the reservoir flashes into a mixture of steam and liquid upon reduction of pressure by the well. After separating the liquid from the steam, the steam is fed to a low pressure turbine. If the reservoir temperature is sufficiently high it may be possible to employ multiple stages of flashing and lower pressure turbines. It appears that there is no significant need for research on surface technology in these two cases.

In the case of hot dry rock systems, the same may be said if the temperature of the formation is high enough to heat the fluids sufficiently. However, it is necessary to conduct research on gaseous emissions, particulate removal, noise control, removal of dissolved salts such as silica and boron, and disposal of brine, condensate, and solids.

There are many geothermal reservoirs (perhaps 80% of all hydrothermal reservoirs) in which the temperature is not high enough to provide fluids that may be used with existing technology to produce economic power. Successful demonstration that these intermediate-temperature and low temperature reservoirs can be utilized economically for generating electric power and for multiple uses would expand the recoverable geothermal resources manyfold.

For the generation of power from such reservoirs, US industry has made some progress in developing the technology of binary fluid systems in which the heat is transferred to a low boiling point fluid (such as Freon or isobutane) which is then used as the working fluid in a closed cycle system. Further research on a variety of binary fluid systems as well as other power cycles deriving energy from hot concentrated brines is needed and demonstration plants should be built. Thermodynamic data is needed on the applicability of various working fluids for geothermal applications.

A great deal of effort must be expended to develop multiple uses of the resource. Desalination pilot plants using geothermal energy are in order. Mineral production will require pilot plants. Many of the utilization problems would be better understood if the physical and chemical properties of brines were better known. Corrosion rates of construction materials exposed to geothermal brines are required.

Immiscible heat transfer systems for extracting geothermal energy should be studied. These systems involve injecting a fluid which will not mix with water into the formation and then producing the combined fluids. Mechanical separation at the surface would provide a clean hot working fluid.

In the case of geopressured resources, research is needed to remove methane from the formation fluid. Direct conversion of geothermal energy to electrical energy, as by thermoelectric techniques, should be studied. As more data become available to support utilization design studies, a great deal of system modeling and analysis work will be necessary as a prerequisite to the design and construction of the several pilot plants.

Research Areas

The research recommended by the utilization technology and economics panel is discussed under the following areas: vapor-dominated systems, liquid-dominated systems, geopressured systems, thermoelectric systems, artifically fractured systems, economics, and demonstration plants.

It should be recorded that the panel specifically did not recommend that the Federal program include demonstration plants. This topic was added to the discussion of utilization technology in preparing the final report. The information was derived from other panel reports (assessment) and supplemented by the editor.

Vapor-Dominated Systems: The technology of utilizing these steam sources, such as at Larderello, Italy and The Geysers, California, is reasonably well advanced. In these systems, the dry steam flows directly from the reservoir and is used as the working fluid in a turbine which drives an electric generator. It is unlikely that this type of resource lends itself to water production.

The non-condensible gases in the steam can vary widely in consistency and quantity. Their removal is necessary to maximize plant efficiency, to minimize corrosion of plant equipment, and for ecological considerations. The Pacific Gas and Electric Company has had a program under way for some time to abate the release of hydrogen sulfide from its plants, and this program should succeed. In other areas, ammonia, or other gases will require attention.

Particulates in steam cause erosion of turbine blading and valves. In order to minimize erosion and to maintain good performance from transfer equipment, it is very desirable to remove as much particulate matter as possible from the steam before it reaches the utilization plant.

Noise control has been a problem in geothermal plants. Considerable improvement has been made in the control of noise from the release of steam during well venting, during release of overpressures in the steam collection system, and during general in-plant operations. Currently, there is technology available to minimize this problem.

Dissolved salts may be carried into the utilization equipment when any wet steam is present. Silica can cause fouling of turbines and decrease their performance; it can restrict the flow in pipes. Boron can be detrimental to plant life. Research should be conducted on preventing dissolved salts from entering the equipment.

Liquid-Dominated Systems: These reservoirs produce hot liquid (largely water) in which a number of dissolved salts exist. Heat is extracted for power generation by either partly flashing the liquid into steam or transferring its heat to a secondary working fluid. Pure water can also be produced by condensing and collecting the steam. Still another use is space heating by conducting the liquid to residences and other buildings such as is done in Iceland. A fourth potential use is the extraction of minerals. Combinations of these uses are sometimes possible and may be more economical then single purpose use.

Non-condensible gases are entrained in the reservoir fluids as they are in vapor-dominated systems. Differences in quantity will exist but the basic techniques for removing and disposing of these gases will be essentially the same as for the vapor-dominated systems.

The noise control problem in the case of liquid-dominated systems is complicated by the fact that flashing produces a two-phase mixture at the exit of the well vent. More energy must be dissipated in the well vent which means that larger and more costly devices are used than on vapor wells. Better muffling is required to solve this problem.

The hot reservoir fluid often contains an abundance of dissolved salts. When the liquid is partially flashed to steam, the dissolved salts remain in the liquid phase which therefore becomes supersaturated with them. Precipitation of the dissolved salts naturally follows.

Identification of the types and concentrations of solutes in a number of reservoirs is necessary to provide design information. Problems of equipment fouling because of precipitates will be serious. These precipitates act to plug the flow passages in the reservoir if flashing occurs underground.

In certain areas which contain geothermal brines, it will be necessary to recharge the reservoirs artifically. The fluids used for recharge generally will have to be treated to make them compatible with the prior formation fluids. Equipment will have to be developed for these tasks.

Environmental considerations may require that the produced brine and the steam condensate from the power plant not be reinjected as a recharge fluid. Alternative disposal schemes will be required which do not cause a nuisance and do not create any dangers to the environment.

Precipitates and separated solids derived from geothermal fluids may contain useful minerals. However, there will be large quantities of minerals which will have little or no value and will simply have to be removed. Burial, sluicing, and injection below ground are potential solutions which require research.

The use of secondary fluid power systems would make it possible to have an isolated working fluid with its characteristics selected for equipment, pressure and temperature requirements of the energy conversion cycle. This would require the transfer of heat from the hot geothermal fluid to the secondary fluid by use of heat exchangers. Research is required on the thermodynamic properties of various appropriate working fluids.

It will be necessary to match the characteristics of the working fluid (perhaps even synthesize a new fluid chemically) to the requirements of the utilization system as revealed by a thorough techno-economic analysis and preliminary design studies.

Turbines will probably have to increase in size and horsepower rating but manufacturers are capable of meeting these research and development requirements and have historically done so out of their own funds as markets developed.

Further research on special fouling problems of heat exchangers utilizing geothermal brines will have to be conducted. As with turbines manufacturers of this equipment have traditionally met these needs using their own funds.

The physical and chemical properties—both static and kinetic—of the geothermal brines are the starting point for the design of utilization equipment. Geothermal brines produced by flowing wells, and their subsequent concentration of solutes in the liquid phase due to flash chamber production of steam, require a study of the chemical kinetics of geothermal brines.

It is anticipated that, in certain cases, multipurpose development of geothermal resources may be more economically advantageous than single-purpose development. Careful study of the potential interface problems for multipurpose development to produce energy, desalinated water, minerals, and low-grade heat is indicated. The development, design, and construction of pilot plants for the desalination of water using geothermal energy is urged.

The extraction of geothermal energy by means of injecting fluids other than water into a geothermal reservoir is possibly feasible. If this fluid is immiscible with the geothermal brine, mechanical separation at the surface will provide a clean, hot working fluid without the necessity for surface heat exchangers.

The severe corrosiveness of geothermal fluids will require investigation of the ability of a number of alloys to withstand, or at least to mitigate, their effect. Design information on rates of corrosion and deposit build-up on materials of construction for turbines, heat exchangers, pipes, valves, and other plant components is required.

The use of geothermal energy for space heating and cooling will be expanded as more geothermal sources are developed. The proximity of the user to the source of supply is very important because of the difficulty of transporting the low-grade energy. Application technology for space heating and cooling, using hot water, have been in existence for many years and no further research is indicated.

Geopressured Systems: Much of the technology that is available, or that will become available as a result of the research which is directed toward utilization of the liquid-dominated system, will be directly applicable to the utilization of geopressured resources.

Technology exists today to pump large quantities of water up to pressures of 3,500 to 4,000 pounds per square inch for supercritical boilers in fossil-fueled steam power plants. This pump experience can be applied to develop a high pressure water turbine which would extract the pressure energy from the geopressured fluid and discharge the fluid at conditions of pressure and temperature such that its thermal energy could be extracted just as if the fluid had originated in the liquid-dominated system discussed above. A demonstration plant will be necessary and is considered below.

Geopressured fluids are normally saturated with methane. At the pressures and temperatures encountered, substantial quantities of methane could be associated with the fluid. This resource would provide an additional incentive for development of geopressured resources. Research is needed to extract the methane from the fluid.

Thermoelectric Systems: Thermoelectric and other approaches to the direct conversion of heat to electrical energy (without using turbines) using geothermal resources should be investigated. It may not even be necessary to go directly to electrical energy; the heat might be used to produce chemicals (such as hydrogen) which could be applied later to generate electricity. Research should be directed toward any such sound approaches to utilizing geothermal resources directly and thereby avoiding some of the problems mentioned above. Proof of concept should precede significant long-range research projects.

Artificially Fractured Systems: Hydrofractured systems (possibly enchanced by thermal fracturing during production) do not appear to pose any unique utilization technology problems that have not already been covered above if it is assumed that injected water is the heat transfer medium. No research specifically allocated to this topic seems needed.

Nuclear fracturing does produce radioactive materials in the formation. Research will be required to identify these materials and to assure that technology is available to prevent their occurrence above ground.

Economics: The current and impending fuel supply situation for electric power generation is presently serious and deteriorating. Supplementary energy sources for this purpose, such as geothermal, are now desirable and soon may be necessary for the economic and social health of the Nation. In this context, the economics of producing electric power by geothermal means may be secondary to its availability.

In dealing directly with the economics of geothermal-electric power production, the lack of definitive cost data makes the usual straightforward cost calculation impossible. The panel, therefore, attacked the problem from the standpoint of what can be afforded in the way of cost for a geothermal system, in order for its energy to be economically competitive with the leading anticipated methods of generation.

Apparently nuclear energy will be the least costly form of generation, among the conventional sources, in the remainder of this century. If the known costs of an electric power plant using geothermal energy are compared with the total cost of producing power from a nuclear plant, the difference will be that margin which is allowable for the unknown geothermal costs. These unknown cost items include reservoir exploration, development, and production; flashing chambers; particulate removal; noise reduction; effluent disposal; transmission differentials; environmental controls; etc.

Employing this technique, the panel has estimated that $200 per kilowatt will be allowable for a 1980 time frame and $250 per kilowatt for a 2000 time frame to cover these unknown geothermal costs.

There is an urgent requirement for comparison of types of energy cycles with the type of resources. System modeling and analysis studies will be very important in matching these elements. Techno-economic comparisons will have to be made.

Demonstration Systems: Historically, the research and development that is needed to make available power facilities has been accomplished by the electric power industry who then installs these facilities.

It is assumed that the manufacturers will continue to make the necessary investments to develop new equipment and that the utilities will invest in the necessary demonstration plants. The exception to this assumption is in the case of a desalination demonstration plant which has already been discussed above. The exception arises because of the traditional governmental role in providing water supplies.

The number and size of demonstration plants to be built is very difficult to predict with any great accuracy at this time, but for calculational purposes, it is assumed that seven demonstration plants will be needed, each of 10,000 kilowatts capacity: two hot water, flashing systems (under different conditions); two hot water, binary fluid systems; one hot brine system; one geopressured system; and one demonstration system for a new approach (unknown as yet).

A test bed laboratory should be installed at a site which has a geothermal fluid supply. This would allow the short term (1-2 years) testing of new equipment as it became available under field conditions without the necessity of building a new plant.

It is most probable that as the dynamics of geothermal fields are better understood, an optimum field management strategy will evolve which will maximize field productivity while meeting demand. To be most effective this should be a fully automated control system.

ENVIRONMENTAL EFFECTS

Objective and Summary

The objectives of the environmental effects panel were to identify the research necessary to define accurately the potential environmental impact which may be associated with the development of the Nation's geothermal resources, and to identify a program of research to alleviate, eliminate, or monitor any possible hazards.

Numerous environmentalists worldwide have reason to believe that geothermal energy may prove to be the cleanest source of convertible power readily available to man.

Geothermal developments are unique in that all activities related to resource production are localized to the immediate vicinity of the power plant or other utilization facility. Support operations such as mines, fuel processing, transportation, and other handling facilities do not exist. For this reason, the environmental effects are solely site dependent in origin.

Certain undesirable effects, however, can extend for several miles from the geothermal field itself and thus introduce environmental problems into the surrounding region. These must be studied, and solutions must be found.

Because the early utilization of geothermal resources will be for the generation of electric power, the panel addressed itself to the effects of this type of development. Potential effects on the environment of most immediate concern are gaseous and particulate emissions, land pollution, subsidence potential, seismic considerations, water pollution, biological effects, and social effects. Research should be directed to an accurate determination and evaluation of their character and magnitude.

Because of the relatively equal weight of the concerns, the panel addressed itself primarily to the problems either known or anticipated to exist in presently operating installations; that is, geothermal plants using dry slightly superheated steam, hot water flashed to steam, and binary fluid systems.

Little consideration was made of multipurpose developments, geopressured systems, direct use of hot waters for heating and cooling, and extractions of energy from hot-dry rock.

Directly concerning the Plowshare proposal, it was the consensus of the panel that the environmental hazards of both peaceful nuclear explosives technology and geothermal technology were combined in this approach and that the panel did not have the expertise to make recommendations concerning the environmental effects of the use of peaceful nuclear explosives technology.

Research Areas

In order to alleviate, eliminate, or monitor any possibly adverse environmental effects associated with the development of the nation's geothermal resources, a number of possible environmental hazards must be investigated and an accurate determination made of their character, magnitude and control. Several of these are briefly described below.

Gaseous and Particulate Emissions: The most publicized noxious material creating air pollution from existing geothermal steam power plants is hydrogen sulfide gas. The majority of this emission comes from the noncondensible gas ejected from direct contact condensates. The remainder is absorbed in the circulated cooling water. A portion of the hydrogen sulfide is oxidized to sulfates and elemental sulfur. The unreacted portion is air-stripped at a cooling tower, creating additional air pollution.

Considerable research is being applied to sulfur emissions, both to sulfur dioxide and hydrogen sulfide, and there are commercially proven processes for bringing those discharges into compliance with emission standards. However, research efforts are still required to resolve the problem of atmospheric pollution from hydrogen sulfide discharged at a cooling tower.

One solution would be to use surface condensers and thereby physically separate the steam condensate from the cooling water. This approach would impose the additional cost and burden of providing surface or ground water for make-up water to the cooling tower.

In addition to the hydrogen sulfide, there are undoubtedly other gaseous chemical substances, e.g., mercuric compounds, radioactive materials such as ^{210}Pb and ^{222}Rn, etc., present in some geothermal reservoirs which could adversely affect the environment if allowed free discharge to the atmosphere.

Research must be applied to identify the chemical composition of the noncondensible gas derived from geothermal reservoirs, and to quantify permissible exposure limits. Based on these findings, methods for containment and safe disposal of these materials must be developed. Concomitant with these research efforts is the need to develop the analytical techniques and instrumentation to monitor and control the discharge of these materials.

Similar research must be directed towards the emission of particulate matter. Particulate emission from an operating geothermal steam plant may not be readily apparent, but it can occur. Plants utilizing direct-contact condensers concentrate the particulates entering with the steam as much as five-fold through evaporation at the cooling tower.

Just as the blowdown from the cooling tower can contain high concentrates of boron, arsenic, fluorides and free ammonia, the entrained water leaving the cooling tower as drift-losses will contain equivalent concentrations in the droplets. Atmospheric dispersion and evaporation of these droplets will leave a dust of these materials on the surrounding lands. The composition of the blowdown has so far been such that it must be reinjected to avoid environmental quality degradation.

Research is required to determine the impact of this dust on the environment. Research can be applied also to mechanical designs which would reduce drift losses. Basic needs for research are to identify, quantify, and regulate the disposal of gaseous and particulate matter from geothermal resources.

Effects on the Land: In relation to the problem of land surface pollution, research is to be directed towards preventing the degradation of usable soil and toward the control of on-site surface deposition of pollutants that may be transported subsequently from the site of production to the surrounding environment. These problems arise in different ways in relation to different resource production processes.

With regard to a vapor dominated system, research is needed to identify and quantify all pollutants (such as Hg, As, Se, ^{210}Pb) in the vapor phase of a geothermal source and at each site proposed for development. Precipitation mechanisms need to be identified to evaluate and quantify fall-out and rain-out problems arising from such pollutants, and emission control methods need to be developed for use at the well head and at the points of power plant and cooling tower exhaust.

In connection with the water dominated geothermal system, the effects of accidental run-off of geothermal fluids from the production site to surrounding land areas need to be assessed, with an identification of surface deposited material that may harm plants, cause soil sterility, or be subject to biological magnification and entrance into food-chains.

Blow-out contingency programs to minimize land pollution by chemical deposition and to control possible erosion are needed, so that they can be applied at each development site. Facilities need to be developed for the adequate and safe storage of both liquid and solid wastes.

Subsidence Studies: Experience has shown that subsidence occurs in some areas when a fluid is removed from the ground, while in other areas the removal of equal quantities of fluid has produced no measurable subsidence. Much knowledge has been accumulated on this problem from studies of both petroleum and ground water reservoirs, but very little has been gathered from the production of geothermal fluids.

The tools and techniques for these studies are presently available and consist first of establishing precise close-spaced leveling data in the present fields known to be capable of producing high enthalpy geothermal fluids. Networks of monitor wells presently exist in some of these areas, but in others they should be established.

The tools, techniques and data from subsidence studies are all applicable to other studies requested below which are related to both water dynamics and water quality.

Seismic Hazards: A fundamental characteristic of many geothermal resource areas is their close association with regions of high geologic activity, which is manifested most commonly as earthquakes. Studies have shown that, if fluid pressures are changed in regions of tectonic stresses, faults can vary their normal patterns of earthquake activity.

Present research is being directed toward resolving many questions regarding seismic activities and some of this information can be applied to geothermal areas. However, seismic monitoring stations should be established near productive geothermal areas to determine if patterns emerge that appear to be related to the removal or injection of fluids from geothermal reservoirs.

Water Studies: Water studies cover a broad area, but the panel felt the subject should be broken into two major groups; fluid dynamics and fluid contamination. At this state of the science of fluid dynamics, the capacity to predict the effects on regional ground-water supply and the effect on land subsidence due to the exploration of geothermal fluids is hampered by inability to distinguish and quantify the sources of the water, to measure reservoir properties, and to define flow systems. New data and cheaper isotopic techniques are needed. Computer modeling with these data could greatly add to the understanding of these complex systems.

A possible risk associated with the development and utilization of geothermal resources is the contamination of surface and ground waters by geothermal fluids. This again is a site dependent phenomenon, since in some regions the high enthalpy fluids carry a heavy load of dissolved solids, while in others the water quality is similar to presently used ground waters.

This is an area where information and tools developed in the related study of subsidence and fluid dynamics could be applied, but specific research is required to identify those chemical constituents, which may have a detrimental effect. Sample collection, analysis, and procedures should be developed where they are presently lacking or are too expensive for widespread field application.

With particular reference to injection of geothermal fluids, chemical and isotopic studies should be undertaken to determine if geothermal fluids will indeed return to the reservoir from which they were produced or if they migrate into other reservoirs.

Biological Effects: Numerous unknowns exist regarding the probable impact of geothermal operations upon the biota native to prospective resource areas as well as areas presently under exploration or development.

Given the delicate balance of a natural environment, damage to many species of plant and animal life can take place through changes of chemical balance in soil and water, through the use of toxic substances in industrial application, through the destruction of such specialized habitats as thermal pools or alpine meadows, through the interruption of migratory patterns, through long term alterations in humidity, and through the introduction of human presence and activity into formerly unaffected regions.

These and other factors are in critical need of study in representatively selected geothermal resource areas to determine necessary procedures for the adequate protection of plant and animal life in regions of development.

Social Effects: Serious social effects arising from geothermal resource development which are needful for research involve the problems of noise and land use. Many phases in the development of a geothermal field can produce high noise levels. Most of the sources of noise can be controlled by present technology, but the blowing of wells in the cleanout and testing phases of the development of the flashed steam production process in a water dominated system presents unique problems for noise attenuation because of the presence of water droplets in the high pressure steam.

The flashed steam process is now in the experimental stage, and the early development of effective muffling equipment is necessary if this potentially important production process is to achieve the degree of public acceptability which will allow its being brought to perfection.

The land use problems connected with geothermal field development involved both conflicts in land use, long term land utility and the possibilities of multiple use. Geothermal operations are industrial in character and geothermal resources areas normally occur in nonindustrial regions. Because of noise, odor and visual impact, geothermal development, under present technology, is not compatible with residential or recreational land use. Problems exist with regard to the possibility of development of known resource areas where different land uses have been previously established.

Sociological, economic and planning studies are greatly needed to determine public policy for the equitable resolution of conflicts of land use arising in these cases. The brief history of geothermal operations in the United States has already generated controversy in this matter, and if it is decided that the development of the resource is to take place at a rate requisite to satisfy the energy demands which it can supply, a means must be found to bring cases of conflict to rapid resolution. Advances in technology which will allow a reduction of the industrialized character of a developed geothermal field will help reduce the land use conflict problem.

The problem of long term land utility arises from the extensive alteration of the physical condition of a developed geothermal resource area, such as by excavation, soil distribution and compaction which take place in the construction of drilling pads, access roads and power plant sites.

Research is needed for the development of resource production methods which will reduce this heavy physical impact. Utilizing the advanced techniques of directional drilling already in existence will make possible the location of several wells on a single drilling pad, thereby considerably confining the physical effects of field development. This method would have the additional benefit of reducing the extensive visual impact arising from the large numbers of drilling pads and piping networks presently utilized in field development.

INSTITUTIONAL CONSIDERATIONS

Objective and Summary

The objective of the institutional panel was to identify the scope of institutional studies, including the organizational, legal, and regulatory considerations, which will permit the development of geothermal resources in a safe, enonomical, environmentally sound, and desirable manner considering the interest of all concerned.

Concerned parties include the general public as ultimate users, industrial and other entities as developers or users, and local, State, and Federal governmental agencies as planners or regulators. The panel recommends that serious studies to weigh the merits and demerits of the different organizational structures be undertaken by a Presidential Study Commission.

The adequacy of the existing body of law should be assessed with the purpose of recommending changes wherever desirable. Governmental planning and regulatory processes should be examined with a clear objective of suggesting amplification, effective coordination, and other improvements. Detailed discussions and recommendations follow.

Research Areas

In order to achieve optimum development of geothermal resources, a number of research problems of an institutional and/or legal nature must be solved. Several of these are described briefly below.

Definition, Ownership, and Rights of Geothermal Resources: Geothermal steam is a resource peculiar unto itself. One may be interested in it for its energy content, for the derivative water, or for the minerals it contains. Thus, it cannot simply be classified either as a water resource or as a mineral resource, and in any event its energy value must not be ignored in establishing its legal position.

Like ground water, geothermal steam or hot fluids may cross property boundaries, carrying with them their energy and mineral content. Exploitation of these entities on one side of a line may affect exploitation on the other side. This poses legal questions and, judging from the current complexity of water laws alone, these questions cannot be easily answered.

The fact that some geothermal steam can be viewed simultaneously as a source of energy, water, and minerals, raises further questions regarding the multiple use of the resource and the allocations among the several resources.

Fragmentation of Research Efforts: Below is a list of the various elements of the Federal, State and private institutions with current interest in some phase of geothermal research, planning, or development. The list reveals the highly fragmented nature of the present national effort.

Although it is possible that wasteful duplication may not be occurring at the present time, duplication almost assuredly is eventually inevitable unless steps are taken to coordinate the national effort. Without effective coordination there can be no realistically achievable programs. In terms of current Federal activity alone, more than a dozen agencies have R&D programs either in the planning stage or already underway.

Some of the questions of the specific responsibilities of the private sector and government (both Federal and State) will have to be worked out and structured to some degree if the private sector and the government are to be meshed in a program that is both effective and efficient in relation to the national need. A listing of Geothermal Research and Development Institutions follows:

- Federal Agencies
  - Department of the Interior
    - US Geological Survey
    - Bureau of Land Management
    - Bureau of Reclamation
    - Bureau of Mines
    - Bureau of Outdoor Recreation
    - Bureau of Sport Fisheries and Wildlife
    - Office of Saline Water
    - Office of Water Resources Research
  - Atomic Energy Commission
    - Division of Applied Technology
    - Division of Research
  - National Science Foundation
    - Earth Sciences Section
    - Energy Research and Technology Program
  - Federal Power Commission
  - Department of the Treasury
  - Department of Defense
    - Advanced Research Project Agency
    - Naval Weapons Center
  - National Aeronautics and Space Administration
  - Environmental Protection Agency
  - Department of State
  - Office of Science and Technology
  - Office of Management and Budget
  - Congress
    - Interior and Insular Affairs Committee (Senate and House)
    - Joint Committee on Atomic Energy

State Agencies
  Variable by state but essentially parallel to those mentioned under Federal Agencies.
Private Sector
  Research Organizations
    Universities
    Industry
    Private research laboratories
  Industry
    Producers (oil companies)
    Consumers (utilities and others)

Regulatory Interfaces: Adequate regulatory jurisdiction for the development of geothermal energy is held by the Federal government and the State and county governments; however, some states and counties with geothermal resources have not as yet developed and promulgated comprehensive regulations. As the geothermal industry develops and each regulatory level becomes more involved, the overlap and, in some cases, the inconsistency of regulations will become very serious. This is an appropriate area for institutional study.

Information Collection and Dissemination: The need for a coordinated effort for the rapid collection and dissemination of current information on geothermal resources research and development activities is critical to the evolution of a sound national program. The information must be validated and qualified, and it must be made available to the interested public.

International Considerations: Several items of international significance should be considered. Among these are the problems of coextensive resources along or near international borders, possible undesirable environmental and exploitational effects resulting from developments near the border, the establishment and implementation of joint international projects, such as is currently proposed under the US-USSR Joint Commission on Scientific and Technical Cooperation, and the need for an information exchange to insure that use is made of the latest available information. The United Nations has done extensive work in this area, and its contributions to the state-of-the-art should be studied thoroughly.

Manpower: Because work on this resource is relatively new in the US, most of the people engaged in geothermal research and development have entered it from other fields. There are few institutions of learning where programs in geothermal resources are sufficiently well developed to provide a ready source of new manpower. The desirability of establishing interdisciplinary programs, specialized curricula, or centers of research should be investigated.

Industry Problems: The private sector will bear the ultimate responsibility for harnessing geothermal energy to generate electrical power and space heating. In doing so, it is faced with a host of problems and decisions that should be considered in designing and implementing a national program.

Among these are: preparing environmental impact statements and obtaining public utility commission certifications as now required on a unit-by-unit basis; seeking governmental cooperation where assistance is needed to establish an economically viable resource; interacting with government agencies in the development of cooperative programs or appropriate leasing regulations for the resource; forming a beneficial industrial association to represent geothermal interests at the national level; considering the economic impact of the lead times required for development; and establishing the lifetime of the resource.

Industry Regulation

In general, the regulation of the geothermal resource industry has been merely a series of prohibitions. This is most natural as the industry is still in its infancy and is little understood. To allow and encourage the development of this resource, within a reasonable time period, many in-depth studies are needed covering the physical, economic and sociological aspects of geothermal resource production.

Land Use: Studies are needed regarding policies for use of geothermal lands. Consideration must be given to beneficial uses other than the dominant one—geothermal resource production (energy, demineralized water, and/or minerals). In the recovery of geothermal resources, regardless of the end product, consideration must be given to regulating well spacing.

For example, should there be a grid system or a clustering of wells in the underground reservoir tapped by slant drilling to minimize surface distance? Should geothermal fluids move in pipelines above ground or in buried ones?

Criteria need to be developed as a basis for forming specific regulations. It will be necessary to reconcile geothermal resource production with other land uses. Studies are needed to formulate policies and methods to reconcile conflicts with grazing, wildlife habitats, watershed, timber production, and so on.

Exploration and Production: Some of the tools of exploration, such as geophysical and geochemical, can have an impact on the land. The impact of such methods and tools must be understood to regulate the exploration for and production of the geothermal resource intelligently.

At the present time, exploration, development, and production of geothermal resources require some forty permits and/or licenses. When power plants, transmission lines, and chemical plants are required, an even greater number of permits will be needed from agencies of the Federal, State, and local governments. Many of these procedures duplicate one another, at least in part. Technology is not in all cases sufficiently developed to implement the present regulations for all levels of government.

Financing and Construction: Studies are needed regarding methods of financing and the relationships between the producer of the geothermal resource and the electric utility or other user. Concerns arise between the electric utility company, which purchases the steam, and the lessee in the areas of lease rights and assignments of those rights to successors. These concerns are compounded when multiple use development of the resource is planned.

For purposes of amortizing capital investments studies are needed to estimate realistically the probable life of a geothermal field. In the case of the Magma

Power and Union Oil holdings in Sonoma County, California, geothermal steam is considered a gas and entitled to the same tax consideration as natural gas. The applicability of that ruling to other geothermal fields should be determined.

Operating Regulations: Existing Federal, State and local regulations cover operations for private investor-owned power plants and those plants that are joint ventures between two entities. Those plants that are publicly owned are essentially regulated by the local governing body. Inconsistencies in this area should be reconciled.

Transportation of Power: Throughout the United States, as the development of electrical energy took place, complex systems of power transmission grids have evolved. Separate lines on the grids may be owned by one company or jointly owned. In some cases an investor-owned public utility and the Federal government jointly own and operate such a facility.

A possible problem exists where a small publicly owned utility has the ability to develop geothermal power but the site is far distant from the utility's operating area. Power transmission becomes a problem since the construction of transmission lines is a costly item for such a utility. Because certain investor-owned public utilities, for a number of reasons, may be legally prevented or reluctant to share transmission facilities with competing utilities, the power transmission facilities are not available to the smaller publicly owned utilities.

Sales of Power: The regulation of the sale of power varies with the ownership of the power plant. A power plant owned by a local public utility essentially sets its own rates. The sale of power from private investor-owned plants and those operated under joint ventures is regulated by the Federal, State and local governments. If the power from such plants is sold wholesale to an entity who plans to resell it, the sales are regulated by the Federal Power Commission, or equivalent, and the local municipality. These regulations need review to assure that the rules are consistent and equitable to the geothermal power source, since geothermal energy as such is a non-transportable resource.

Regulation Interface Problems: Adequate regulatory jurisdiction for the development of geothermal energy is held by the Federal Government and the State and County governments; however, some states and counties with geothermal resources have not as yet developed and promulgated comprehensive regulations.

As the geothermal industry develops and each regulatory level becomes more involved, the overlap, and in some cases the duplication, of regulatory duties will become more serious. The county government is the smallest unit and it essentially regulates the land use of the county through zoning. The counties can also regulate those actions which will have an environmental effect, and they can move into those areas where they have not been preempted by the State.

The State and Federal governments, as possible land holders in the counties, are faced with the need to comply with county regulations. Where not preempted by Federal regulations, the State government should regulate the geothermal industry statewide, except in those areas that are considered under county control, such as land use planning. The Federal geothermal regulations for Federal lands should take into consideration the applicable State regulations where State regulatory programs exist.

Legislative Considerations

The Geothermal Steam Act of 1970 (Public Law 91-581; 84 Stat. 1566) was enacted some 21 months ago, but because of problems growing out of the environmental protection laws and interpretative court decisions, as well as administrative difficulties, as of November 1972 there are no regulations to implement the Geothermal Steam Act. Therefore, no geothermal leases have been issued on any Federal lands and the statute has not been subject to any practical test of workability and effectiveness. However, developments in the industry since the Act was passed and further study and analysis indicate that consideration might well be given to amendatory or new legislation on the following matters, among others.

Application of Mineral or Water Laws: Clarification of whether geothermal steam is, as a matter of law, a mineral resource or a water resource is needed. The Act is not clear on this point and, while it does provide for the leasing of "geothermal steam and associated geothermal resources," as such, the answers to a number of associated problems arising out of geothermal steam development depend upon whether water law or mineral law is applicable.

For example, who has the rights to the water remaining after the energy-producing potential has been exhausted and the by-products extracted? Section 9 of the Geothermal Steam Act provides that State water law govern production and use, but does not settle the question of rights to the waters.

The Act itself recognizes that this basic question is unanswered. Section 21 (b) requires that, where the title to geothermal leased lands has passed from Federal ownership, but the minerals have been reserved to the United States, the Attorney General must bring a quiet title action in Federal Court to determine whether the reservation of minerals to the United States included the geothermal resources. Such action shall be taken upon a report by the Secretary of the Interior when he finds that development of geothermal leasing under the Act is imminent.

The action need be taken only once. Also, in Section 22, the uncertainty as to whether geothermal steam is water or mineral is recognized again where provision is made that nothing in the Act shall constitute an express or implied claim or denial on the part of the Federal government as to its exemption from State water laws.

Continuity of Resources: Whether geothermal steam is a renewable or depletable resource is of extreme importance with respect to taxation. The Internal Revenue Service as noted above treats geothermal steam from The Geysers field in California as a gas and thus the producers are entitled to the 22-1/2 percent depletion allowance. However, it would be well to have the eligibility of geothermal steam for such depletion allowance provided by legislation.

Unitization: Section 18 of the Act authorizes voluntary unitization by lessees for conservation, and gives the Secretary of the Interior power to require unitization by provision in the lease. However, such forced unitization applies, of course, only to Federal lands. Unitization of Federal, State, and private lands is authorized, but not compulsory. Required unitization of lands within the same field appears desirable in some instances for conservation and efficient production. State cooperation will be required.

Incentives to Industry: The Geothermal Steam Act of 1970 is based on the concept of the development of the geothermal resources of Federally owned lands by and through private enterprise. However, the industry is a new one, with very little economic data on which to base decisions. Initial investments required are extremely high and the risks great at the present stage of knowledge and experience.

Therefore, it is highly desirable that ways be found to give private industry incentive to engage in development of this gravely needed natural resource. Such incentives might include a sliding scale of royalties, instead of the bonus system.

For competitive bidding, sliding royalties would have the three-fold advantage of requiring less immediate capital outlay by the private investor; it would have the government share in the success of development; and it would enable the so-called "little man," persons of limited capital, to have the opportunity to participate in the publicly-owned wealth of our national lands.

Another incentive might be an arrangement whereby the government guarantees a private entrepreneur up to 50 per cent against loss on his investment. A third incentive might be a credit for expenditures by the lessee for research and development against royalties, term of use, and acreage limitations. These and other avenues of providing incentives to private industry to carry out the purpose of the Act should be explored.

Program Development

Prime consideration in getting the nation's geothermal resources research program under way should go to the selection of the institutions required to assess exploration methods, resource appraisal, reservoir development and production, utilization technology and economics, and environmental effects.

There exists today no less an urgency for an early solution of the impending national energy shortage than existed in the areas of nuclear research in the 1940's, or space exploration and space communication in the 1960's, when it was agreed that means to meet these national needs required the establishment of new and unique federal organizational entities.

Organizational structures should be studied and a suitable institution selected which could implement geothermal resources research as its sole or principal mission. It is envisioned that such a mission-oriented organization would be separate from, but function in consonance with, the existing Federal, State, and local regulating bodies.

Currently there are no less than nine executive departments and independent Federal agencies concerned, in some part, with what is developing as the national geothermal program. These organizations should be studied to determine if improvements should be made in institutional and technical competence, intra-agency priorities, planning, budgeting, administration, and coordination of geothermal activities. It is almost trite to observe that in order to realize a coordinated, well managed and effective national program, a solution to these fragmented responsibilities needs to be found.

While the actual performance of the geothermal resources research effort as outlined in this report may well be carried out by existing competence in private companies, educational institutions, national laboratories, and newly created geothermal field research centers, the funding, direction, management, and support functions may well best be carried out by one, rather than several, Federal department or agency.

A partial listing of such common support functions is as follows: budget and program planning, cost effective analysis, audit and accounting, technical information collection and dissemination, intellectual property arbitration and review, public information and acceptance programs, environmental impact assessment and National Environmental Policy Act compliance, joint Federal-private industry cooperative project policy and administration, Federal-State coordination and cooperation, international information exchange and cooperative projects, manpower development (scientific and technical training programs), legislative programs regulating coordination and advisory functions, Congressional relations, operation of geothermal field research and experimental centers, technical advisory support for regulatory agencies, and coordination with other Federal agencies with related responsibilities.

Varying degrees of centralization are possible to embody these common functions and support services. Consideration should be given to any and all possible institutional opportunities, some of which are listed below:

Geothermal Energy Commission: A new independent agency, established by legislation, reporting to the President and combining the function and responsibilities of all other Federal agencies presently concerned with geothermal resources research and development.

Geothermal Resources Corporation: A quasi-governmental corporation, established by legislation and funded jointly by appropriations and private capital or operating revenues. Other generally similar organizations are the Tennessee Valley Authority (TVA), the Communications Satellite Corporation (COMSAT), and the Bonneville Power Administration (BPA).

Joint Office of Geothermal Research and Development: A separate office jointly funded and staffed by participating Federal agencies and existing by formal agreement of these organizations or by executive order. The Office would have all the responsibilities of a separate agency with budgets and funding received through participating agencies. Examples of such offices are: the Space Nuclear Propulsion Office (SNPO), the Office of Naval Reactors, and the Office for International Decade for Ocean Exploration.

National Geothermal Coordinating Council: A council providing a forum for formal coordination of Federal programs which exist by executive order. The Council would have no programmatic nor mission oriented responsibility.

Ad Hoc Interagency Coordinating Committee: The continuation of the existing ad hoc Federal group which assisted in preparation of the report entitled *Assessment of Geothermal Energy Resources.* It is a voluntary participation group with no programmatic function nor responsibilities of its own.

National Council for Geothermal Resource Development: A Presidential council to set policy, coordinate and oversee the conduct of a national geothermal resources research program carried out by the several agencies involved. Similar councils have been employed for national security affairs, as well as domestic and foreign affairs.

National Geothermal Field Research Laboratories: A small set of laboratories organized around a small set of different types of geothermal resources. A large number of techniques and different technologies must be employed in finding, appraising, mapping (in three dimensions), and experimenting with each typical geothermal resource in order to develop and validate the methodology and technology for finding, appraising and exploiting similar resources.

If all the research is concentrated on one each of all different types of geothermal resources, then cooperation between research projects will reduce both the time and money required to complete the research. Laboratories would be available for research and development projects being conducted by industry, government, and university investigators subject to the scientific approval of a Geothermal Resources Advisory Council. Laboratories should be jointly operated by USGS and NSF.

In addition to a Federal institutional forum for geothermal development, there is a need for a national association of geothermal industries. This would conceptually be a privately sponsored association not to be substituted for any of the above, but highly desirable to provide effective industry representation at the national level and to provide a responsible point of liaison with the Federal geothermal institution or institutions.

Recommendations

It is recommended that a Study Commission be established by, and report directly to the President of the United States. A study group, given a broad scope and composed of highly qualified individuals, is required to establish a Commission of sufficient status to deal effectively with the authorities and jurisdictional prerogatives of cabinet level departments.

The recommended charter is to examine existing Federal structures and missions responsible for aspects of geothermal resources and their development, to assess the capabilities and effectiveness of such components, and to recommend institutional and programmatic solutions to mount a vigorous national geothermal resources program.

ENERGY FROM SOLID WASTE 1974

by Frederick R. Jackson

Pollution Technology Review No. 8

Energy Technology Review No. 1

The solid waste disposal problem is reaching alarming proportions everywhere. The United States alone produces close to 300 million tons of solid waste per year, which is equivalent to about one ton per person.

At the present time the prevalent methods of disposal are dumping and sanitary landfill. Municipal incineration disposes of a small portion only, with attendant high capital and operating costs.

Many methods have been proposed for coping with the problem, such as source separation, source reduction, or material recovery. However, with the energy crisis descending upon us, producing energy from waste is becoming more and more attractive. An estimate currently making the rounds in financial circles is that when the price of crude oil reaches $7.00 a barrel, alternate sources of energy become practicable.

This book is based primarily upon information from studies conducted under the auspices of the EPA. Its foremost topic is burning of solid wastes to create steam directly. The air pollution problem created by burning can be solved quite easily with known technology. Solid wastes are low in sulfur, consequently there is no SO_2 removal problem.

Another technique that may assume more importance in the future is the controlled pyrolysis of wastes, yielding so-called pyrolysis gas or oil. Chapter seven is devoted to this. The final chapter discusses European practice, which is historically far more extensive than that of the U.S. A condensed table of contents follows here:

ISBN 0-8155-0528-0 **163 pages**

UTILIZATION OF WASTE HEAT FROM POWER PLANTS 1974

by David Rimberg

Pollution Technology Review No. 14

Energy Technology Review No. 3

Present-day steam-driven turbo-electric power plants in the United States discharge as waste heat an amount of energy roughly equivalent to twice their total electricity-generating capacity. This energy is most difficult to utilize, because it is degraded in temperature. The effluent water is warm, but it is far from being near the boiling point. It constitutes a necessary, but unwanted, by-product of the energy conversion process for generating electricity.

The growing quantities of waste heat discharged, and the increasing ecological anxieties about undesirable growths, energy utilization and thermal discharge problems, have stimulated an examination of methods for productively using energy presently wasted on the environment.

Part I of this book discusses the reasons for present-day ineffective utilization, but it also shows ways and means to promote more efficient energy usage.

Part II assesses the cause, magnitude, and possible effects of heat discharges into water from steam electric power plants and related condenser cooling systems. Also contained in this section is a discussion of the thermodynamics of the electric power generation cycle detailing the reasons for the "inefficiencies" in by-product heat generation.

Ultimately all the accessible waste energy appears as low temperature heat, and the term "waste heat utilization" refers to the performance of useful functions with this heat before it is discharged into the environment. Within the framework of this book, the minimum temperature of the effluent water considered for subsequent use is about 38°C or 100°F. These uses are discussed in **Part III:** Food production in agriculture, hydroponics and aquaculture (fish farming) and the use of heat in wastewater treatment.

In the past the dissipation of waste heat was accomplished by wet and dry cooling towers and cooling ponds, lakes and streams. These methods are now being challenged by some sectors of society, and industry is being forced to consider their environmental impact. In this regard **Part IV** discusses the research needs necessary to equate the complicated interactions of the physical, engineering, biological, and social aspects of this waste heat problem.

This Pollution Technology Review is based on studies conducted by industrial and engineering firms or university research teams under the auspices of various governmental agencies, e.g. The National Water Commission, Oak Ridge National Laboratory Federal Water Pollution Control Administration, Office of Water Resources Research, Environmental Protection Agency, National Science Foundation, and the Department of Commerce.

A partial and condensed table of contents follows here:

ISBN 0-8155-0555-8 **175 pages**

ENERGY IN THE CITY ENVIRONMENT 1973

Edited by Robert N. Rickles

This book is based upon material supplied by Robert Rickles, the Institute for Public Transportation, and the New York Board of Trade.

I: CRISES IN ENERGY SOURCES AND PRODUCTION

1. OVERVIEW (1972-1985)
   - Primary Reliance on Conventional Fuel Sources
   - Problems in Research and Technology
   - Lead Time for Breeder Reactors
   - Fusion
   - Solar Energy
   - Coal Gasification

2. OIL
   - NEPA and Other Environmental Issues
   - Accelerated Government Leasing Programs on Outer Continental Shelf
   - Oil Imports
   - Costs and Risks: Economic, National Security and Environmental Problems
   - Improved Exploration Incentives
   - Product Prices versus Foreign Imports
   - Tax Incentives

3. GAS
   - Accelerated Government Leasing of Outer Continental Shelf
   - Amendments to Natural Gas Act
   - Security of Contract Terms
   - Competitive Pricing
   - LNG/SNG — Imported and Domestic
   - Interstate and Intrastate Marketing Policies

4. ELECTRICITY
   - Generating Plant Licensing and Regulations
   - Nuclear Technology
   - Pumped Storage
   - Availability of Low Sulfur Fuel Oil and Gas
   - Efficiency of Conventional Sources
   - Purchasing Power (Imports)

II: CRISES IN ENERGY DEMANDS

1. OVERVIEW
   - Are Our Priorities and Objectives Compatible and Ready for Action?
   - Competition for Available Supplies: National and International Demands Constrain City Energy Resources
   - Effects of Differing Demand Breakdowns
   - Institutional Alternatives for Energy Management

2. REALISTIC APPRAISAL OF DEMAND GROWTH RATES (through 1985)
   - Population Growth (Inevitable Modest Demand Expansion)
   - Social Side Effects of Energy Constraint Policy
   - Additional Energy of Environmental Protection
   - Air Pollution and Energy

3. CONSERVATION OF ENERGY RESOURCES
   - Rationing and End Use Controls
   - Taxing Certain Energy Uses
   - Changes in Life Style
   - Improved Energy Efficiency
   - New Buildings
   - Mass Transit
   - Appliances

III: ELEMENTS IN THE ENERGY-ENVIRONMENT BALANCE

1. OVERVIEW
   - Need to Develop New Energy Reserves Without Unreasonable Environmental Delays
   - Ten Year Planning Lead
   - New Explorations to Preclude Shortages

2. IMPACT OF INADEQUATE ENERGY SUPPLIES
   - Economy-GNP-Unemployment-Inhibition of Investments
   - Insecurity and Reduced Standard of Living
   - Slowdown and Abandonment of Public Sector Programs
   - National Security
   - Environmental Protection: Air Pollution, Water and Waste Management

3. ENERGY-ENVIRONMENT BALANCING
   - Air Pollution
   - Offshore Drilling
   - Increased Tanker Traffic
   - Reliance on Fuel Imports
   - Oil and LNG
   - Nuclear Power
   - Radiation and Waste Disposal
   - Interim Nuclear Licensing
   - Selection and Nature of Plant Sites
   - Air Quality
   - Thermal Pollution

IV. NYS POWER PLANT SITING ACT

The important **charts** in the **appendix** present a list of energy conservation measures for the short-term (1972-1975), mid-term (1976-1980) and the long-term (beyond 1980) for the Transportation, Residential/Commercial, Industrial and Electric Utility sectors. The charts also indicate estimated maximum attainable energy savings, possible means for implementing each conservation measure, and pros and cons.

ISBN 0-8155-5019-7 **173 pages**